Atmospheric modeling, data assimilation and predictability

This comprehensive text and reference work on numerical weather prediction covers for the first time, not only methods for numerical modeling, but also the important related areas of data assimilation and predictability.

It incorporates all aspects of environmental computer modeling including an historical overview of the subject, equations of motion and their approximations, a modern and clear description of numerical methods, and the determination of initial conditions using weather observations (an important new science known as data assimilation). Finally, this book provides a clear discussion of the problems of predictability and chaos in dynamical systems and how they can be applied to atmospheric and oceanic systems. This includes discussions of ensemble forecasting, El Niño events, and how various methods contribute to improved weather and climate prediction. In each of these areas the emphasis is on clear and intuitive explanations of all the fundamental concepts, followed by a complete and sound development of the theory and applications.

Professors and students in meteorology, atmospheric science, oceanography, hydrology and environmental science will find much to interest them in this book which can also form the basis of one or more graduate-level courses. It will appeal to professionals modeling the atmosphere, weather and climate, and to researchers working on chaos, dynamical systems, ensemble forecasting and problems of predictability.

Eugenia Kalnay was awarded a PhD in Meteorology from the Massachusetts Institute of Technology in 1971 (Jule Charney, advisor). Following a position as Associate Professor in the same department, she became Chief of the Global Modeling and Simulation Branch at the NASA Goddard Space Flight Center (1983–7). From 1987 to 1997 she was Director of the Environmental Modeling Center (US National Weather Service) and in 1998 was awarded the Robert E. Lowry endowed chair at the University of Oklahoma. In 1999 she became the Chair of the Department of Meteorology at the University of Maryland. Professor Kalnay is a member of the US National Academy of Engineering and of the Academia Europaea, is the recipient of two gold medals from the US Department of Commerce and the NASA Medal for Exceptional Scientific Achievement, and has received the Jule Charney Award from the American Meteorological Society. The author of more than 100 peer reviewed papers on numerical weather prediction, data assimilation and predictability, Professor Kalnay is a key figure in this field and has pioneered many of the essential techniques.

Atmospheric modeling, data assimilation and predictability

Eugenia Kalnay
University of Maryland

CAMBRIDGE
UNIVERSITY PRESS

CAMBRIDGE UNIVERSITY PRESS
Cambridge, New York, Melbourne, Madrid, Cape Town,
Singapore, São Paulo, Delhi, Mexico City

Cambridge University Press
The Edinburgh Building, Cambridge CB2 8RU, UK

Published in the United States of America by Cambridge University Press, New York

www.cambridge.org
Information on this title: www.cambridge.org/9780521796293

First published 2003
Third printing with corrections 2006
7th printing 2012

A catalogue record for this publication is available from the British Library

Library of Congress Cataloguing in Publication Data

Kalnay, Eugenia, 1942–
Atmospheric modeling, data assimilation and predictability / Eugenia Kalnay.
p. cm.
Includes bibliographical references and index.
ISBN 0-521-79179-0 – ISBN 0-521-79629-6 (pbk.)
1. Numerical weather forecasting. I. Title.
QC996 .K35 2002 551.63′4–dc21 2001052687

ISBN 978-0-521-79179-3 Hardback
ISBN 978-0-521-79629-3 Paperback

I dedicate this book to the Grandmothers of Plaza de Mayo for their tremendous courage and leadership in defense of human rights and democracy

Contents

Foreword

During the 50 years of numerical weather prediction the number of textbooks dealing with the subject has been very small, the latest being the 1980 book by Haltiner and Williams. As you will soon realize, the intervening years have seen impressive development and success. Eugenia Kalnay has contributed significantly to this expansion, and the meteorological community is fortunate that she has applied her knowledge and insight to writing this book.

Eugenia was born in Argentina, where she had exceptionally good teachers. She had planned to study physics, but was introduced to meteorology by a stroke of fate; her mother simply entered her in a competition for a scholarship from the Argentine National Weather Service! But a military coup took place in Argentina in 1966 when Eugenia was a student, and the College of Sciences was invaded by military forces. Rolando Garcia, then Dean of the College of Sciences, was able to obtain for her an assistantship with Jule Charney at the Massachusetts Institute of Technology. She was the first female doctoral candidate in the Department and an outstanding student. In 1971, under Charney's supervision, she finished an excellent thesis on the circulation of Venus. She recalls that an important lesson she learned from Charney at that time was that if her numerical results did not agree with accepted theory it might be because the theory was wrong.

What has she written in this book? She covers many aspects of numerical weather prediction and related areas in considerable detail, on which her own experience enables her to write with relish and authority. The first chapter is an overview that introduces all the major concepts discussed later in the book. Chapter 2 is a presentation of the standard equations used in atmospheric modeling, with a concise but complete discussion of filtering approximations. Chapter 3 is a roadmap to numerical methods providing a student without background in the subject with all the tools needed to develop a new model. Chapter 4 is an introduction to the parameterization of subgrid-scale physical processes, with references to specialized textbooks and papers. I found her explanations in Chapter 5 of data assimilation methods and in Chapter 6 on predictability and ensemble forecasting to be not only inclusive but thorough and well presented, with good attention to historical developments. These chapters, however, contain many definitions and equations. (I take this wealth as a healthy sign of the technical maturity of the subject.) This complexity may be daunting for many readers, but this has obviously been recognized by Eugenia. In response she has devised many simple graphical sketches that illustrate the important relations and definitions. An added bonus is the description in an appendix of the use of *Model Output Statistics* by the National Weather Service, its successes, and the rigid constraints that it imposes on the forecast model. She also includes in the appendices a simple adaptive regression scheme based on Kalman filtering and an introduction to the generation of linear tangent and adjoint model codes.

Before leaving the National Centers for Environmental Prediction in 1998 as Director of the Environmental Modeling Center, Eugenia directed the *Reanalysis Project*, with Robert Kistler as Technical Manager. This work used a 1995 state-of-the-art analysis and forecast system to reanalyze and reforecast meteorological events from past years. The results for November 1950 were astonishing. On November 24 of that year an intense snowstorm developed over the Appalachians that had not been operationally predicted even 24 hours in advance. This striking event formed a test situation for the emerging art of numerical weather prediction in the years immediately following the first computations in 1950 on the ENIAC computer discussed in Chapter 1. In 1953, employing his baroclinic model, and with considerable "tuning" Jule Charney finally succeeded in making a 24-hour forecast starting on November 23 1950 of a cyclonic development, which, however, was still located some 400 kilometers northeast of the actual location of the storm. This "prediction" played a major role in justifying the creation of the Joint Numerical Weather Prediction Unit in 1955 (Chapter 1). By contrast, in the Reanalysis Project, this event was forecast extremely well, both in intensity and location – as much as three days in advance. (Earlier than this the associated vorticity center at 500 mbs had been located over the Pacific Ocean, even though at that time there was no satellite data!) This is a remarkable demonstration of the achievements of the numerical weather prediction community in the past decades, achievements that include many by our author.

After leaving NCEP in 1998, Eugenia was appointed Lowry Chair in the School of Meteorology at the University of Oklahoma, where she started writing her book. She returned to Maryland in 1999 to chair the Department of Meteorology, where she continues to do research on a range of topics, including applications of chaos to ensemble forecasting and data assimilation. We look forward to future contributions by Professor Kalnay.

Norman Phillips

Acknowledgements

I drafted about two thirds of this book while teaching the subject for the first time at Oklahoma University, during the fall of 1998. OU provided me with a supportive environment that made it possible to write the first draft. I made major revisions and finished the book while teaching the course again in the fall in 1999 through 2001 at the University of Maryland. The students that took the course at UM and OU gave me essential feedback, and helped me find many (hopefully most) of the errors in the drafts.

In addition, several people helped to substantially revise one or more of the manuscript chapters, and their suggestions and corrections have been invaluable. Norm Phillips read an early draft of Chapter 1 and made important historical comments. Anders Persson wrote the notes on the early history of numerical weather prediction, especially in Europe, reproduced in an appendix. Alfredo Ruiz Barradas reviewed Chapter 2. Will Sawyer reviewed and made major suggestions for improvements for Chapter 3. Hua-lu Pan influenced Chapter 4. Jim Geiger reviewed Chapter 5 and pointed out sections that were obscure. Jim Purser also reviewed this chapter and not only made very helpful suggestions but also provided an elegant demonstration of the equivalence of the 3D-Var and OI formulations. Discussions with Peter Lyster on this chapter were also very helpful. D. J. Patil suggested many improvements to Chapter 6, and Bill Martin pointed out the story by Ray Bradbury concerning the "butterfly effect". Joaquim Ballabrera substantially improved the appendix on model output post-processing. Shu-Chih Yang and Matteo Corazza carefully reviewed the complete book, including the appendices, and suggested many clarifications and corrections.

I am grateful to Malaquias Peña, who wrote the abbreviation list and helped with many figures and corrected references. Dick Wobus created the beautiful 6-day ensemble forecast figure shown on the cover. Seon Ki Park provided the linear tangent and adjoint code in Appendix B. The help and guidance of Matt Lloyd and Susan Francis of Cambridge University Press, the editing of the text by Maureen Storey, and the kind foreword by Norm Phillips are also very gratefully acknowledged.

I began to learn numerical weather prediction (NWP) in the late 1960s from professors at the University of Buenos Aires, especially Rolando Garcia and Ruben Norscini, and from the inspiring book of P. D. Thompson. At MIT, my thesis advisor, Jule Charney, and the lectures of Norm Phillips and Ed Lorenz, influenced me more than I can describe. The NWP class notes of Akio Arakawa at UCLA and the NCAR text on numerical methods by John Gary helped me teach the subject at MIT. Over the last 30 years I have continued learning from numerous colleagues at other institutions where I had the privilege of working. They include the University of Montevideo, MIT, NASA/GSFC, OU, and UM. However, my most important experience came from a decade I spent as Director of the Environmental Modeling Center at the National Centers for Environmental Prediction, where my extremely dedicated colleagues and I learned together how to best transition from research ideas to operational improvements.

Finally, I would like to express my gratitude for the tremendous support, patience and encouragement that my husband, Malise Dick, my son, Jorge Rivas, and my sisters Patricia and Susana Kalnay have given me, and for the love for education that my parents instilled in me.

Abbreviations

3D-Var	Three-dimensional variational analysis
4DDA	Four-dimensional data assimilation
4D-Var	Four-dimensional variational analysis
AC	Anomaly correlation
ADI	Alternating direction implicit
AMIP	Atmospheric Models Intercomparison Project (frequently refers to long model runs in which the observed SST is used instead of climatology)
AO	Arctic oscillation
AP	Arrival point in semi-Lagrangian schemes
ARPS	Advanced Regional Prediction System
AVN	NCEP's aviation (global) spectral model
CAPS	Center for Analysis and Prediction of Storms
CFL	Courant–Friedrichs–Lewy
COAMPS	US Navy's coupled ocean/atmosphere mesoscale prediction system
CONUS	Continental USA
CPC	Climate Prediction Center (NCEP)
CSI	Critical success index (same as threat score)
DP	Departure point in semi-Lagrangian schemes
DWD	German Weather Service
ECMWF	European Centre for Medium-Range Weather Forecasts
EDAS	Eta data assimilation system
ENIAC	Electronic numerical integrator and computer

ENSO	El Niño–Southern Oscillation
FASTEX	Fronts and Storm Track Experiment
FDE	Finite difference equation
FDR	Frequency dispersion relationship
FFSL	Flux-form-semi-Lagrangian scheme
GLE	Global Lyapunov exponents
GPS	Global positioning system
hPa	hecto Pascals (also known as millibars)
HPC	Hydrometeorological Prediction Center (NCEP)
JMA	Japan Meteorological Agency
JNWPU	Joint Numerical Weather Prediction Unit
LFM	Limited fine mesh
LLV	Local Lyapunov vectors
MCC	Mesoscale compressible community (model)
MeteoFrance	National Meteorological Service for France
MJO	Madden and Julian oscillation
MM5	Penn State/NCAR mesoscale model, version 5
MOS	Model output statistics
NAO	North Atlantic oscillation
NASA	National Aeronautics and Space Administration
NCAR	National Center for Atmospheric Research
NCEP	National Centers for Environmental Prediction (US National Weather Service)
NCI	Nonlinear computational instability
NGM	Nested grid model
NLNMI	Nonlinear normal mode initialization
NMC	National Meteorological Center
NOAA	National Oceanic and Atmospheric Administration
NORPEX	North Pacific Experiment
NWP	Numerical weather prediction
NWS	National Weather Service
OI	Optimal interpolation
PDE	Partial differential equation
PDO	Pacific decadal oscillation
PIRCS	Project to Intercompare Regional Climate Systems
PQPF	Probabilistic quantitative precipitation forecast
PSAS	Physical space analysis scheme
PVE	Potential vorticity equation
RAFS	Regional analysis and forecasting system
RAOB	Rawinsonde observation
RDAS	Regional data assimilation system
RSM	NCEP's Regional Spectral Model

RUC	NCEP's rapid update cycle
SAC	Standardized anomaly correction
SCM	Successive correction method
SOR	Successive overrelaxation
SST	Sea surface temperature
SWE	Shallow water equations
TOGA	Tropical ocean, global atmosphere
TOVS	TIROS-N operational vertical sounder
TS	Threat score
UKMO	United Kingdom Meteorological Office
UTC	Universal time or Greenwich time, e.g. 1200 UTC. Frequently abbreviated as 1200Z
WMO	World Meteorological Organization

Variables

a	radius of the Earth
$\mathbf{A}$	analysis error covariance matrix
$\mathbf{B}$	background error covariance matrix
$\mathbf{C}$	covariance matrix
C_p, C_v	specific heat at constant pressure, constant volume
$\mathbf{d}$	innovation or observational increments vector
D	fluid depth
$E(\,)$	expected value
f	Coriolis parameter
g	gravitational constant
$\mathbf{H}$	linear observation operator matrix
H	observational operator, scale height of the atmosphere
$\mathbf{I}$	identity matrix
J	cost function
JM	maximum number of grid points j
$\mathbf{K}$	Kalman gain matrix
$L(t_0, t)$	resolvent or propagator of TLM
$\mathbf{M}$	TLM matrix
N	Brunt–Vaïsälä frequency
$\mathbf{P}$	projection matrix
p	pressure, probability, distribution function
q	mixing ratio of water vapor and dry air mass
$\mathbf{Q}$	forecast model error covariance
$\mathbf{r}$	position vector

$\mathbf{R}$	observations error covariance matrix
R	root mean square error, gas constant
R_d	Rossby radius of deformation
R_0	Rossby number
RE	relative error
T	temperature
TS	threat score
u,v	eastward and northward wind components
$\mathbf{W}$	weight matrix
W	vertical wind component, optimal weight
x,y	horizontal coordinates
δ_{ij}	Kronecker delta
ε_a	analysis error
ε_b	background error
η	absolute vorticity
Φ	geopotential height
φ	geopotential, latitude
λ	longitude
λ_i	global Lyapunov exponent
ρ_{ij}	element *i,j* of the correlation matrix C
σ	standard deviation
σ^2	variance
ψ	streamfunction
ω	vertical velocity in pressure coordinates, spectral frequency
ζ	relative vorticity

1

Historical overview of numerical weather prediction

1.1 Introduction

In general, the public is not aware that our daily weather forecasts start out as initial-value problems on the major national weather services supercomputers. Numerical weather prediction provides the basic guidance for weather forecasting beyond the first few hours. For example, in the USA, computer weather forecasts issued by the National Center for Environmental Prediction (NCEP) in Washington, DC, guide forecasts from the US National Weather Service (NWS). NCEP forecasts are performed by running (integrating in time) computer models of the atmosphere that can simulate, given one day's weather observations, the evolution of the atmosphere in the next few days.[1] Because the time integration of an atmospheric model is an *initial-value problem*, the ability to make a skillful forecast requires both that *the computer model be a realistic representation of the atmosphere*, and that *the initial conditions be known accurately*.

NCEP (formerly the National Meteorological Center or NMC) has performed operational computer weather forecasts since the 1950s. From 1955 to 1973, the forecasts included only the Northern Hemisphere; they have been global since 1973. Over the years, the quality of the models and methods for using atmospheric observations has improved continuously, resulting in major forecast improvements.

1 In this book we will provide many examples mostly drawn from the US operational numerical center (NCEP), because of the availability of long records, and because the author's experience in this center facilitates obtaining such examples. However, these operational NCEP examples are only given for illustration purposes, and are simply representative of the evolution of operational weather forecasting in all major operational centers.

Figure 1.1.1(a) shows the longest available record of the skill of numerical weather prediction. The "$S1$" score (Teweles and Wobus, 1954) measures the relative error in the horizontal gradient of the height of the constant pressure surface of 500 hPa (in the middle of the atmosphere, since the surface pressure is about 1000 hPa) for 36-h forecasts over North America. Empirical experience at NMC indicated that a score of 70% or more corresponds to a useless forecast, and a score of 20% or less corresponds to an essentially perfect forecast. This was found from the fact that 20% was the average $S1$ score obtained when comparing analyses hand-made by several experienced forecasters fitting the same observations over the data-rich North American region.

Figure 1.1.1(a) shows that current 36-h 500-hPa forecasts over North America are close to what was considered essentially "perfect" 40 years ago: the computer forecasts are able to locate generally very well the position and intensity of the large-scale atmospheric waves, major centers of high and low pressure that determine the general evolution of the weather in the 36-h forecast. The sea level pressure forecasts contain smaller-scale atmospheric structures, such as fronts, mesoscale convective systems that dominate summer precipitation, etc., and are still difficult to forecast in detail (although their prediction has also improved very significantly over the years) so their $S1$ score is still well above 20% (Fig. 1.1.1(b)). Fig. 1.1.1(a) also shows that the 72-h forecasts of today are as accurate as the 36-h forecasts were 10–20 years ago. This doubling (or better) of skill in the forecasts is observed for other forecast variables, such as precipitation. Similarly, 5-day forecasts, which had no useful skill 15 years ago, are now moderately skillful, and during the winter of 1997–8, ensemble forecasts for the second week average showed useful skill (defined as anomaly correlation close to 60% or higher).

The improvement in skill of numerical weather prediction over the last 40 years apparent in Fig.1.1.1 is due to four factors:

- the increased power of supercomputers, allowing much finer numerical resolution and fewer approximations in the operational atmospheric models;
- the improved representation of small-scale physical processes (clouds, precipitation, turbulent transfers of heat, moisture, momentum, and radiation) within the models;
- the use of more accurate methods of data assimilation, which result in improved initial conditions for the models; and
- the increased availability of data, especially satellite and aircraft data over the oceans and the Southern Hemisphere.

In the USA, research on numerical weather prediction takes place in the national laboratories of the National Oceanic and Atmospheric Administration (NOAA), the National Aeronautics and Space Administration (NASA) and the National Center for Atmospheric Research (NCAR), and in universities and centers such as the

NCEP operational S1 scores at 36 and 72 hr over North America (500 hPa)

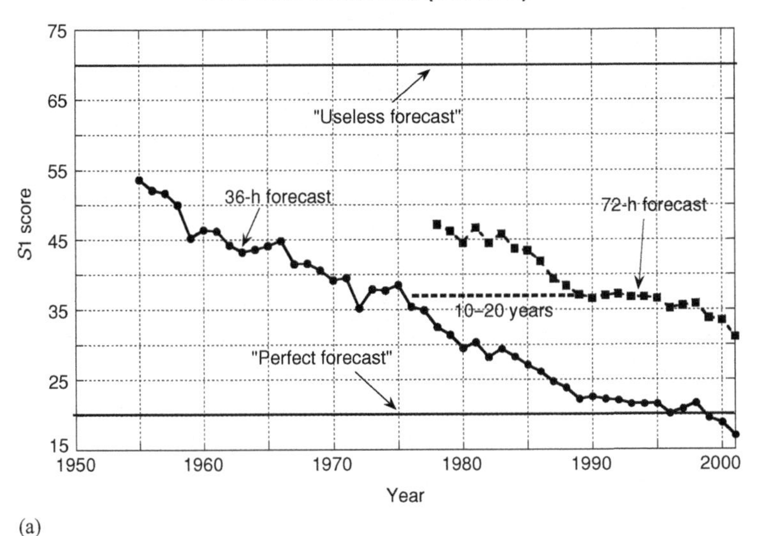

(a)

NCEP operational models S1 scores: Mean Sea Level Pressure over North America

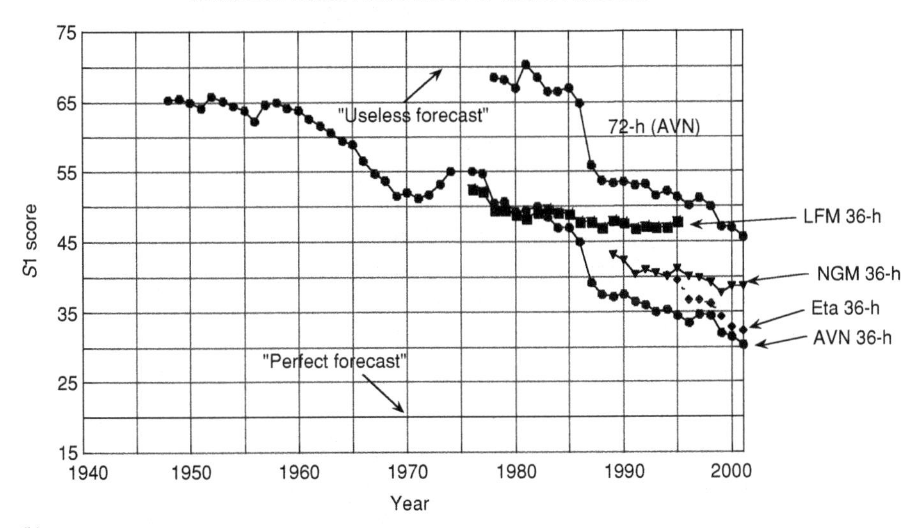

(b)

Figure 1.1.1: (a) Historic evolution of the operational forecast skill of the NCEP (formerly NMC) models over North America (500 hPa). The $S1$ score measures the relative error in the horizontal pressure gradient, averaged over the region of interest. The values $S1 = 70\%$ and $S1 = 20\%$ were empirically determined to correspond respectively to a "useless" and a "perfect" forecast when the score was designed. Note that the 72-h forecasts are currently as skillful as the 36-h were 10–20 years ago (data courtesy C.Vlcek, NCEP). (b) Same as (a) but showing $S1$ scores for sea level pressure forecasts over North America (data courtesy C.Vlcek, NCEP). It shows results from global (AVN) and regional (LFM, NGM and Eta) forecasts. The LFM model development was "frozen" in 1986 and the NGM was frozen in 1991.

Center for Prediction of Storms (CAPS). Internationally, major research takes place in large operational national and international centers (such as the European Center for Medium Range Weather Forecasts (ECMWF), NCEP, and the weather services of the UK, France, Germany, Scandinavian and other European countries, Canada, Japan, Australia, and others). In meteorology there has been a long tradition of sharing both data and research improvements, with the result that progress in the science of forecasting has taken place on many fronts, and all countries have benefited from this progress.

In this introductory chapter, we give an overview of the major components and milestones in numerical forecasting. They will be discussed in detail in the following chapters.

1.2 Early developments

Jule G. Charney (1917–1981) was one of the giants in the history of numerical weather prediction. In his 1951 paper "Dynamical forecasting by numerical process", he introduced the subject of this book as well as it could be introduced today. We reproduce here parts of the paper (with emphasis added):

> As meteorologists have long known, *the atmosphere exhibits no periodicities of the kind that enable one to predict the weather in the same way one predicts the tides*. No simple set of causal relationships can be found which relate the state of the atmosphere at one instant of time to its state at another. It was this realization that *led V. Bjerknes (1904) to define the problem of prognosis as nothing less than the integration of the equations of motion of the atmosphere.*[2] But it remained for *Richardson (1922)* to suggest the practical means for the solution of this problem. *He proposed to integrate the equations of motion numerically and showed exactly how this might be done. That the actual forecast used to test his method was unsuccessful was in no way a measure of the value of his work.* In retrospect it

2 The importance of the Bjerknes (1904) paper is clearly described by Thompson (1990), another pioneer of NWP, and the author of a very inspiring text on NWP (Thompson, 1961a). His paper "Charney and the revival of NWP" contains extremely interesting material on the history of NWP as well as on early computers:

> It was not until 1904 that Vilhelm Bjerknes – in a remarkable manifesto and testament of deterministic faith – stated the central problem of NWP. This was the first explicit, coherent recognition that the future state of the atmosphere is, *in principle*, completely determined by its detailed initial state and known boundary conditions, together with Newton's equations of motion, the Boyle–Charles–Dalton equation of state, the equation of mass continuity, and the thermodynamic energy equation. Bjerknes went further: he outlined an ambitious, but logical program of observation, graphical analysis of meteorological data and graphical solution of the governing equations. He succeeded in persuading the Norwegians to support an expanded network of surface observation stations, founded the famous Bergen School of synoptic and dynamic meteorology, and ushered in the famous polar front theory of cyclone formation. Beyond providing a clear goal and a sound physical approach to dynamical weather prediction, V. Bjerknes instilled his ideas in the minds of his students and their students in Bergen and in Oslo, three of whom were later to write important chapters in the development of NWP in the US (Rossby, Eliassen and Fjörtoft).

> becomes obvious that the inadequacies of observation alone would have doomed any attempt, however well conceived, a circumstance of which Richardson was aware. The real value of his work lay in the fact that it crystallized once and for all the essential problems that would have to be faced by future workers in the field and it laid down a thorough groundwork for their solution.
>
> For a long time no one ventured to follow in Richardson's footsteps. The paucity of the observational network and the enormity of the computational task stood as apparently insurmountable barriers to the realization of his dream that one day it might be possible to advance the computation faster than the weather. But with the increase in the density and extent of the surface and upper-air observational network on the one hand, and the development of large-capacity high-speed computing machines on the other, interest has revived in Richardson's problem, and attempts have been made to attack it anew.
>
> These efforts have been characterized by a devotion to objectives more limited than Richardson's. Instead of attempting to deal with the atmosphere in all its complexity, one tries to be satisfied with *simplified models* approximating the actual motions to a greater or lesser degree. By *starting with models incorporating only what it is thought to be the most important of the atmospheric influences*, and by gradually bringing in others, one is able to proceed inductively and thereby to avoid the pitfalls inevitably encountered when a great many poorly understood factors are introduced all at once.
>
> A necessary condition for the success of this stepwise method is, of course, that the first approximations bear a recognizable resemblance to the actual motions. Fortunately, the science of meteorology has progressed to the point where one feels that at least the main factors governing the large-scale atmospheric motions are well known. *Thus integrations of even the linearized barotropic and thermally inactive baroclinic equations have yielded solutions bearing a marked resemblance to reality.* At any rate, it seems clear that the models embodying the collective experience and the positive skill of the forecast cannot fail utterly. This conviction has served as the guiding principle in the work of the meteorology project at The Institute for Advanced Study [at Princeton University] with which the writer has been connected.

As indicated by Charney, Richardson performed a remarkably comprehensive numerical integration of the full primitive equations of motion (Chapter 2). He used a horizontal grid of about 200 km, and four vertical layers of approximately 200 hPa, centered over Germany. Using the observations at 7 UTC (Universal Coordinate Time) on 20 May 1910, he computed the time derivative of the pressure in central Germany between 4 and 10 UTC. *The predicted 6-h change was 146 hPa, whereas in reality there was essentially no change observed in the surface pressure.* This huge error was discouraging, but it was due mostly to the fact that the initial conditions were *not balanced*, and therefore included fast-moving gravity waves which masked the *initial rate of change* of the meteorological signal in the forecast (Fig. 1.2.1). Moreover, if the integration had been continued, it would have suffered "computational blow-up" due to the violation of the Courant–Friedricks–Lewy (CFL) condition

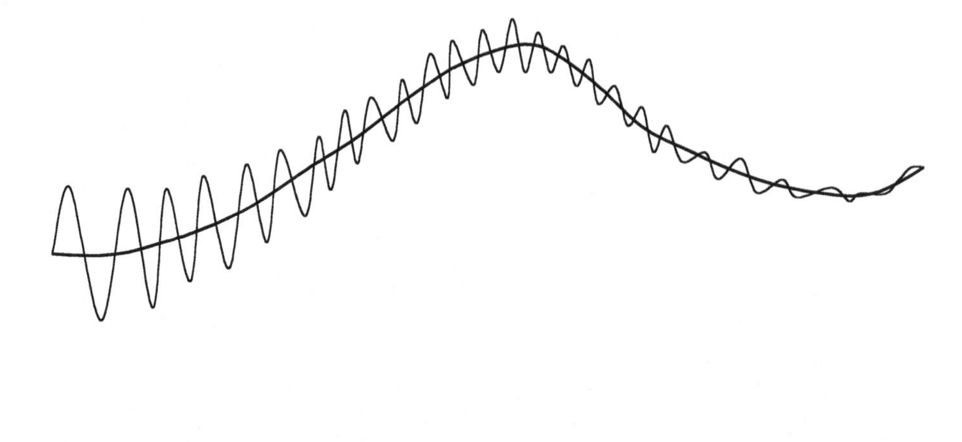

Figure 1.2.1: Schematic of a forecast with slowly varying weather-related variations and superimposed high-frequency gravity waves. Note that even though the forecast of the slow waves is essentially unaffected by the presence of gravity waves, the initial time derivative is much larger in magnitude, as obtained in the Richardson (1922) experiment.

(Chapter 3) which requires that the time step should be smaller than the grid size divided by the speed of the fastest traveling signal (in this case horizontally moving sound waves, traveling at about 300 m/s).

Charney (1948, 1949) and Eliassen (1949) solved both of these problems by deriving "filtered" equations of motion, based on quasi-geostrophic (slowly varying) balance, which filtered out (i.e., did not include) gravity and sound waves, and were based on pressure fields alone. Charney points out that this approach was justified by the fact that forecasters' experience was that they were able to predict tomorrow's weather from pressure charts alone:

> In the selection of a suitable first approximation, Richardson's discovery that the horizontal divergence was an unmeasurable quantity had to be taken into account. Here a consideration of forecasting practice gave rise to the belief that this difficulty could be surmounted: forecasts were made by means of geostrophic reasoning from the pressure field alone – forecasts in which the concept of horizontal divergence played no role.

In order to understand better Charney's comment, we quote an anecdote from Lorenz (1990) on his interactions with Jule Charney:

> On another[3] occasion when our conversations had turned closer to scientific matters, Jule was talking again about the early days of NWP. For a proper

3 The previous occasion was a story about an invitation Charney received to appear on the "Today" show, to talk about how computers were going to forecast the weather. Since the show was at 7 am, Charney, a late riser, had never watched it. "He told us that he felt that he ought to see the show at least once before agreeing to appear on it, and so, one morning, he managed to pull himself out of bed and turn on the TV set, and the first person he saw was a chimpanzee.

> perspective, we should recall that at the time when Charney was a student, pressure was king. The centers of weather activity were acknowledged to be the highs and lows. A good prognostic chart was one that had the isobars in the right locations. Naturally, then, the thing that was responsible for the weather changes was the thing that made the pressure change. This was readily shown to be the divergence of the wind field. The divergence could not be very accurately measured, and a corollary deduced by some meteorologists, including some of Charney's advisors, was that the dynamic equations could not be used to forecast the weather.
>
> Such reasoning simply did not make sense to Jule. The idea that the wind field might serve instead of the pressure field as a basis for dynamical forecasting, proposed by Rossby, gave Jule a route to follow.[4] He told us, however, that what really inspired him to develop the equations that later became the basis for NWP was a determination to prove, to those who had assured him that the task was impossible, that they were wrong.

Charney, R. Fjørtoft, and J. von Neuman (1950) computed a historic first one-day weather forecast using a barotropic (one-layer) filtered model. The work took place in 1948–9. They used one of the first electronic computers (the Electronic Numerical Integrator and Computer, ENIAC), housed at the Aberdeen Proving Grounds of the US Army in Maryland. It incorporated von Neuman's idea of "stored programming" (i.e., the ability to perform arithmetic operations over different operands (loops) without having to repeat the code). The results of the first forecasts were quite encouraging: Fig. 1.2.2, reproduced from Charney (1951) shows the 24-h forecast and verification for 30 January 1949. Unlike Richardson's results, the forecast remains meteorological, and there is a pattern correlation between the predicted and the observed pressure field 24-h changes.

It is remarkable that in his 1951 paper, just after the triumph of performing the first successful forecasts with filtered models, Charney already saw that much more progress would come from the use of the primitive (unfiltered) equations of motion as Richardson had originally attempted:

> The discussion so far has dealt exclusively with the quasi-geostrophic equations as the basis for numerical forecasting. Yet there has been no intention to exclude the possibility that the primitive Eulerian equations can also be used for this purpose. *The outlook for numerical forecasting would be indeed dismal if the quasi-geostrophic approximation represented the upper limit of attainable accuracy, for it is known that it applies only indifferently, if at all, to many of the small-scale but meteorologically significant motions.* We have merely indicated two obstacles that stand in the way of the applications of the primitive equations:

He decided he could never compete with a chimpanzee for the public's favor, and so he gracefully declined to appear, much to the dismay of the computer company that had engineered the invitation in the first place" (Lorenz, 1990).

4 The development of the "Rossby waves" phase speed equation $c = U - \beta L^2/\pi^2$ based on the linearized, non-divergent vorticity equation (Rossby *et al.*, 1939, Rossby, 1940), and its success in predicting the motion of the large-scale atmospheric waves, was an essential stimulus to Charney's development of the filtered equations (Phillips, 1990b, 1998).

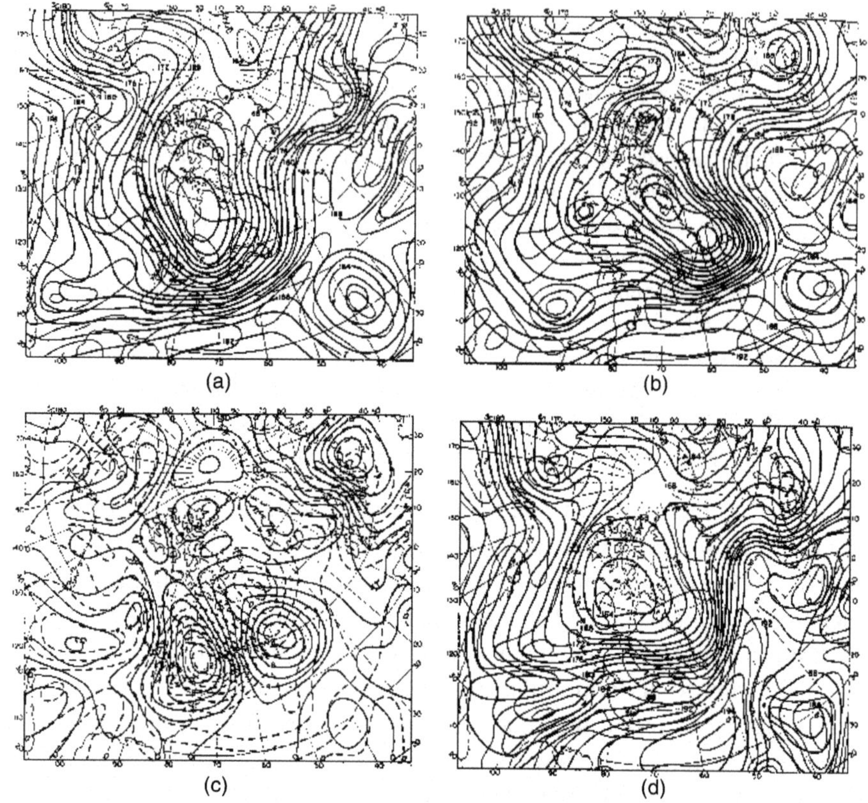

Figure 1.2.2: Forecast of 30 January 1949, 0300 GMT: (a) contours of observed z and $\zeta + f$ at $t = 0$; (b) observed z and $\zeta + f$ at $t = 24$ h; (c) observed (continuous lines) and computed (broken lines) 24-h height change; (d) computed z and $\zeta + f$ at $t = 24$ h. The height unit is 100 ft and the unit of vorticity is $1/3 \times 10^{-4}$ s^{-1}. (Reproduced from the *Compendium of Meteorology*, with permission of the American Meteorological Society.)

> First, there is the difficulty raised by Richardson that *the horizontal divergence cannot be measured with sufficient accuracy. Moreover, the horizontal divergence is only one of a class of meteorological unobservables which also includes the horizontal acceleration. And second, if the primitive Eulerian equations are employed, a stringent and seemingly artificial bound is imposed on the size of the time interval for the finite difference equations. The first obstacle is the most formidable, for the second only means that the integration must proceed in steps of the order of fifteen minutes rather than two hours.* Yet the first does not seem insurmountable, as the following considerations will indicate.

He proceeded to describe an unpublished study in which he and J.C. Freeman integrated barotropic primitive equations (i.e., shallow water equations, Chapter 2) which include not only the slowly varying quasi-geostrophic solution, but also fast gravity waves. They initialized the forecast assuming zero initial divergence, and compared the result with a barotropic forecast (with gravity waves filtered out). The results were similar to those shown schematically in Fig. 1.2.1: they observed that over a day or so the gravity waves subsided (through a process that we call geostrophic adjustment) and did not otherwise affect the forecast of the slow waves. From this result Charney concluded that numerical forecasting could indeed use the full primitive equations (as eventually happened in operational practice). He listed in the paper the complete primitive equations in pressure coordinates, essentially as they are used in current operational weather prediction, but without heating (nonadiabatic) and frictional terms, which he expected to have minor effects in one- or two-day forecasts. Charney concluded this remarkable paper with the following discussion, which includes a list of the physical processes that take place at scales too small to be resolved, and are incorporated in present models through "parameterizations of the subgrid-scale physics" (condensation, radiation, and turbulent fluxes of heat, momentum and moisture, Chapter 4):

> *Nonadiabatic and frictional terms have been ignored in the body of the discussion because it was thought that one should first seek to determine how much of the motion could be explained without them. Ultimately they will have to be taken into account, particularly if the forecast period is to be extended to three or more days.*
>
> Condensational phenomena appear to be the simplest to introduce: one has only to add the equation of continuity for water vapor and to replace the dry by the moist adiabatic equation. Long-wave radiational effects can also be provided for, since our knowledge of the absorptive properties of water vapor and carbon dioxide has progressed to a point where quantitative estimates of radiational cooling can be made, although the presence of clouds will complicate the problem considerably.
>
> The most difficult phenomena to include have to do with the turbulent transfer of momentum and heat. A great deal of research remains to be done before enough is known about these effects to permit the assignment of even rough values to the eddy coefficients of viscosity and heat conduction. Owing to their statistically indeterminate nature, the turbulent properties of the atmosphere *place an upper limit to the accuracy obtainable by dynamical methods of forecasting*, beyond which we shall have to rely upon statistical methods. But it seems certain that much progress can be made before these limits can be reached.

This paper, which although written in 1951 has not become dated, predicted with almost supernatural vision the path that numerical weather forecasting was to follow over the next five decades. It described the need for objective analysis of meteorological data in order to replace the laborious hand analyses. We now refer to this process as data assimilation (Chapter 5), which uses both observations and short forecasts to estimate initial conditions. Note that at a time at which only one-day forecasts had ever been attempted, Charney already had the intuition that there was an *upper limit* to weather predictability, which Lorenz (1965) later estimated to be about two weeks. However, Charney attributed the expected limit to model deficiencies (such as the parameterization of turbulent processes), rather than to the chaotic nature of the atmosphere, which imposes a limit of predictability even if the model is perfect

(Lorenz, 1963b; Chapter 6). Charney was right in assuming that in practice model deficiencies, as well as errors in the initial conditions, would limit predictability. At the present time, however, the state of the art in numerical forecasting has advanced enough that, when the atmosphere is highly predictable, the theoretically estimated limit for weather forecasting (about two weeks) is occasionally reached and even exceeded through techniques such as ensemble forecasting (Chapter 6).

Following the success of Charney *et al.* (1950), Rossby moved back to Sweden, and was able to direct a group that reproduced similar experiments on a powerful Swedish computer known as BESK. As a result, the first operational (real time) numerical weather forecasts started in Sweden in September 1954, six months before the start-up of the US operational forecasts[5] (Döös and Eaton, 1957, Wiin-Nielsen, 1991, Bolin, 1999).

1.3 Primitive equations, global and regional models, and nonhydrostatic models

As envisioned by Charney (1951, 1962) the filtered (quasi-geostrophic) equations, although very useful for understanding of the large-scale extratropical dynamics of the atmosphere, were not accurate enough to allow continued progress in NWP, and were eventually replaced by primitive equation models (Chapter 2). The primitive equations are conservation laws applied to individual parcels of air: conservation of the three-dimensional momentum (equations of motion), conservation of energy (first law of thermodynamics), conservation of dry air mass (continuity equation), and equations for the conservation of moisture in all its phases, as well as the equation of state for perfect gases. They include in their solution fast gravity and sound waves, and therefore in their space and time discretization they require the use of smaller time steps, or alternative techniques that slow them down (Chapter 3). For models with a horizontal grid size larger than 10 km, it is customary to replace the vertical component of the equation of motion with its hydrostatic approximation, in which the vertical acceleration is considered negligible compared with the gravitational acceleration (buoyancy). With this approximation, it is convenient to use atmospheric pressure, instead of height, as a vertical coordinate.

The continuous equations of motions are solved by discretization in space and in time using, for example, finite differences (Chapter 3). It has been found that the accuracy of a model is very strongly influenced by the spatial resolution: in general, the higher the resolution, the more accurate the model. Increasing resolution, however, is extremely costly. For example, doubling the resolution in the three space dimensions also requires halving the time step in order to satisfy conditions for computational

5 Anders Persson (1999 personal communication) kindly provided the notes on the historical development of NWP in the USA and Sweden reproduced in Appendix A.

stability. Therefore, the computational cost of doubling the resolution is a factor of 2^4 (three space and one time dimensions). Modern methods of discretization attempt to make the increase in accuracy less onerous by the use of semi-implicit and semi-Lagrangian time schemes. These schemes (pioneered by Canadian scientists under the leadership of Andre Robert) have less stringent stability conditions on the time step, and more accurate space discretization. Nevertheless, there is a constant need for higher resolution in order to improve forecasts, and as a result running atmospheric models has always been a major application of the fastest supercomputers available.

When the "conservation" equations are discretized over a given grid size (typically from a few to several hundred kilometers) it is necessary to add "sources and sinks" terms due to small-scale physical processes that occur at scales that cannot be explicitly resolved by the models. As an example, the equation for water vapor conservation on pressure coordinates is typically written as

$$\frac{\partial \overline{q}}{\partial t} + \overline{u}\frac{\partial \overline{q}}{\partial x} + \overline{v}\frac{\partial \overline{q}}{\partial y} + \overline{\omega}\frac{\partial \overline{q}}{\partial p} = \overline{E} - \overline{C} + \frac{\partial \overline{\omega' q'}}{\partial p} \tag{1.3.1}$$

where q is the ratio between water vapor and dry air mass, x and y are horizontal coordinates with appropriate map projections, p is pressure, t is time, u and v are the horizontal air velocity (wind) components, $\omega = dp/dt$ is the vertical velocity in pressure coordinates, and the product of primed variables represents turbulent transports of moisture on scales unresolved by the grid used in the discretization, with the overbar indicating a spatial average over the grid of the model. It is customary to call the left-hand side of the equation, the "dynamics" of the model, which is computed explicitly (Chapter 3).

The right-hand side represents the so-called "physics" of the model. For the moisture equation, it includes the effects of physical processes such as evaporation and condensation $\overline{E} - \overline{C}$, and turbulent transfers of moisture which take place at small scales that cannot be explicitly resolved by the "dynamics". These *subgrid-scale physical processes*, which are sources and sinks for the equations, are then "parameterized" in terms of the variables explicitly represented in the atmospheric dynamics (Chapter 4).

Two types of models are in use for NWP: global and regional models (Chapter 5). Global models are generally used for guidance in medium-range forecasts (more than 2 d), and for climate simulations. At NCEP, for example, the global models are run through 16 d every day. Because the horizontal domain of global models is the whole earth, they usually cannot be run at high resolution. For more detailed forecasts it is necessary to increase the resolution, and this can only be done over limited regions of interest.

Regional models are used for shorter-range forecasts (typically 1–3 d), and are run with a resolution two or more times higher than global models. For example, the NCEP global model in 1997 was run with 28 vertical levels, and a horizontal resolution of 100 km for the first week, and 200 km for the second week. The regional

(Eta) model was run with a horizontal resolution of 48 km and 38 levels, and later in the day with 29 km and 50 levels. Because of their higher resolution, regional models have the advantage of higher accuracy and the ability to reproduce smaller-scale phenomena such as fronts, squall lines, and much better orographic forcing than global models. On the other hand, regional models have the disadvantage that, unlike global models, they are not "self-contained" because they require lateral boundary conditions at the borders of the horizontal domain. These boundary conditions must be as accurate as possible, because otherwise the interior solution of the regional models quickly deteriorates. Therefore it is customary to "nest" the regional models within another model with coarser resolution, whose forecast provides the boundary conditions. For this reason, regional models are used only for short-range forecasts. After a certain period, which is proportional to the size of the model, the information contained in the high-resolution initial conditions is "swept away" by the influence of the boundary conditions, and the regional model becomes merely a "magnifying glass" for the coarser model forecast in the regional domain. This can still be useful, for example, in climate simulations performed for long periods (seasons to multiyears), and which therefore tend to be run at coarser resolution. A "regional climate model" can provide a more detailed version of the coarse climate simulation in a region of interest. Several other major NWP centers in Europe (United Kingdom (http://www.met-office.gov.uk/), France (http://www.meteo.fr/), Germany (http://www.dwd.de/)), Japan (http://www.kishou.go.jp/), Australia (http://www.bom.gov.au/nmoc/ab_nmc_op.shtml), and Canada (http://www.ec.gc.ca/) also have similar global and regional models, whose details can be obtained at their web sites.

More recently the resolution of some regional models has been increased to just a few kilometers in order to resolve better storm-scale phenomena. Storm-resolving models such as the Advanced Regional Prediction System (ARPS) cannot use the hydrostatic approximation which ceases to be accurate for horizontal scales of the order of 10 km or smaller. Several major nonhydrostatic models have been developed and are routinely used for mesoscale forecasting. In the USA the most widely used are the ARPS, the MM5 (Penn State/NCAR Mesoscale Model, Version 5), the RSM (NCEP Regional Spectral Model) and the COAMPS (US Navy's Coupled Ocean/Atmosphere Mesoscale Prediction System). There is a tendency towards the use of nonhydrostatic models that can be used globally as well.

1.4 Data assimilation: determination of the initial conditions for the computer forecasts

As indicated previously, NWP is an initial-value problem: given an estimate of the present state of the atmosphere, the model simulates (forecasts) its evolution. The problem of determination of the initial conditions for a forecast model is very important and complex, and has become a science in itself (Daley, 1991). In this section we introduce methods that have been used for this purpose (successive corrections method or SCM, optimal interpolation or OI, variational methods in three and four dimensions, 3D-Var and 4D-Var, and Kalman filtering or KF). We discuss this subject in more detail in Chapter 5, and refer the reader to Daley (1991) as a much more comprehensive text on atmospheric data analysis.

In the early experiments, Richardson (1922) and Charney *et al.* (1950) performed hand interpolations of the available observations to grid points, and these fields of initial conditions were manually digitized, which was a very time consuming procedure. The need for an automatic "objective analysis" quickly became apparent (Charney, 1951), and interpolation methods fitting data to grids were developed (e.g., Panofsky, 1949, Gilchrist and Cressman, 1954, Barnes, 1964, 1978). However, there is an even more important problem than spatial interpolation of observations to gridded fields: the data available are not enough to initialize current models. Modern primitive equations models have a number of degrees of freedom of the order of 10^7. For example, a latitude–longitude model with a typical resolution of 1° and 20 vertical levels would have $360 \times 180 \times 20 = 1.3 \times 10^6$ grid points. At each grid point we have to carry the values of at least four prognostic variables (two horizontal wind components, temperature, moisture), and the surface pressure for each column, giving over 5 million variables that need to be given an initial value. For any given time window of ± 3 hours, there are typically 10–100 thousand observations of the atmosphere, two orders of magnitude less than the number of degrees of freedom of the model. Moreover, their distribution in space and time is very nonuniform (Fig. 1.4.1), with regions like North America and Eurasia which are relatively data-rich, while others much more poorly observed.

For this reason, it became obvious rather early that it was necessary to use additional information (denoted *background*, *first guess* or *prior information*) to prepare initial conditions for the forecasts (Bergthorsson and Döös, 1955). Initially climatology was used as a first guess (e.g., Gandin, 1963), but as the forecasts became better, a short-range forecast was chosen as the first guess in the operational data assimilation systems or "analysis cycles". The intermittent data assimilation cycle shown schematically in Fig. 1.4.2 is continued in present-day operational systems, which typically use a 6-h cycle performed four times a day.

In the 6-h data assimilation cycle for a global model, the background field is a model 6-h forecast x^b (a three-dimensional array). To obtain the background or first guess "observations", the model forecast is interpolated to the observation location, and if they are different, converted from model variables to observed variables y^o (such as satellite radiances or radar reflectivities). The first guess of the observations is therefore $H(x^b)$, where H is the observation operator that performs the necessary interpolation and transformation from model variables to observation space. The difference between the observations and the model first guess $y^o - H(x^b)$ is denoted "observational increments" or "innovations". The analysis x^a is obtained by

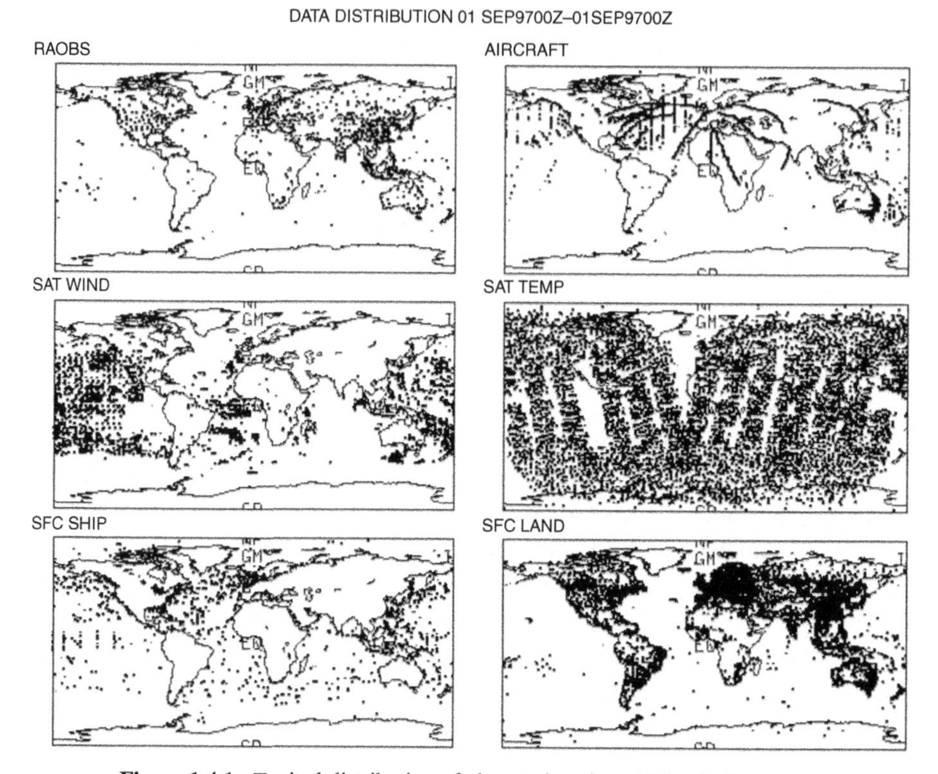

Figure 1.4.1: Typical distribution of observations in a ±3-h window.

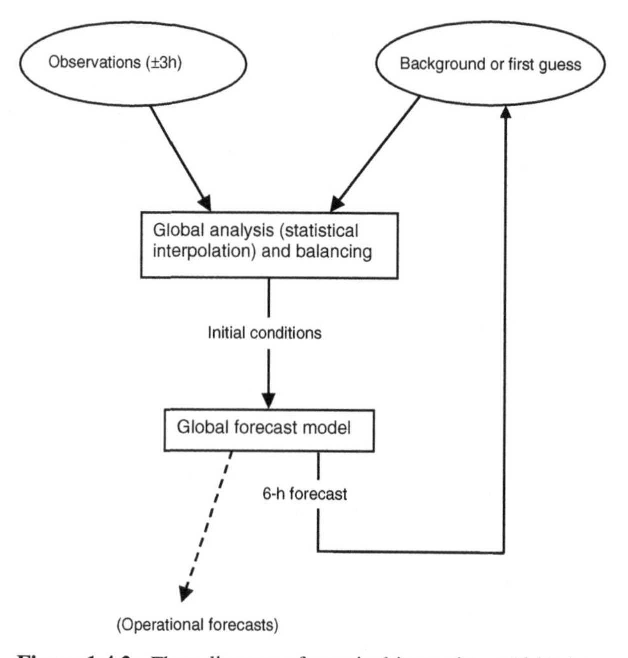

Figure 1.4.2: Flow diagram of a typical intermittent (6-h) data assimilation cycle.

adding the innovations to the model forecast (first guess) with weights W that are determined based on the estimated statistical error covariances of the forecast and the observations:

$$\mathbf{x}^a = \mathbf{x}^b + \mathbf{W}[\mathbf{y}^o - \mathbf{H}(\mathbf{x}^b)] \tag{1.4.1}$$

Different analysis schemes (SCM, OI, 3D-Var, and KF) are based on (1.4.1) but differ by the approach taken to combine the background and the observations to produce the analysis. Earlier methods such as the SCM (Bergthorsson and Döös, 1955, Cressman, 1959, Barnes, 1964) were of a form similar to (1.4.1), with weights determined empirically. The weights are a function of the distance between the observation and the grid point, and the analysis is iterated several times. In OI (Gandin, 1963) the matrix of weights W is determined from the minimization of the analysis errors at each grid point. In the 3D-Var approach one defines a cost function proportional to the square of the distance between the analysis and both the background and the observations (Sasaki, 1970). The cost function is minimized directly to obtain the analysis. Lorenc (1986) showed that OI and the 3D-Var approach are equivalent if the cost function is defined as:

$$J = \frac{1}{2}\{[\mathbf{y}^o - H(\mathbf{x})]^T R^{-1}[\mathbf{y}^o - H(\mathbf{x})] + (\mathbf{x} - \mathbf{x}^b)^T B^{-1}(\mathbf{x} - \mathbf{x}^b)\} \tag{1.4.2}$$

The cost function J in (1.4.2) measures the distance of a field x to the observations (the first term in the cost function) and the distance to the first guess or background x^b (the second term in the cost function). The distances are scaled by the observation error covariance R and by the background error covariance B respectively. The minimum of the cost function is obtained for $x = x^a$, which is defined as the "analysis". The analysis obtained in (1.4.1) and (1.4.2) is the same if the weight matrix in (1.4.1) is given by

$$\mathbf{W} = \mathbf{B}\mathbf{H}^T(\mathbf{H}\mathbf{B}\mathbf{H}^T + \mathbf{R}^{-1})^{-1} \tag{1.4.3}$$

The difference between OI (1.4.1) and the 3D-Var approach (1.3) is in the method of solution: in OI, the weights W are obtained for each grid point or grid volume, using suitable simplifications. In 3D-Var, the minimization of (1.4.2) is performed directly, allowing for additional flexibility and a simultaneous global use of the data (Chapter 5).

More recently, the variational approach has been extended to four dimensions, by including within the cost function the distance to observations over a time interval (assimilation window). A first version of this considerably more expensive method was implemented at ECMWF at the end of 1997 (Bouttier and Rabier, 1997). Research on the even more advanced and computationally expensive KF (e.g., Ghil *et al.*, 1981), and ensemble KF (Evensen, 1994, Houtekamer and Mitchell, 1998) is discussed in Chapter 5. That chapter also includes a discussion about the problem of enforcing a balance in the analysis so that the presence of gravity waves does not mask the meteorological signal, as happened to Richardson (1922) (Fig. 1.2.1). The method used for many years to solve this "initialization" problem was "nonlinear normal mode initialization" (Machenhauer, 1977, Baer and Tribbia, 1977). The balance in the initial conditions is usually obtained by either adding a constraint to the cost function (1.4.2) (Parrish and Derber, 1992), or through the use of a digital filter (Lynch and Huang, 1992, Chapter 5).

In the analysis cycle, no matter which analysis scheme is employed, the use of the model forecast is essential in achieving "four-dimensional data assimilation" (4DDA). This means that the data assimilation cycle is like a long model integration, in which the model is "nudged" by the observational increments in such a way that it remains close to the real atmosphere. The importance of the model cannot be overemphasized: it transports information from data-rich to data-poor regions, and it provides a complete estimation of the four-dimensional state of the atmosphere. Figure 1.4.3 presents the rms difference between the 6-h forecast (used as a first guess) and the rawinsonde observations from 1978 to the present (in other words, the rms of the observational increments for 500-hPa heights). It should be noted that the rms differences are not necessarily forecast errors, since the observations also contain errors. In the Northern Hemisphere the rms differences have been halved from about 30 m in the late 1970s, to about 13 m in 2000, equivalent to a mean temperature error of about 0.65 K, similar to rawinsonde observational errors. In the Southern Hemisphere the improvements are even larger, with the differences decreasing from about 47 m to about 12 m. The improvements in these short-range forecasts are a

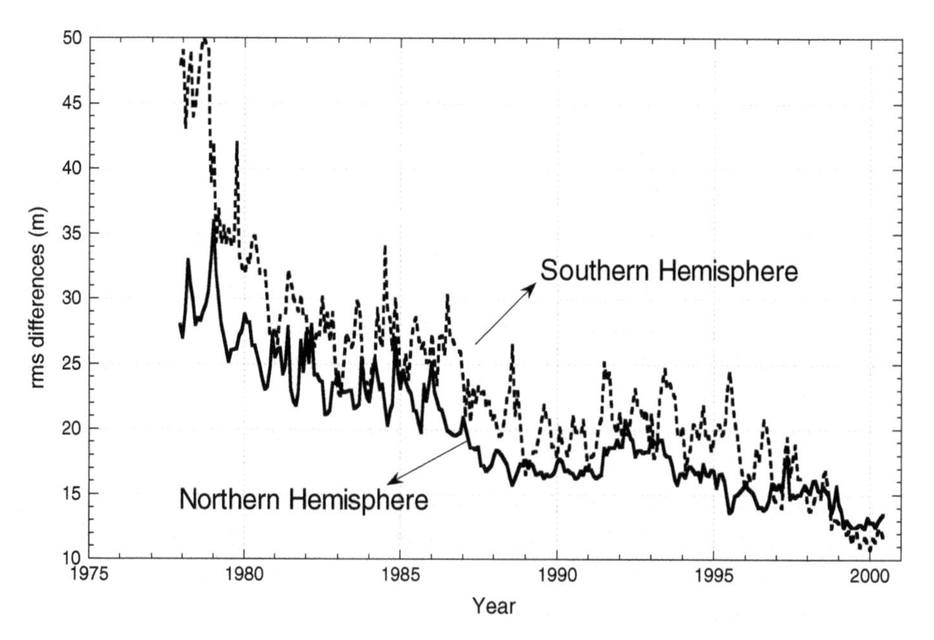

Figure 1.4.3: Rms observational increments (differences between 6-h forecast and rawinsonde observations) for 500-hPa heights (data courtesy of Steve Lilly, NCEP).

reflection of improvements in the model, the analysis scheme used to assimilate the data, and the quality and quality control of the data (Chapter 5).

1.5 Operational NWP and the evolution of forecast skill

Major milestones of operational numerical weather forecasting include the paper by Charney *et al.* (1950) with the first successful forecast based on the primitive equations, and the first operational forecasts performed in Sweden in September 1954, followed 6 months later by the first operational (real time) forecasts in the USA. We describe in what follows the evolution of NWP at NCEP, but as mentioned before, similar developments took place at several major operational NWP centers: in the UK, France, Germany, Japan, Australia and Canada.

The history of operational NWP at the NMC (now NCEP) has been reviewed by Shuman (1989) and Kalnay *et al.* (1998). It started with the organization of the Joint Numerical Weather Prediction Unit (JNWPU) on 1 July 1954, staffed by members of the US Weather Bureau (later the National Weather Service, NWS), the Air Weather Service of the US Air Force, and the Naval Weather Service.[6] Shuman pointed out that in the first few years, numerical predictions could *not* compete with those produced manually. They had several serious flaws, among them overprediction of cyclone development. Far too many cyclones were predicted to deepen into storms. With time, and with the joint work of modelers and practising synopticians, major sources of model errors were identified, and operational NWP became the central guidance for operational weather forecasts.

Shuman (1989) included a chart with the evolution of the $S1$ score (Teweles and Wobus, 1954), the first measure of error in a forecast weather chart which, according to Shuman (1989), was designed, tested, and modified to correlate well with expert forecasters' opinions on the quality of a forecast. The $S1$ score measures the average relative error in the pressure gradient (compared to a verifying analysis chart). Experiments comparing two independent subjective analyses of the same data-rich North American region made by two experienced analysts suggested that a "perfect" forecast would have an $S1$ score of about 20%. It was also found empirically that forecasts with an $S1$ score of 70% or more were useless as synoptic guidance.

Shuman pointed out some of the major system improvements that enabled NWP forecasts to overtake and surpass subjective forecasts. The first major improvement took place in 1958 with the implementation of a barotropic (one-level) model, which was actually a reduction from the three-level model first tried, but which included better finite differences and initial conditions derived from an objective analysis scheme (Bergthorsson and Döös, 1955, Cressman, 1959). It also extended the domain of the

6 In 1960 the JNWPU reverted to three separate organizations: the National Meteorological Center (National Weather Service), the Global Weather Central (US Air Force) and the Fleet Numerical Oceanography Center (US Navy).

model to an octagonal grid covering the Northern Hemisphere down to 9–15° N. These changes resulted in numerical forecasts that for the first time were competitive with subjective forecasts, but in order to implement them JNWPU had to wait for the acquisition of a more powerful supercomputer, an IBM 704, to replace the previous IBM 701. This pattern of forecast improvements which depend on a combination of the better use of the data and better models, and would require more powerful supercomputers in order to be executed in a timely manner has been repeated throughout the history of operational NWP. Table 1.5.1 (adapted from Shuman (1989)) summarizes the major improvements in the first 30 years of operational numerical forecasts at the NWS. The first primitive equations model (Shuman and Hovermale, 1968) was implemented in 1966. The first regional system (Limited Fine Mesh or LFM model, Howcroft, 1971) was implemented in 1971. It was remarkable because it remained in use for over 20 years, and it was the basis for Model Output Statistics (MOS). Its development was frozen in 1986. A more advanced model and data assimilation system, the Regional Analysis and Forecasting System (RAFS) was implemented as the main guidance for North America in 1982. The RAFS was based on the multiple Nested Grid Model (NGM, Phillips, 1979) and on a regional OI scheme (DiMego, 1988). The global spectral model (Sela, 1980) was implemented in 1980.

Table 1.5.2 (from Kalnay *et al.*, 1998 and P. Caplan, personal communication, 2000) summarizes the major improvements implemented in the global system starting

Table 1.5.1. *Major operational implementations and computer acquisitions at NMC between 1955 and 1985 (adapted from Shuman, 1989)*

Year	Operational model	Computer
1955	Princeton three-level quasi-geostrophic model (Charney, 1954). Not used by the forecasters	IBM 701
1958	Barotropic model with improved numerics, objective analysis initial conditions, and octagonal domain.	IBM 704
1962	Three-level quasi-geostrophic model with improved numerics	IBM 7090 (1960) IBM 7094 (1963)
1966	Six-layer primitive equations model (Shuman and Hovermale, 1968)	CDC 6600
1971	LFM model (Howcroft, 1971) (first regional model at NMC)	
1974	Hough functions analysis (Flattery, 1971)	IBM 360/195
1978	Seven-layer primitive equation model (hemispheric)	
1978	OI (Bergman, 1979)	Cyber 205
Aug 1980	Global spectral model, R30/12 layers (Sela, 1980)	
March 1985	Regional Analysis and Forecast System based on the NGM (Phillips, 1979) and OI (DiMego, 1988)	

Table 1.5.2. *Major changes in the NMC/NCEP global model and data assimilation system since 1985 (adapted from Kalnay et al. 1998 and P. Caplan, pers. comm., 2000)*

Year	Operational model	Computer
April 1985	GFDL physics implemented on the global spectral model with silhouette orography, R40/18 layers	
Dec 1986	New OI code with new statistics	
1987		2nd Cyber 205
Aug 1987	Increased resolution to T80/18 layers, Penman–Montieth evapotranspiration and other improved physics (Caplan and White, 1989, Pan, 1990)	
Dec 1988	Implementation of hydrostatic complex quality control (CQC) (Gandin, 1988)	
1990		Cray YMP/8cpu/ 32 megawords
Mar 1991	Increased resolution to T126 L18 and improved physics, mean orography. (Kanamitsu *et al.*, 1991)	
June 1991	New 3D-Var (Parrish and Derber, 1992, Derber *et al.*, 1991)	
Nov 1991	Addition of increments, horizontal and vertical OI checks to the CQC (Collins and Gandin, 1990)	
7 Dec 1992	First ensemble system: one pair of bred forecasts at 00Z to 10 days, extension of AVN to 10 days (Toth and Kalnay, 1993, Tracton and Kalnay, 1993)	
Aug 1993	Simplified Arakawa–Schubert cumulus convection (Pan and Wu, 1995). Resolution T126/28 layers	
Jan 1994		Cray C90/16cpu/ 128 megawords
March 1994	Second ensemble system: five pairs of bred forecasts at 00Z, two pairs at 12Z, extension of AVN, a total of 17 global forecasts every day to 16 days	
10 Jan 1995	New soil hydrology (Pan and Mahrt, 1987), radiation, clouds, improved data assimilation. Reanalysis model	
25 Oct 1995	Direct assimilation of TOVS cloud-cleared radiances (Derber and Wu, 1998). New planetary boundary layer (PBL) based on nonlocal diffusion (Hong and Pan, 1996). Improved CQC	Cray C90/16cpu/ 256 megawords

Table 1.5.2. *(cont.)*

Year	Operational model	Computer
5 Nov 1997	New observational error statistics. Changes to assimilation of TOVS radiances and addition of other data sources	
13 Jan 1998	Assimilation of noncloud-cleared radiances (Derber *et al.*, pers.comm.). Improved physics.	
June 1998	Resolution increased to T170/40 layers (to 3.5 days). Improved physics. 3D ozone data assimilation and forecast. Nonlinear increments in 3D-Var. Resolution reduced to T62/28levels on Oct. 1998 and upgraded back in Jan. 2000	IBM SV2 256 processors
June 2000	Ensemble resolution increased to T126 for the first 60 h	
July 2000	Tropical cyclones relocated to observed position every 6 h	

in 1985 with the implementation of the first comprehensive package of physical parameterizations from GFDL (Geophysical Fluid Dynamics Laboratory). Other major improvements in the physical parameterizations were made in 1991, 1993, and 1995. The most important changes in the data assimilation were an improved OI formulation in 1986, the first operational 3D-Var in 1991, the replacement of the satellite retrievals of temperature with the direct assimilation of cloud-cleared radiances in 1995, and the use of "raw" (not cloud-cleared) radiances in 1998. The model resolution was increased in 1987, 1991, and 1998. The first operational ensemble system was implemented in 1992 and enlarged in 1994. The resolution of the ensembles was increased in 2000.

Table 1.5.3 contains a summary of the regional systems used for short-range forecasts (up to 48 h). The RAFS (triple nested NGM and OI) were implemented in 1985. The Eta model, designed with advanced finite differences, step-mountain coordinates, and physical parameterizations, was implemented in 1993, with the same 80-km horizontal resolution as the NGM. It was denoted "early" because of a short data cut-off. The resolution was increased to 48 km, and a first "mesoscale" version with 29 km and reduced coverage was implemented in 1995. A cloud prognostic scheme was implemented in 1995, and a new land-surface parameterization in 1996. The OI data assimilation was replaced by a 3D-Var in 1998, and at this time the early and meso-Eta models were unified into a 32-km/45-level version. Many other less significant changes were also introduced into the global and regional operational systems and are not listed here for the sake of brevity. The Rapid Update Cycle (RUC), which provides frequent updates of the analysis and very-short-range forecasts over

Table 1.5.3. *Major changes in the NMC/NCEP regional modeling and data assimilation since 1985 (from compilations by Fedor Mesinger and Geoffrey DiMego, pers. comm., 1998)*

Year	Operational model	Computer
March 1985	RAFS based on triply NGM (Phillips, 1979) and OI (DiMego, 1988). Resolution: 80 km/16 layers.	Cyber 205
August 1991	RAFS upgraded for the last time: NGM run with only two grids with inner grid domain doubled in size. Implemented Regional Data Assimilation System (RDAS) with three-hourly updates using an improved OI analysis using all off-time data including Profiler and Aircraft Communication Addressing and Reporting System (ACARS) wind reports (DiMego *et al.*, 1992) and CQC procedures (Gandin *et al.*, 1993).	Cray YMP 8 processors 32 megawords
June 1993	First operational implementation of the Eta model in the 00Z & 12Z early run for North America at 80-km and 38-layer resolution (Mesinger *et al.*, 1988, Janjic, 1994, Black *et al.*, 1993)	
September 1994	The RUC (Benjamin *et al.*, 1996) was implemented for CONUS domain with three-hourly OI updates at 60-km resolution on 25 hybrid (sigma-theta) vertical levels.	Cray C-90 16 processors 128 megawords
September 1994	Early Eta analysis upgrades (Rogers *et al.*, 1995)	
August 1995	A mesoscale version of the Eta model (Black, 1994) was implemented at 03Z and 15Z for an extended CONUS domain, with 29-km and 50-layer resolution and with NMC's first predictive cloud scheme (Zhao and Black, 1994) and new coupled land-surface–atmosphere package (two-layer soil).	Cray C-90 16 processors 256 megawords
October 1995	Major upgrade of early Eta runs: 48-km resolution, cloud scheme and Eta Data Assimilation System (EDAS) using three-hourly OI updates (Rogers *et al.*, 1996)	
January 1996	New coupled land-surface–atmosphere scheme put into early Eta runs (Chen *et al.*, 1997, Mesinger, 1997)	
July–August 1996	Nested capability demonstrated with twice-daily support runs for Atlanta Olympic Games with 10-km 60-layer version of Meso Eta.	

Table 1.5.3. *(cont.)*

Year	Operational model	Computer
February 1997	Upgrade package implemented in the early and Meso Eta runs.	
February 1998	Early Eta runs upgraded to 32 km and 45 levels with four soil layers. OI analysis replaced by 3D-Var with new data sources. EDAS now partially cycled (soil moisture, soil temperature, cloud water/ice & turbulent kinetic energy).	
April 1998	RUC (three-hourly) replaced by hourly RUC II system with extended CONUS domain, 40-km and 40-level resolution, additional data sources and extensive physics upgrades.	
June 1998	Meso runs connected to early runs as a single 4/day system for North American domain at 32-km and 45-level resolution, 15Z run moved to 18Z, added new snow analysis. All runs connected with EDAS, which is fully cycled for all variables.	IBM SV2 256 processors

continental USA (CONUS), developed at NOAA's Forecast System Laboratory, was implemented in 1994 and upgraded in 1998 (Benjamin *et al.*, 1996).

The 36-h *S*1 forecast verification scores constitute the longest record of forecast verification available anywhere. They were started in the late 1940s for subjective surface forecasts, before operational computer forecast guidance, and for 500 hPa in 1954, with the first numerical forecasts. Figure 1.1.1(a) includes the forecast scores for 500 hPa from 1954 until the present, as well as the scores for the 72-h forecasts. It is clear that the forecast skill has improved substantially over the years, and that the current 36-h 500-hPa forecasts are close to a level that in the 1950s would have been considered "perfect" (Shuman, 1989). The 72-h forecasts have also improved, and are now as accurate as the 36-h forecasts were about 15 years ago. This doubling of the skill over 10–20 years can be observed in other types of forecasts verifications as well.

As indicated at the beginning of this chapter, the 36-h forecasts of 500 hPa showing the position and intensity of the large-scale atmospheric waves and centers of high and low pressure are generally excellent, as suggested by the nearly "perfect" *S*1 score. However, sea level pressure maps are more affected by mesoscale structures, such as fronts and convective systems which are still difficult to forecast in detail, and hence they have a poorer *S*1 score (Fig. 1.1.1(b)). The solid line with circles starts in 1947 with scores from subjectively made surface forecasts, then barotropic and baroclinic quasi-geostrophic models (Table 1.5.1), the LFM model and since 1983, the global

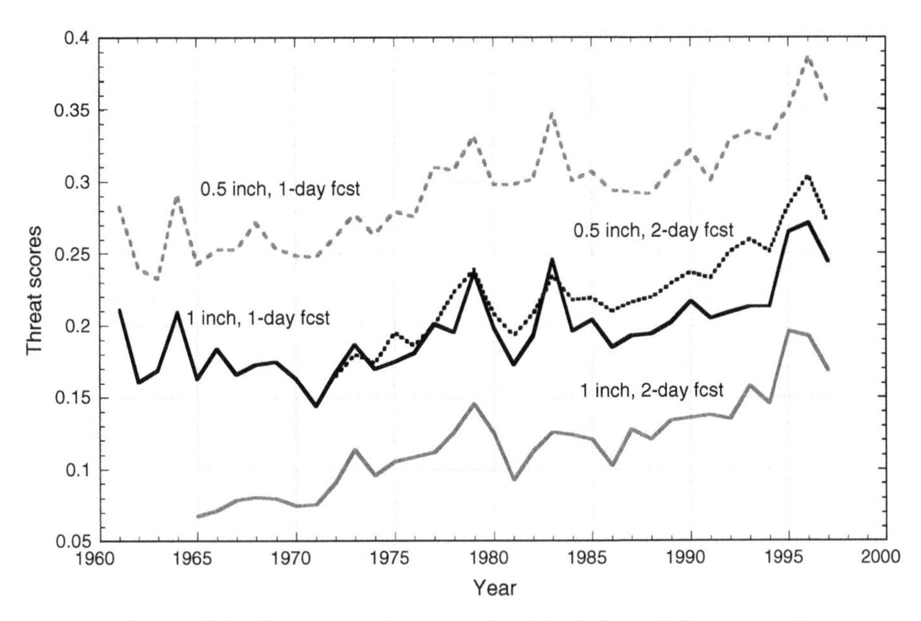

Figure 1.5.1: Threat scores (day 1 and day 2 for 0.5′ and 1′ 24-h accumulation, annual average) of human forecasters at NCEP (data courtesy of J. Hoke).

spectral model (denoted Aviation or AVN). Other model forecasts are also presented separately on Fig. 1.1.1(b). Note that the AVN model and the Eta model, which continue to be developed, show the most improvement. The development of the LFM was "frozen" in 1986, and that of the NGM in 1991, when more advanced systems were implemented, and therefore their forecasts show no further improvement with time (except for the effect of improved global forecasts used as a first guess for the LFM).

Fig. 1.5.1 shows threat scores for precipitation predictions made by expert forecasters from the NCEP Hydrometeorological Prediction Center (HPC, the Meteorological Operations Division of the former NMC). The threat score (*TS*) is defined as the intersection of the predicted area of precipitation exceeding a particular threshold (*P*), in this case 0.5 inches in 24 h, and the observed area (*O*), divided by the union of the two areas: $TS = (P \cap O)/(P \cup O)$. The bias (not shown) is defined by P/O. The *TS*, also known as critical success index (CSI) is a particularly useful score for quantities that are relatively rare. Fig. 1.4.2 indicates that the forecasters skill in predicting accumulated precipitation has been increasing with time, and that the current average skill in the 2-d forecast is as good as the 1-d forecasts were in the 1970s. Beyond the first 6–12 h, the forecasts are based mostly on numerical guidance, so that the improvement reflects to a large extent improvements of the numerical forecasts, which the human forecasters in turn improve upon based on their knowledge and expertise. The forecasters also have access to several model forecasts, and they use their judgment in assessing which one is more accurate in each case. This constitutes a major source of the "value-added" by the human forecasters.

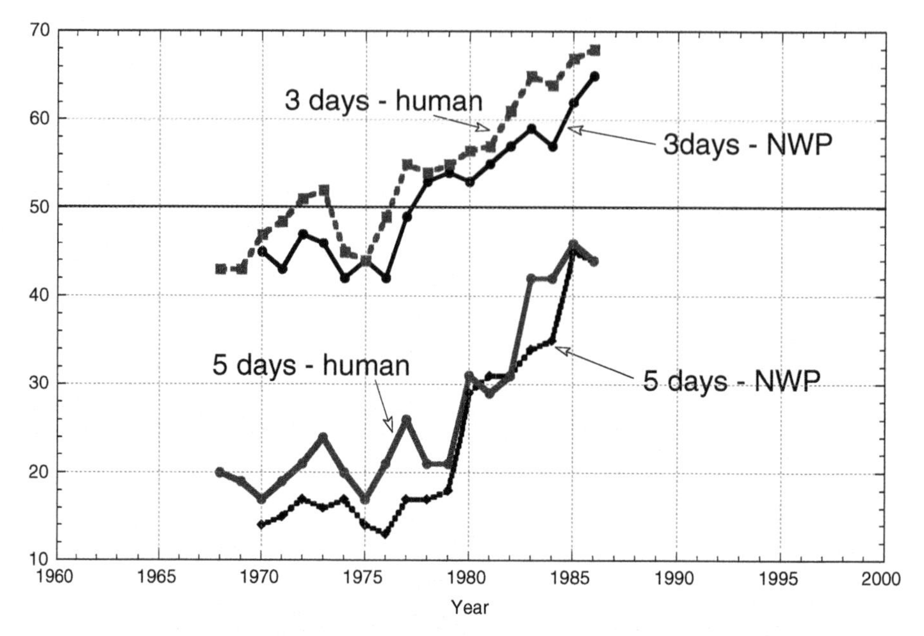

Figure 1.5.2: Hughes data: comparison of the forecast skill in the medium-range from NWP guidance and from human forecasters.

The relationship between the evolution of human and numerical forecasts is clearly shown in a record compiled by the late F. Hughes (1987), reproduced in Fig. 1.5.2. It is the first operational score maintained for the "medium-range" (beyond the first two days of the forecasts). The score used by Hughes was a standardized anomaly correlation (SAC), which accounted for the larger variability of sea level pressure at higher latitudes compared to lower latitudes. Unfortunately the SAC is not directly comparable to other scores such as the anomaly correlation (discussed in the next section). The fact that until 1976 the 3-day forecast scores from the model were essentially constant is an indication that their rather low skill was more based on synoptic experience than on model guidance. The forecast skill started to improve after 1977 for the 3-day forecast, and after 1980 for the 5-day forecast. Note that the human forecasts are on the average significantly more skillful than the numerical guidance, but it is the improvement in NWP forecasts that drives the improvements in the subjective forecasts.

1.6 Nonhydrostatic mesoscale models

The hydrostatic approximation involves neglecting vertical accelerations in the vertical equation of motion, compared to gravitational acceleration. This is a very good approximation, even in stratified fluids, as long as horizontal scales of motion are larger than the vertical scales. The main advantage of the hydrostatic equation (Chapter 2) is that it filters sound waves (except those propagating horizontally, or Lamb waves). Because of the problem of computational instability, the absence of sound waves allows the use of larger time steps (the Lamb waves are handled generally with semi-implicit time schemes, discussed in Section 3.3).

The hydrostatic approximation is very accurate if the horizontal scales are much larger than the vertical scales. For atmospheric models with horizontal grid sizes of the order of 100 km, the hydrostatic equation is very accurate and convenient. Furthermore, for quasi-geostrophic (slow) motion, the hydrostatic equation is accurate even if the horizontal scales are of the same order as the vertical scales, i.e., the hydrostatic approximation can be used even in mesoscale models with grid sizes of the order of 10 km or larger without introducing large errors.

However, in order to represent smaller-scale phenomena such as storms or convective clouds which have vertical accelerations that are not negligible compared to buoyancy forces, it is necessary to use the equations of motion without the hydrostatic approximation. In the last decade a number of nonhydrostatic models have been developed in order to simulate mesoscale phenomena in North America. They include the Penn State/NCAR Mesoscale Model (e.g., Dudhia, 1993), the CAPS Advanced Regional Prediction System (Xue *et al.*, 1995), NCEP's Regional Spectral Model (Juang *et al.*, 1997), the Mesoscale Compressible Community (MCC) model (Laprise *et al.*, 1997), the CSU RAMS (Tripoli and Cotton 1980), the US Navy COAMPS (Hodur, 1997). In Europe and Japan several other nonhydrostatic models have been developed as well.

Sound waves, which are generally of no consequence for atmospheric flow but would require the use of very small steps, require a special approach in nonhydrostatic models in order to maintain a reasonable computational efficiency. Sound waves depend on compressibility (three-dimensional divergence) for their propagation. For this reason, some nonhydrostatic models use the quasi-Boussinesq or "anelastic" equations, where the atmosphere is assumed to be separated into a hydrostatic basic state and perturbations, and where the density perturbations are neglected everywhere except in the buoyancy terms (Ogura and Phillips, 1962, Klemp and Wilhelmson, 1978). Other approaches are the use of artificial "divergence damping" in the pressure gradient terms (e.g., Xue *et al.*, 1995, Skamarock and Klemp, 1992), and the use of implicit time schemes for the terms affecting sound waves that are unconditionally stable (Durran and Klemp, 1983, Laprise *et al.*, 1997).

Nonhydrostatic models with an efficient (e.g., semi-implicit) treatment of sound waves are computationally competitive with hydrostatic models, and future generations of models may become nonhydrostatic even in the global domain.

1.7 Weather predictability, ensemble forecasting, and seasonal to interannual prediction

In a series of remarkable papers, Lorenz (1963a,b, 1965, 1968) made the fundamental discovery that *even with perfect models and perfect observations, the chaotic nature of the atmosphere would impose a finite limit of about two weeks to the predictability of the weather.* He proved this by running a simple atmospheric model, introducing (by mistake) exceedingly small perturbations in the initial conditions, and running the model again. With time, the small difference between the two forecasts became larger and larger, until after about two weeks, the forecasts were as different as two randomly chosen states of the model. In the 1960s Lorenz's discovery, which started the theory of chaos, was "only of academic interest" and not relevant to operational weather forecasting, since at that time the skill of even two-day operational forecasts was low. Since then, however, computer-based forecasts have improved so much that Lorenz's limit of predictability is starting to become attainable in practice, especially with ensemble forecasting. Furthermore, skillful prediction of longer lasting phenomena such as El Niño is becoming feasible (Chapter 6).

Because the skill of the forecasts decreases with time, Epstein (1969) and Leith (1974) suggested that instead of performing "deterministic" forecasts, stochastic forecasts providing an estimate of the skill of the prediction should be made. The only computationally feasible approach in order to achieve this goal is through "ensemble forecasting" in which several model forecasts are performed by introducing perturbations in the initial conditions or in the models themselves.

After considerable research on how to most effectively perturb the initial conditions, ensemble forecasting was implemented operationally in December 1992 at both NCEP and ECMWF (Tracton and Kalnay, 1993, Toth and Kalnay, 1993, Palmer *et al.*, 1993, Molteni *et al.*, 1996, Toth and Kalnay, 1997). Since 1994 NCEP has been running 17 global forecasts per day, each out to 16 days, with initial perturbations obtained using the method of *breeding growing perturbations*. This ensures that the initial perturbations contain naturally growing dynamical perturbations in the atmosphere, which are also present in the analysis errors. The length of the forecasts allows the generation of "outlooks" for the second week. The NCEP ensemble forecasts can be accessed through the world-wide web at the EMC home page (nic.fb4.noaa.gov:8000), and linking to the ensemble home page. At ECMWF, the perturbation method is based on the use of *singular vectors*, which grow even faster than the *bred* or *Lyapunov* vector perturbations. The ECMWF ensemble contains 50 members (Chapter 6).

Ensemble forecasting has accomplished two main goals: the first one is to provide an ensemble average forecast that beyond the first few days is more accurate than individual forecasts, because *the components of the forecast that are most uncertain tend to be averaged out*. The second and more important goal is *to provide forecasters with an estimation of the reliability of the forecast*, which because of changes in atmospheric predictability, varies from day to day and from region to region.

The first goal is illustrated in Fig. 1.7.1, prepared at the Climate Prediction Center (CPC, the Climate Analysis Center of the former NMC) for the verification of the NCEP ensemble during the winter of 1997–8. This was an El Niño winter with major anomalies in the atmospheric circulation, and the operational forecasts had excellent

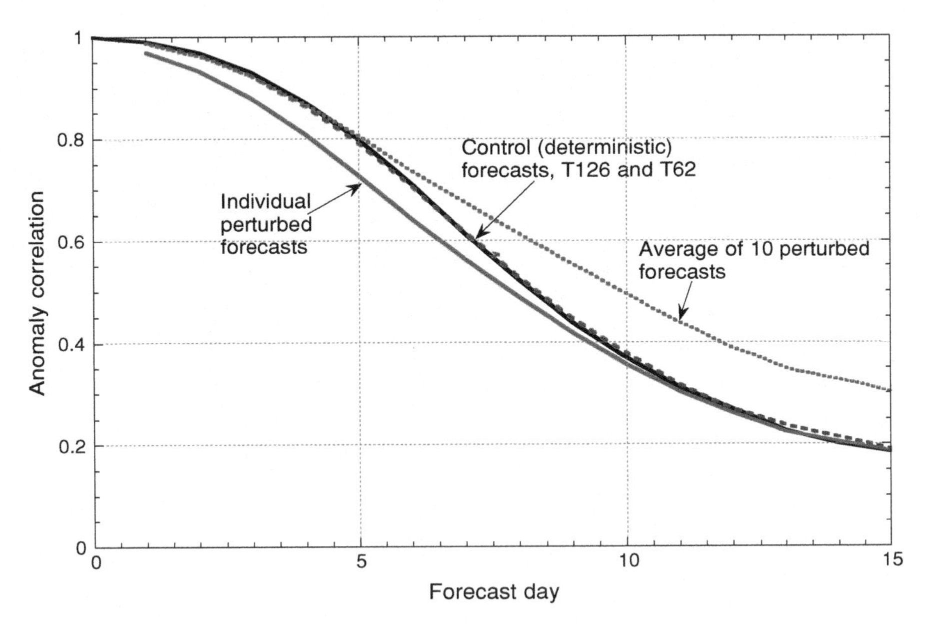

Figure 1.7.1: Anomaly correlation of the ensembles during the winter of 1997–8 (controls, T_{126} and T_{62}, and ten perturbed ensemble forecasts). (Data courtesy Jae Schemm, of NCEP.)

skill. The control "deterministic" forecast (circles) had an "anomaly correlation" (AC, pattern correlation between predicted and analyzed anomalies) in the 5-day forecast of 80%, which is quite good. The ten perturbed ensemble members have individually a poorer verification with an average AC of about 73% at 5 days. This is because, in the initial conditions, the control starts from the best estimate of the state of the atmosphere (the analysis), but growing perturbations are added to this analysis for each additional ensemble member. However, the ensemble average forecast tends to average out uncertain components, and as a result, it has better skill than the control forecast starting at day 5. Note that the ensemble extends by one day the length of the useful forecast (defined as an AC greater than 60%) from about 7 days in the control to about 8 days in the ensemble average.

The second goal of the ensemble forecasting, to provide guidance on the uncertainty of each forecast, is accomplished best by the use of two types of plots. The "spaghetti" plots show a single contour line for all 17 forecasts, and the probabilistic plots show, for example, what percentage of the ensemble predicts 24-h accumulated precipitation of more than 1 inch at each grid point (for probabilistic Quantitative Precipitation Forecasts or pQPF). Both of them provide guidance on the reliability of the forecasts in an easy-to-understand way. The use of the ensembles has provided the US NWS forecasters with the confidence to issue storm forecasts 5–7 days in advance when the spaghetti plots indicate good agreement in the ensemble. Conversely, the spaghetti plots also indicate when a short-range development may be particularly

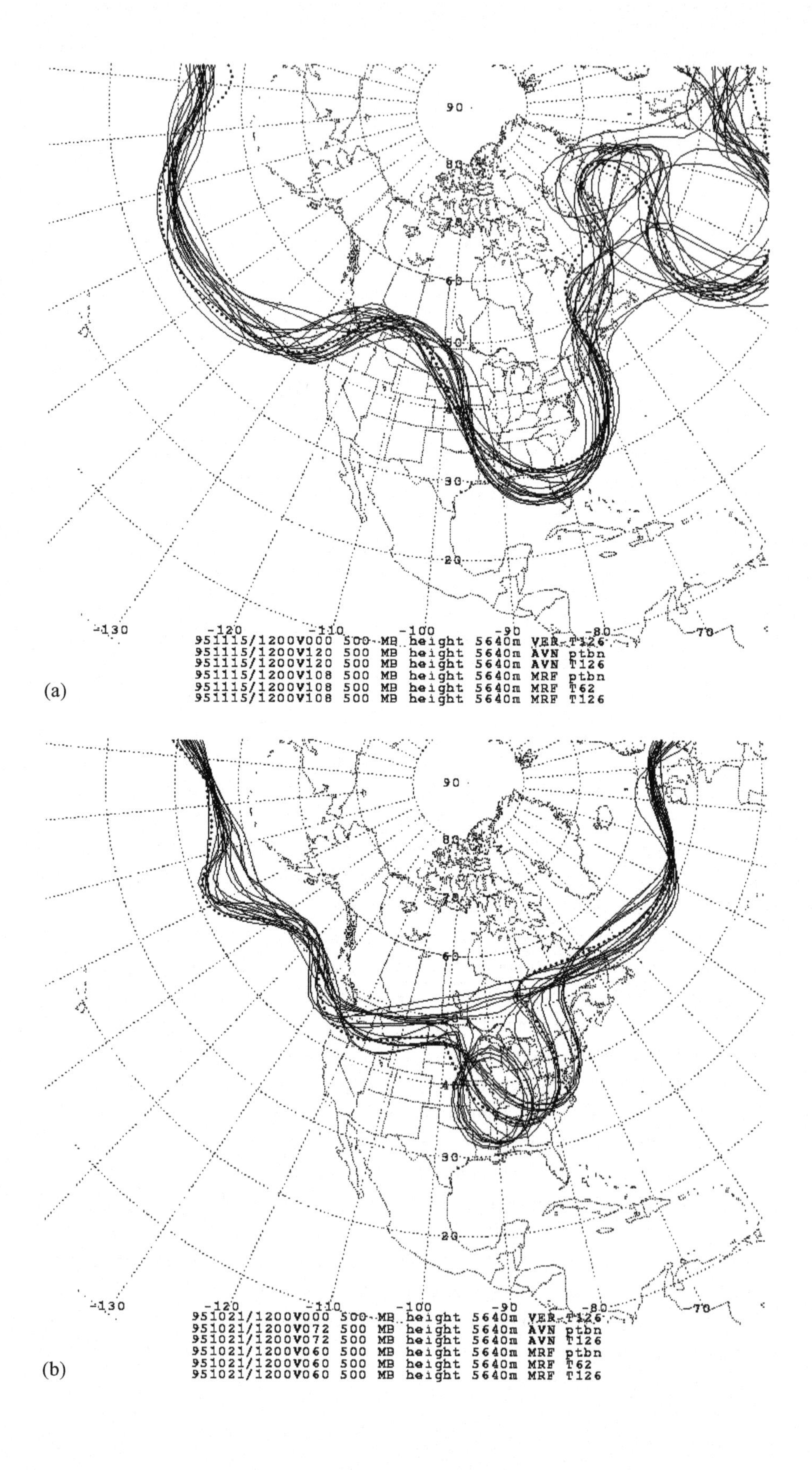

90
80
70
60
50
40
30
20
-130
-120
-110
-100
-90
-80
-70
951115/1200V000 500 MB height 5640m VER T126
951115/1200V120 500 MB height 5640m AVN ptbn
951115/1200V120 500 MB height 5640m AVN T126
951115/1200V108 500 MB height 5640m MRF ptbn
951115/1200V108 500 MB height 5640m MRF T62
951115/1200V108 500 MB height 5640m MRF T126
(a)
951021/1200V000 500 MB height 5640m VER T126
951021/1200V072 500 MB height 5640m AVN ptbn
951021/1200V072 500 MB height 5640m AVN T126
951021/1200V060 500 MB height 5640m MRF ptbn
951021/1200V060 500 MB height 5640m MRF T62
951021/1200V060 500 MB height 5640m MRF T126
(b)

difficult to predict, so that the users should be made aware of the uncertainty of the forecast. Fig. 1.7.2(a) shows an example of the 5-day forecast for 15 November 1995, the first East Coast winter storm of 1995–6: the fact that the ensemble showed good agreement provided the forecasters with the confidence to issue a storm forecast these many days in advance. By contrast, Fig. 1.7.2(b) shows a 2.5-day forecast for a storm with verification time 21 October 1995, and it is clear that even at this shorter range, the atmosphere is much less predictable and there is much more uncertainty about the location of the storm.

The use of ensembles has also led to another major development, the possibility of an *adaptive* or *targeted* observing system. As an example, consider a case in which the lack of agreement among the ensemble members indicates that a 3-day forecast in a certain region is exceedingly uncertain, as in Fig. 1.7.2(b). Several new techniques have been developed to trace such a region of uncertainty backward in time, for example 2 days. These techniques will point to a region or regions where additional observations would be especially useful. The additional observations could be dropwinsondes launched from a reconnaissance or a pilotless airplane, additional rawinsondes, or especially intensive use of satellite data such as a Doppler Wind Lidar. If additional observations are available 24 h after the start of the originally critically uncertain 3-day forecast, they can increase substantially the usefulness of the 2-day forecast. Similarly, a few additional rawinsondes could be launched where short-range ensemble forecasts (12–24 h) indicate that they are most needed. Preliminary tests of this approach of targeted observations have been successfully performed within an international Fronts and Storm Track Experiment (FASTEX) in the North Atlantic during January and February 1997, and in the North Pacific Experiment (NORPEX) in January and February 1998 (Szunyogh *et al.*, 2000).

Ensemble forecasting also provides the basic tool to extend forecasts beyond Lorenz's 2-week limit of weather predictability (Chapter 6). Slowly varying surface forcing, especially from the tropical ocean and from land-surface anomalies, can produce atmospheric anomalies that are longer lasting and more predictable than individual weather patterns. The most notable of these is the El Niño–Southern Oscillation (ENSO) produced by unstable oscillations of the coupled ocean–atmosphere system, with a frequency of 3–7 years. Because of their long time scale, the ENSO oscillations should be predictable a year or more in advance (in agreement with the chaos theory). The first successful experiments in this area were made by Cane *et al.* (1986) with a simple coupled atmosphere–ocean model. The warm phases of ENSO (El Niño episodes) are associated with warm sea surface temperature (SST) anomalies

Caption for Figure 1.7.2: (a) Spaghetti plot for the 5-day forecast for 15 Nov 1995, a case of a very predictable storm over eastern USA. (Figure courtesy of R. Wobus, NCEP.) (b) Spaghetti plot for the 2.5-day forecast for 21 Oct 1995, the case of a very unpredictable storm over the USA. (Courtesy of R. Wobus, NCEP.) Dashes indicate the control forecast.

in the equatorial central and eastern Pacific Ocean, and cold phases (La Niña episodes) with cold anomalies. NCEP started performing multiseasonal predictions with coupled comprehensive atmosphere–ocean models in 1995, and ECMWF did so in 1997.

A single atmospheric forecast forced with the SST anomalies would not be useful beyond the first week or so, when unpredictable weather variability would mask the forced atmospheric anomalies. Ensemble averaging many forecasts made with atmospheric models forced by SST anomalies (and by other slowly varying anomalies over land such as soil moisture and snow cover) allows the filtering out of the unpredictable components of the forecast, and the retention of more of the forced predictable components. This filtering is reflected in the fact that the ensemble average for the second week of the forecasts for the winter of 1997–8 (Fig. 1.7.1) had a high AC of 57%, much higher than previously obtained. Researchers at the Japanese Meteorological Agency have performed forecasts for the 28-day average and also found that ensemble averaging substantially increased the information on the second week and the last 2 weeks of the forecast. The very successful operational forecasts of the ENSO episode of 1997–8 performed at both NCEP and ECMWF have been substantially based on the use of ensembles to extract the useful information on the impact of El Niño from the "weather noise".

1.8 The future

The last decades have seen the expectations of Charney (1951) fulfilled, and an amazing improvement in the quality of the forecasts based on NWP guidance. From the active research taking place, one can envision that the next decade will continue to bring improvements, especially in the following areas:

- detailed short-range forecasts, using storm-scale models able to provide skillful predictions of severe weather;
- more sophisticated methods of data assimilation able to extract the maximum possible information from observing systems, especially remote sensors such as satellites and radars;
- development of adaptive observing systems, in which additional observations are placed where ensembles indicate that there is rapid error growth (low predictability);
- improvement in the usefulness of medium-range forecasts, especially through the use of ensemble forecasting;
- fully coupled atmospheric–hydrological systems, where the atmospheric model precipitation is appropriately downscaled and used to extend the length of river flow prediction;
- more use of detailed atmosphere–ocean–land coupled models, in which long-lasting coupled anomalies such as SST and soil moisture anomalies lead

to more skillful predictions of anomalies in weather patterns beyond the limit of weather predictability (about two weeks);

- more guidance to governments and the public on subjects such as air pollution, ultraviolet radiation and transport of contaminants, which affect health;
- an explosive growth of systems with emphasis on commercial applications of NWP, from guidance on the state of highways to air pollution, flood prediction, guidance to agriculture, construction, etc.

2

The continuous equations

2.1 Governing equations

V. Bjerknes (1904) pointed out for the first time that there is a complete set of seven equations with seven unknowns that governs the evolution of the atmosphere:

- Newton's second law or conservation of momentum (three equations for the three velocity components);
- the continuity equation or conservation of mass;
- the equation of state for ideal gases;
- the first law of thermodynamics or conservation of energy;
- a conservation equation for water mass.

To these equations we have to add appropriate boundary conditions at the bottom and top of the atmosphere.

In this section we briefly derive the governing equations. The reader may refer to other texts, such as Haltiner and Williams (1980), or James (1994) for more details.

Newton's second law *or conservation of momentum:*
On an inertial frame of reference, the absolute acceleration of a parcel of air in three dimensions is given by

$$\frac{d_a \mathbf{v}_a}{dt} = \mathbf{F}/m \tag{2.1.1}$$

On a rotating frame of reference centered at the center of the earth, the absolute velocity $\mathbf{v}_a$ is given by the sum of the relative velocity $\mathbf{v}$ plus the velocity due to the rotation with angular velocity $\boldsymbol{\Omega}$:

$$\mathbf{v}_a = \mathbf{v} + \boldsymbol{\Omega} \times \mathbf{r} \tag{2.1.2}$$

where $\mathbf{r}$ is the position vector of the parcel. This is a particular case (for $\mathbf{A} = \mathbf{r}$) of the general formula relating the total time derivative of any vector on a rotating frame $d\mathbf{A}/dt$ to its total derivative in an inertial frame $d_a\mathbf{A}/dt$:

$$\frac{d_a\mathbf{A}}{dt} = \frac{d\mathbf{A}}{dt} + \boldsymbol{\Omega} \times \mathbf{A} \tag{2.1.3}$$

We can also apply this formula to $\mathbf{A} = \mathbf{v}_a$, giving

$$\frac{d_a\mathbf{v}_a}{dt} = \frac{d\mathbf{v}_a}{dt} + \boldsymbol{\Omega} \times \mathbf{v}_a \tag{2.1.4}$$

Substituting (2.1.2) into (2.1.4) we obtain that the accelerations in an inertial (absolute) and a rotating frame of reference are related by

$$\frac{d_a\mathbf{v}_a}{dt} = \frac{d\mathbf{v}}{dt} + 2\boldsymbol{\Omega} \times \mathbf{v} + \boldsymbol{\Omega} \times (\boldsymbol{\Omega} \times \mathbf{r}) \tag{2.1.5}$$

This equation indicates that on a rotating frame of reference there are two *apparent* forces per unit mass: the Coriolis force (second term on the right-hand side) and the centrifugal force (third term).

The left-hand side of (2.1.5) represents the *real* forces acting on a parcel of air, i.e., the pressure gradient force $-\alpha\boldsymbol{\nabla} p$, the gravitational acceleration $\mathbf{g}_e = -\boldsymbol{\nabla}\phi_e$, and the frictional force $\mathbf{F}$. Therefore in a rotating frame of reference moving with the earth, the apparent acceleration is given by

$$\frac{d\mathbf{v}}{dt} = -\alpha\boldsymbol{\nabla} p - \boldsymbol{\nabla}\phi_e + \mathbf{F} - 2\boldsymbol{\Omega} \times \mathbf{v} - \boldsymbol{\Omega} \times (\boldsymbol{\Omega} \times \mathbf{r}) \tag{2.1.6}$$

Here $\alpha = 1/\rho$ is the specific volume (the inverse of the density ρ), p is the pressure, ϕ_e is the Newtonian gravitational potential of the earth, and, as indicated before, the last two terms are the apparent accelerations, denoted the Coriolis force and centrifugal force respectively. We have not included the *tidal potential*, whose effects are negligible below about 100 km.

We can now combine the centrifugal force with the gravitational force, since $-\boldsymbol{\Omega} \times (\boldsymbol{\Omega} \times \mathbf{r}) = \boldsymbol{\Omega}^2\mathbf{l} = \boldsymbol{\nabla}(\Omega^2 l^2/2)$, where $\mathbf{l}$ is the position vector from the axis of rotation to the parcel. Therefore we can define as the "geopotential" $\phi = \phi_e - \Omega^2 l^2/2$, and the apparent gravity is given by

$$-\boldsymbol{\nabla}\phi = \mathbf{g} = \mathbf{g}_e + \Omega^2\mathbf{l} \tag{2.1.7}$$

We define the geographic latitude φ to be perpendicular to the geopotential ϕ. At the surface of the earth, the geographic latitude and the geocentric latitude differ by less than 10 minutes of a degree of latitude. Therefore, Newton's law on the rotating frame of the earth is written as

$$\frac{d\mathbf{v}}{dt} = -\alpha\nabla p - \nabla\phi + \mathbf{F} - 2\boldsymbol{\Omega} \times \mathbf{v} \tag{2.1.8}$$

Continuity equation *or equation of conservation of mass*

This can be derived as follows: Consider the mass of a parcel of air of density ρ

$$M = \rho\Delta x\Delta y\Delta z \tag{2.1.9}$$

If we follow the parcel in time, it conserves its mass, i.e., the total time derivative (also called the substantial, individual or Lagrangian time derivative) is equal to zero: $dM/dt = 0$. If we take a logarithmic derivative of the mass

$$\frac{1}{M}\frac{dM}{dt} = 0$$

in (2.1.9) we obtain the continuity equation:

$$\frac{1}{\rho}\frac{d\rho}{dt} + \boldsymbol{\nabla}_3 \cdot \mathbf{v} = 0 \tag{2.1.10}$$

since

$$\frac{1}{\Delta x}\frac{d\Delta x}{dt} = \frac{\partial u}{\partial x}$$

and similarly for the other directions $y,\ z$.

Now, the total derivative of any function $f(x, y, z, t)$, following a parcel, can be expanded as

$$\frac{df}{dt} = \frac{\partial f}{\partial t} + \frac{\partial f}{\partial x}\frac{dx}{dt} + \frac{\partial f}{\partial y}\frac{dy}{dt} + \frac{\partial f}{\partial z}\frac{dz}{dt} = \frac{\partial f}{\partial t} + \mathbf{v} \cdot \boldsymbol{\nabla} f \tag{2.1.11}$$

Equation (1.11) indicates that the total (or Lagrangian or individual) time derivative of a property is given by the local (partial, Eulerian) time derivative (at a fixed point) plus the changes due to advection. If we expand $d\rho/dt$ in (2.1.10) using (2.1.11) we obtain an alternative form of the continuity equation, usually referred to as "in flux form":

$$\frac{\partial\rho}{\partial t} = -\boldsymbol{\nabla} \cdot (\rho\mathbf{v}) \tag{2.1.12}$$

Equation of state for perfect gases

The atmosphere can be assumed to be a perfect gas, for which the pressure p, specific volume α (or its inverse ρ, density), and temperature T are related by

$$p\alpha = RT \tag{2.1.13}$$

where R is the gas constant for dry air. For moist air this has to take into account the partial pressure of moist air, usually done by defining the virtual temperature $Tv = 1 + 0.6q)$, i.e., the dry temperature having the same density as moist air at the same pressure. This equation indicates that, given two thermodynamic variables, the others are determined.

Thermodynamic energy equation *or conservation of energy equation*
This equation expresses that if heat is applied to a parcel at a rate of Q per unit mass, this heat can be used to increase the internal energy $C_v T$ and/or to produce work of expansion:

$$Q = C_v \frac{dT}{dt} + p \frac{d\alpha}{dt} \tag{2.1.14}$$

The coefficients of specific heat at constant volume C_v and at constant pressure C_p are related by $C_p = C_v + R$. We can use the equation of state (2.1.13) to derive another form of the thermodynamic equation:

$$Q = C_p \frac{dT}{dt} - \alpha \frac{dp}{dt} \tag{2.1.15}$$

The rate of change of the specific entropy s of a parcel is given by $ds/dt = Q/T$, i.e., the diabatic heating divided by the absolute temperature. We now define potential temperature by $\theta = T\,(p_0/p)^{R/C_p}$, where p_0 is a reference pressure (1000 hPa). With this definition, it is easy to show that the potential temperature and the specific entropy are related by

$$\frac{ds}{dt} = C_p \frac{1}{\theta} \frac{d\theta}{dt} = \frac{Q}{T} \tag{2.1.16}$$

This shows that *potential temperature is individually conserved* in the absence of diabatic heating.

Equation for conservation of water vapor mixing ratio q
This equation simply indicates that the total amount of water vapor in a parcel is conserved as the parcel moves around, except when there are *sources* (evaporation E) and *sinks* (condensation C):

$$\frac{dq}{dt} = E - C \tag{2.1.17}$$

Conservation equations for other atmospheric constituents can be similarly written in terms of their corresponding sources and sinks. If we multiply (2.1.17) by ρ, expand the total derivative $dq/dt = \partial q/\partial t + \mathbf{v} \cdot \nabla q$, and add the continuity equation (2.1.12) multiplied by q, we can write the conservation of water in an alternative "flux form":

$$\frac{\partial \rho q}{\partial t} = -\nabla \cdot (\rho \mathbf{v} q) + \rho(E - C) \tag{2.1.18}$$

The flux form of the time derivative is very useful in the construction of models. The first term of the right-hand side of (2.1.18) is the convergence of the flux of q. Note that we can include similar conservation equations for additional tracers such as liquid water, ozone, etc., as long as we also include their corresponding sources and sinks.

We now have seven equations with seven unknowns: $\mathbf{v} = (u, v, w)$, T, p, ρ or α, and q. For convenience we repeat the governing equations, which (when written without friction $\mathbf{F}$) are sometimes referred to as "*the Euler equations*":

$$\frac{d\mathbf{v}}{dt} = -\alpha\nabla p - \nabla\phi + \mathbf{F} - 2\boldsymbol{\Omega}\times\mathbf{v} \tag{2.1.19}$$

$$\frac{\partial\rho}{\partial t} = -\nabla\cdot(\rho\mathbf{v}) \tag{2.1.20}$$

$$p\alpha = RT \tag{2.1.21}$$

$$Q = C_p\frac{dT}{dt} - \alpha\frac{dp}{dt} \tag{2.1.22}$$

$$\frac{\partial\rho q}{\partial t} = -\nabla\cdot(\rho\mathbf{v}q) + \rho(E - C) \tag{2.1.23}$$

2.2 Atmospheric equations of motion on spherical coordinates

Since the earth is nearly spherical, it is natural to use spherical coordinates. Near the earth, gravity is almost constant, and the ellipticity of the earth is very small, so that one can accurately approximate scale factors by those appropriate for true spherical coordinates (Phillips, 1966, 1973, 1990a). The three velocity components are then

$$\left.\begin{aligned} u &= \text{zonal (positive eastward)} = r\cos\varphi\frac{d\lambda}{dt} \\ v &= \text{meridional (positive northward)} = r\frac{d\varphi}{dt} \\ w &= \text{vertical (positive up)} = \frac{dr}{dt} \end{aligned}\right\} \tag{2.2.1}$$

Note that $\mathbf{v} = u\mathbf{i} + v\mathbf{j} + w\mathbf{k}$, where $\mathbf{i}$, $\mathbf{j}$, $\mathbf{k}$ are the unit vectors in the three orthogonal spherical coordinates. When the acceleration (total derivative of the velocity vector) is calculated, the rate of change of the unit vectors has to be included. For example, geometrical considerations show that

$$\frac{d\mathbf{k}}{dt} = \frac{u}{r\cos\varphi}\frac{\partial\mathbf{k}}{\partial\lambda} + \frac{v}{r}\frac{\partial\mathbf{k}}{\partial\varphi} = \frac{u\mathbf{i}}{r} + \frac{v\mathbf{j}}{r} \tag{2.2.1a}$$

Exercise 2.2.1: Use spherical geometry to show (2.2.1a). Derive

$$\frac{d\mathbf{i}}{dt} = \frac{u}{r\cos\varphi}(\mathbf{j}\sin\varphi - \mathbf{k}\cos\varphi) \quad \text{and} \quad \frac{d\mathbf{j}}{dt} = \frac{1}{r\cos\varphi}(-u\mathbf{i}\sin\varphi - v\mathbf{k}\cos\varphi) \tag{2.2.2}$$

When we include these time derivatives, take into account that $\boldsymbol{\Omega} = \Omega\sin\varphi\mathbf{k} + \Omega\cos\varphi\mathbf{j}$, and expand the momentum equation (2.1.19) into its three components, we obtain

$$\left.\begin{aligned}\frac{du}{dt} &= -\frac{\alpha}{r\cos\varphi}\frac{\partial p}{\partial\lambda} + F_\lambda + \left(2\Omega + \frac{u}{r\cos\varphi}\right)(v\sin\varphi - w\cos\varphi)\\ \frac{dv}{dt} &= -\frac{\alpha}{r}\frac{\partial p}{\partial\varphi} + F_\varphi - \left(2\Omega + \frac{u}{r\cos\varphi}\right)u\sin\varphi - \frac{vw}{r}\\ \frac{dw}{dt} &= -\alpha\frac{\partial p}{\partial r} - g + F_r + \left(2\Omega + \frac{u}{r\cos\varphi}\right)u\cos\varphi + \frac{v^2}{r}\end{aligned}\right\} \tag{2.2.1c}$$

The terms proportional to $u/r\cos\varphi$ are known as "metric terms".

A "*traditional approximation*" (Phillips, 1966) has been routinely made in NWP, since most of the atmospheric mass is confined to a few tens of kilometers. This suggests that in considering the distance of a point to the center of the earth $r = a + z$, one can neglect z and replace r by the radius of the earth $a = 6371$ km, replace $\partial/\partial r$ by $\partial/\partial z$, and neglect the metric and Coriolis terms proportional to $\cos\varphi$. Then the equations of motion in spherical coordinates become

$$\left.\begin{aligned}\frac{du}{dt} &= -\frac{\alpha}{a\cos\varphi}\frac{\partial p}{\partial\lambda} + F_\lambda + \left(2\Omega + \frac{u}{a\cos\varphi}\right)v\sin\varphi\\ \frac{dv}{dt} &= -\frac{\alpha}{a}\frac{\partial p}{\partial\varphi} + F_\varphi - \left(2\Omega + \frac{u}{a\cos\varphi}\right)u\sin\varphi\\ \frac{dw}{dt} &= -\alpha\frac{\partial p}{\partial z} - g + F_z\end{aligned}\right\} \tag{2.2.1d}$$

which possess the angular momentum conservation principle

$$\frac{d}{dt}[(u + \Omega a\cos\varphi)a\cos\varphi] = a\cos\varphi\left(-\frac{\alpha}{a\cos\varphi}\frac{\partial p}{\partial\lambda} + F_\lambda\right) \tag{2.2.1e}$$

With the "traditional approximation" the total time derivative operator in spherical coordinates is given by

$$\frac{d()}{dt} = \frac{\partial()}{\partial t} + \frac{u}{a\cos\varphi}\frac{\partial()}{\partial\lambda} + \frac{v}{a}\frac{\partial()}{\partial\varphi} + w\frac{\partial()}{\partial z} \tag{2.2.1f}$$

and the three-dimensional divergence that appears in the continuity equation by

$$\nabla_3\cdot\mathbf{v} = \frac{1}{a\cos\varphi}\left(\frac{\partial u}{\partial\lambda} + \frac{\partial v\cos\varphi}{\partial\varphi}\right) + \frac{\partial w}{\partial z} \tag{2.2.1g}$$

2.3 Basic wave oscillations in the atmosphere

In order to understand the problems in Richardson's result in 1922 (Fig. 1.2.1) and the effect of the filtering approximations introduced by Charney *et al.* (1950), we need to have a basic understanding of the characteristics of the different types of waves present in the atmosphere. The characteristics of these waves, (sound, gravity, and slower weather waves) have also profound implications for the present use of hydrostatic and nonhydrostatic models. The three types of waves are present in the solutions of the governing equations, and different approximations such as the hydrostatic, the quasi-geostrophic, and the anelastic approximations are designed to filter out some of them.

To simplify the analysis we make a tangent plane or "f-plane" approximation. We consider motions with horizontal scales L smaller than the radius of the earth. On this tangent plane we can approximate the spherical coordinates (Section 2.2) by

$$\frac{1}{a\cos\varphi_0}\frac{\partial}{\partial\lambda} \approx \frac{\partial}{\partial x} \qquad \frac{1}{a}\frac{\partial}{\partial\varphi} \approx \frac{\partial}{\partial y}, \qquad f \approx 2\Omega\sin\varphi_0$$

and ignore the metric terms, since $u/(a\tan\varphi)$ is small compared with Ω.

The governing equations on an f-plane (rotating with the local vertical component of the earth rotation) are:

$$\frac{du}{dt} = +fv - \frac{1}{\rho}\frac{\partial p}{\partial x} \tag{2.3.1a}$$

$$\frac{dv}{dt} = -fu - \frac{1}{\rho}\frac{\partial p}{\partial y} \tag{2.3.1b}$$

$$\frac{dw}{dt} = -\frac{1}{\rho}\frac{\partial p}{\partial z} - g \tag{2.3.1c}$$

$$\frac{d\rho}{dt} = -\rho\left(\frac{\partial u}{\partial x} + \frac{\partial v}{\partial y} + \frac{\partial w}{\partial z}\right) \tag{2.3.1d}$$

$$\frac{ds}{dt} = \frac{Q}{T}; \qquad s = C_p \ln\theta \tag{2.3.1e}$$

$$p = \rho RT \tag{2.3.1f}$$

Consider a *basic state* at rest $u_0 = v_0 = w_0 = 0$. From (2.3.1a) and (2.3.1b), we see that p_0 does not depend on x, y, $p_0 = p_0(z)$. From (2.3.1c), ρ_0 and therefore the other basic state thermodynamic variables also depend on z only.

Assume that the motion is adiabatic and frictionless, $Q = 0$, $\mathbf{F} = 0$. Consider *small perturbations* $p = p_0 + p'$, etc. so that we can linearize the equations (neglect terms which are products of perturbations). For convenience, we define $u^* = \rho_0 u'$; $v^* = \rho_0 v'$; $w^* = \rho_0 w'$; $s^* = \rho_0 s'$. The perturbation equations are then

$$\frac{\partial u^*}{\partial t} = +fv^* - \frac{\partial p'}{\partial x} \tag{2.3.2a}$$

$$\frac{\partial v^*}{\partial t} = -fu^* - \frac{\partial p'}{\partial y} \tag{2.3.2b}$$

$$\frac{\partial w^*}{\partial t} = -\frac{\partial p'}{\partial z} - \rho' g \tag{2.3.2c}$$

$$\frac{\partial \rho'}{\partial t} = -\left(\frac{\partial u^*}{\partial x} + \frac{\partial v^*}{\partial y} + \frac{\partial w^*}{\partial z}\right) \tag{2.3.2d}$$

$$\frac{\partial s^*}{\partial t} = -w^* \frac{ds_0}{dz} \tag{2.3.2e}$$

$$\frac{p'}{p_0} = \frac{\rho'}{\rho_0} + \frac{T'}{T_0} \tag{2.3.2f}$$

where

$$s^* = \rho_0 C_p \frac{\theta'}{\theta_0} = \rho_0 C_p \left(\frac{T'}{T_0} - \frac{R}{C_p} \frac{p'}{p_0} \right) = C_p \left(\frac{p'}{\gamma R T_0} - \rho' \right) \tag{2.3.2g}$$

Exercise 2.3.1: Derive (2.3.2a)–(2.3.2g), recalling that $p = \rho RT, \theta = T\,(1000hP_a/p)^{R/C_p}$, $C_p = R + C_v$, $\gamma = C_p/C_v = 1.4$, and $c_s^2 = \gamma RT \approx (320\text{ m/s})^2$ is the square of the speed of sound.

2.3.1 Pure types of plane wave solutions

We first consider *special cases with pure wave type solutions*. They exist in their pure form only under very simplified assumptions. However, if we understand their basic characteristics, we will understand their role in the full nonlinear models, and the methodology used for filtering some of the waves out. We will be assuming plane wave solutions aligning the x-axis along the horizontal direction of propagation:

$$(u^*, v^*, w^*, p') = (U, V, W, P)e^{i(kx+mz-\nu t)} \tag{2.3.3}$$

Here $k = 2\pi/L_x$ and $m = 2\pi/L_z$ are horizontal and vertical wavenumbers, respectively, $\nu = 2\pi/T$ is the frequency, and $U, V, W,$ and P are constant amplitudes. We will aim to derive the *frequency dispersion relationship (FDR)* $\nu = f(k, m, parameters)$ for each type of wave by substituting the plane wave formulation (2.3.3) into the linear equation, and eliminating variables. The FDR gives us not only the frequency, but also the *phase speed* components $(\nu/k, \nu/m)$ as well as the *group velocity* components $(\partial\nu/\partial k, \partial\nu/\partial m)$. The phase speed is the speed of individual wave crests and valleys, and the group velocity is the speed at which wave energy propagates in the horizontal and vertical directions. A pure type of wave occurs under idealized conditions, such as no rotation, no stratification for sound waves, but its basic characteristics are retained even if the ideal conditions are not valid (sound waves are still present but slightly modified in the presence of rotation and stratification).

2.3.1.1 Pure sound waves

We neglect rotation, stratification and gravity: $f = 0$, $g = 0$, $ds_0/dz = 0$. From (2.3.2e), we have $s^* = 0$ (recall that s^* is a perturbation, and if it was constant, we would have included its value into the basic state s_0). Therefore $p' = c_s^2\rho'$, and (2.3.2) reduce to

$$\frac{\partial u^*}{\partial t} = -\frac{\partial p'}{\partial x} \tag{2.3.4a}$$

$$\frac{\partial v^*}{\partial t} = -\frac{\partial p'}{\partial y} \tag{2.3.4b}$$

$$\frac{\partial w^*}{\partial t} = -\frac{\partial p'}{\partial z} \tag{2.3.4c}$$

$$\frac{1}{c_s^2}\frac{\partial p'}{\partial t} = -\left(\frac{\partial u^*}{\partial x} + \frac{\partial v^*}{\partial y} + \frac{\partial w^*}{\partial z}\right) \tag{2.3.4d}$$

These show that sound waves occur through adiabatic expansion and contraction (three-dimensional divergence), and that the pressure perturbation is proportional to the density perturbation.

Assuming plane wave solutions (2.3.3), with the x-axis along the horizontal direction of the waves, and substituting into (2.3.4), we get

$$-i\nu U = -ikP \tag{2.3.5a}$$

$$-i\nu V = 0 \tag{2.3.5b}$$

$$-i\nu W = -imP \tag{2.3.5c}$$

$$-i\nu P = -c_s^2(ikU + imW) \tag{2.3.5d}$$

From (2.3.5b) $V = 0$, and substituting U and W from (2.3.5a) and (2.3.5c) into (2.3.5d), we get the FDR:

$$\nu^2 = c_s^2(k^2 + m^2) \tag{2.3.6}$$

These are sound waves that propagate through air compression or three-dimensional divergence. The components of the phase velocity are $(\nu/k, \nu/m)$ and the total phase velocity is

$$\frac{\nu}{\sqrt{k^2 + m^2}} = \pm c_s$$

2.3.1.2 Lamb waves (horizontally propagating sound waves)

We now neglect rotation and assume that there is only horizontal propagation (no vertical velocity), but we allow for the fluid to be gravitationally stratified. With $f = 0$ and $w^* = 0$, we again have $s^* = 0$, and from (2.3.2f) $p' = c_s^2\rho'$, but from (2.3.2c) the flow is now hydrostatic: $\partial p'/\partial z = -\rho' g$. If we insert the same type of plane wave solutions (2.3.3) into (2.3.2), we find that $p' = Pe^{-(g/c_s^2)z}e^{i(kx-\nu t)}$, i.e., the vertical wavenumber is imaginary $m = ig/c_s^2$, and the phase speed is $\nu^2/k^2 = c_s^2$. Since the vertical wavenumber is imaginary, there is no vertical propagation, and the waves are *external*.

Therefore, a Lamb wave is a type of external horizontal sound wave, which is present in the solutions of models even when the hydrostatic approximation is made.

This is very important because it means primitive equation models (which make the hydrostatic approximation) contain these fast moving horizontal sound waves. We will see that Lamb waves are also equivalent to the gravity waves in a shallow water model. Note also that the FDR is such that $\nu/k = \pm c_s$, so that the phase speed does not depend on the wavenumber. This implies that the group velocity $\partial\nu/\partial k = \pm c_s$. It is also independent of the wavenumber, and as a result Lamb waves without rotation are nondispersive, so that a package of waves will move together and not disperse.

2.3.1.3 Vertical gravitational oscillations

Now we neglect rotation and pressure perturbations, $f = p' = 0$, so that there is no horizontal motion, but allow for vertical stratification. Equations (2.3.2) become

$$\frac{\partial w^*}{\partial t} = -\rho' g \tag{2.3.7a}$$

$$\frac{\partial \rho'}{\partial t} = \frac{w^*}{C_p}\frac{ds_0}{dz} = w^*\frac{d\ln\theta_0}{dz} \tag{2.3.7b}$$

From these two equations we get

$$\frac{\partial^2 w^*}{\partial t^2} + N^2 w^* = 0 \tag{2.3.8}$$

and from the continuity equation we obtain

$$\frac{\partial \rho'}{\partial t} = -\frac{\partial w^*}{\partial z} \tag{2.3.9}$$

Substituting the plane wave solution (2.3.3) into (2.3.8) we obtain $\nu^2 = N^2$, where $N^2 = gd\ln\theta_0/dz$ is the square of the *Brunt–Vaïsälä frequency*. A typical value of N for the atmosphere is $N \sim 10^{-2}\ \mathrm{s}^{-1}$. A parcel displaced in a stable atmosphere will oscillate vertically with frequency N. Equations (2.3.7b) and (2.3.9) show that the amplitude of w^* will decrease with height as $e^{-(d\ln\theta/dz)z}$.

2.3.1.4 Inertia oscillations

Inertia oscillations are horizontal and are due to the basic rotation. We now assume that $p' = 0$, $ds_0/dz = 0$, and there are no pressure perturbations and no stratification. Then $s^* = 0$, and, therefore, $\rho' = 0$ and the horizontal equations of motion become

$$\frac{\partial \mathbf{v}^*}{\partial t} = -f\mathbf{k}\times\mathbf{v}^* \quad \text{or} \quad \frac{\partial^2 \mathbf{v}^*}{\partial t^2} = f\mathbf{k}\times(f\mathbf{k}\times\mathbf{v}^*) = -f^2\mathbf{v}^* \tag{2.3.10}$$

As indicated by (2.3.10), the frequency of inertia oscillations is $\nu = \pm f$, with the acceleration perpendicular to the wind, corresponding to a circular wind oscillation. In the presence of a basic flow, there is also a translation, and the trajectories look like Fig. 2.3.1.

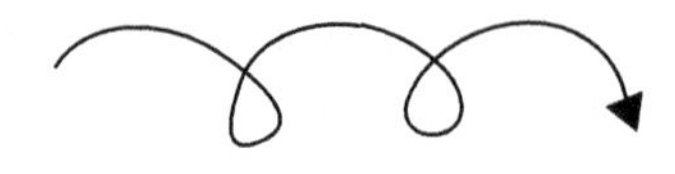

Figure 2.3.1: Schematic of an inertial oscillation in the presence of a basic flow to the right.

2.3.1.5 Lamb waves in the presence of rotation and geostrophic modes:

We now consider the same case as in Section 2.3.1.2 of horizontally propagating Lamb waves, but without neglecting rotation, i.e., $f \neq 0$, but the vertical velocity is still zero. From $w^* = 0$ and (2.3.2c) we have again $p' = c_s^2 \rho'$, and the hydrostatic balance in (2.3.2g) then implies $\partial p'/\partial z = -p' g/c^2$. Therefore the three-dimensional perturbations can be written as $p'(x, y, z, t) = p'(x, y, 0, t)e^{-z/\gamma H}$, where $\gamma H = c_s^2/g$.

The system of equations (2.3.2) becomes

$$\left.\begin{aligned}\frac{\partial \mathbf{v}^*}{\partial t} &= -f\mathbf{k} \times \mathbf{v}^* - \nabla p' \\ \frac{\partial p'}{\partial t} &= -c_s^2 \nabla \cdot \mathbf{v}^*\end{aligned}\right\} \tag{2.3.11}$$

This system is completely analogous to the linearized *shallow water equations* (SWE) which are widely used in NWP as the simplest primitive equations model:

$$\left.\begin{aligned}\frac{\partial \mathbf{v}}{\partial t} &= -f\mathbf{k} \times \mathbf{v} - \nabla \phi' \\ \frac{\partial \phi'}{\partial t} &= -\Phi \nabla \cdot \mathbf{v} \\ \text{where} \quad \phi &= \Phi + \phi'\end{aligned}\right\} \tag{2.3.12}$$

If we assume plane wave solutions of the form $(u^*, v^*, p') = (U, V, P)e^{-i(kx-\nu t)}$, and substitute in (2.3.11) we obtain:

$$\left.\begin{aligned}-i\nu U &= fV - ikP \\ -i\nu V &= -fU \\ -i\nu P &= -c_s^2 ikU\end{aligned}\right\} \tag{2.3.13}$$

Therefore the FDR is

$$\nu(\nu^2 - f^2 - c_s^2 k^2) = 0 \tag{2.3.14}$$

Note that this FDR contains *two* types of solution: one type is $\nu^2 = f^2 + c_s^2 k^2$, Lamb waves modified by inertia (rotation), or inertia Lamb waves. In the SWE analog, these are inertia-gravity waves (external gravity waves modified by inertia), $\nu^2 = f^2 + \Phi k^2$. Note that in the presence of rotation the phase speed and group velocity depend on the wavenumber: rotation makes Lamb waves dispersive (and this helps with the problem of getting rid of noise in the initial conditions as in Fig. 1.2.1).

The *second type of solution* (and for us the more important!) is *the steady state solution* $\nu = 0$. This means that $\partial()/\partial t = -i\nu() = 0$ for all variables. Without the presence of rotation, this steady state solution would be trivial: $u^* = v^* = w^* = p' = 0$. But *with rotation*, an examination of (2.3.13) or (2.3.12) shows that this is *the geostrophic mode*: $U = 0$, $\nabla \cdot \mathbf{v}^* = \partial U/\partial x = 0$, *but* $V = ikP/f$, i.e.,

$$v^* = \frac{1}{f}\frac{\partial p'}{\partial x}$$

This is a steady state, but nontrivial, geostrophic solution. If we add a dependence of f on latitude, the geostrophic solution becomes the Rossby waves solution, which is not steady state, but is still much *slower* than gravity waves or sound waves.

2.3.2 General wave solution of the perturbation equations in a resting, isothermal atmosphere

So far we have been making drastic approximations to obtain "pure" elementary waves (sound, inertia and gravity oscillations). We now consider a more general case, including all waves simultaneously. We consider again the equations for small perturbations (2.3.2), and assume a resting, isothermal basic state in the atmosphere: $T_0(z) = T_{00}$, a constant. Then

$$N^2 = g\frac{d\ln\theta}{dz} = -g\kappa\frac{d\ln p_0}{dz} \tag{2.3.15}$$

where $\kappa = R/C_p = 0.4$. Since the basic state is hydrostatic,

$$N^2 = g\kappa\frac{\rho_0 g}{p_0} = g\kappa\frac{g}{RT_0} = \frac{g\kappa}{H} \tag{2.3.16}$$

These equations show that for an isothermal atmosphere, both N^2 and the scale height $H = RT/g$ are constant.

We continue considering an f-plane, a reasonable approximation for horizontal scales L small compared to the radius of the earth: $L << a$. If L were not small compared with the radius of the earth, we would have to take into account the variation of the Coriolis parameter with latitude, and spherical geometry. With some manipulation, assuming that the waves propagate along the x-axis, and there is no y-dependence, the perturbation equations (2.3.2) become

$$\frac{\partial u^*}{\partial t} = +fv^* - \frac{\partial p'}{\partial x} \tag{2.3.17a}$$

$$\frac{\partial v^*}{\partial t} = -fu^* \tag{2.3.17b}$$

$$\alpha\frac{\partial w^*}{\partial t} = -\frac{\partial p'}{\partial z} - \rho' g \tag{2.3.17c}$$

$$\beta\frac{\partial \rho'}{\partial t} = -\left(\frac{\partial u^*}{\partial x} + \frac{\partial w^*}{\partial z}\right) \tag{2.3.17d}$$

$$\frac{g}{C_p}\frac{\partial s^*}{\partial t} = -w^* N^2 \tag{2.3.17e}$$

$$s^* = C_p\left(\frac{p'}{c_s^2} - \rho'\right) \tag{2.3.17f}$$

In these equations we have introduced two constants α and β as *markers* for the hydrostatic and the quasi-Boussinesq approximations respectively. They can take the value 1 or 0. If we make $\alpha = 0$, it indicates that we are making the *hydrostatic* approximation, i.e., neglecting the vertical acceleration in (2.3.17c). If we make $\beta = 0$, it indicates that we are making the *anelastic or quasi-Boussinesq* approximation, i.e., assuming that the mass weighted three-dimensional divergence is zero. Otherwise the markers take the value 1. These markers will be used in the next section, where we discuss filtering approximations.

We now try plane wave solutions, where the basic state is a function of z of the form

$$(u^*, v^*, w^*, p', \rho') = (U(z), V(z), W(z), P(z), R(z))e^{i(kx-\nu t)} \tag{2.3.18}$$

Instead of assuming a z-dependence of the form $e^{i(mz)}$, we will determine it explicitly. If the horizontal scale is not small compared with the radius of the earth, $L \sim a$, then the solutions are of the form $(u^*, v^*, w^*, p', \rho') = (U(z), V(z), W(z), P(z), R(z))A(\varphi)e^{i(s\lambda-\nu t)}$, and the equation obtained for $A(\varphi)$ is *the Laplace tidal equation.*

Substituting the assumed form of the solution (2.3.18) into (2.3.17) we get

$$-i\nu U = -ikP + fV \tag{2.3.19a}$$

$$-i\nu V = -fU \tag{2.3.19b}$$

$$-i\nu\alpha W = -Rg - \frac{dP}{dz} \tag{2.3.19c}$$

$$-i\nu\beta R = -ikU - \frac{dW}{dz} \tag{2.3.19d}$$

$$-i\nu\left(\frac{P}{c_s^2} - R\right) = -W\frac{N^2}{g} \tag{2.3.19e}$$

From (2.3.19a) and (2.3.19b)

$$U = \frac{k\nu}{\nu^2 - f^2}P \tag{2.3.19f}$$

From (2.3.19d) and (2.3.19f)

$$\beta R = \frac{k^2}{\nu^2 - f^2}P - \frac{i}{\nu}\frac{dW}{dz} \tag{2.3.19g}$$

From (2.3.19c) and (2.3.19e)

$$\frac{dP}{dz} + \frac{g}{c_s^2}P = \frac{i}{\nu}(\nu^2\alpha - N^2)W \tag{2.3.19h}$$

From (2.3.19e) and (2.3.19g)

$$\frac{dW}{dz} + \beta\frac{N^2}{g}W = \frac{i\nu}{c_s^2}\left[\frac{\beta(\nu^2 - f^2) - c_s^2k^2}{\nu^2 - f^2}\right]P \tag{2.3.19i}$$

From (2.3.19h) and (2.3.19i)

$$\left(\frac{d}{dz} + \frac{g}{c_s^2}\right)\left(\frac{d}{dz} + \beta\frac{N^2}{g}\right)W = -\frac{1}{c_s^2}\left[\frac{(\beta(\nu^2 - f^2) - c_s^2k^2)(\nu^2\alpha - N^2)}{\nu^2 - f^2}\right]W \tag{2.3.20}$$

or a similar equation for P. This last equation is of the form

$$\frac{d^2W}{dz^2} + A\frac{dW}{dz} + BW = 0$$

In order to eliminate the first derivative, we try a substitution of the form $W = e^{\delta z}\Omega$, and obtain $d^2\Omega/dz^2 + C\Omega = 0$. This requires that we choose $\delta = -A/2$, and in that case $C = B - A^2/4$.

From (2.3.20), the variable substitution, and additional sweat, we finally obtain

$$\frac{d^2\Omega}{dz^2} + n^2\Omega = 0 \tag{2.3.21}$$

where

$$n^2 = \frac{(\beta(\nu^2 - f^2) - c_s^2k^2)(\nu^2\alpha - N^2)}{c_s^2(\nu^2 - f^2)} - \frac{1}{4}\left(\beta\frac{N^2}{g} + \frac{g}{c_s^2}\right)^2 \tag{2.3.22}$$

This is the frequency dispersion relationship for waves in an atmosphere with an isothermal basic state. Given a horizontal structure of the wave (k), and its frequency (ν), (2.3.22) determines the vertical structure (n) of Ω (and W), and vice versa. The same FDR would have been obtained making the substitution $Q = e^{-\delta z}P$, and solving for Q.

Equation (2.3.22) indicates that depending on the sign of n^2 we can have either external or internal wave solutions.

2.3.2.1 External waves

If $n^2 < 0$, the vertical wavenumber n is imaginary, $n = im$. The solution of (2.3.21) is then $\Omega = Ae^{mz} + Be^{-mz}$, or, going back to the vertical velocity,

$$w^*(x, z, t) = e^{i(kx-\nu t)}e^{-\frac{1}{2}\left(\beta\frac{N^2}{g}+\frac{g}{c_s^2}\right)z}\left(Ae^{mz} + Be^{-mz}\right) \tag{2.3.23}$$

These are external waves (the waves do not oscillate in the vertical, and therefore do not propagate vertically). If the boundary condition at the ground is that the vertical velocity is zero, then $\Omega = Ae^{mz} + Be^{-mz} = 0$ at $z = 0$, so that $A + B = 0$, and

$$w^*(x, z, t) = e^{i(kx-\nu t)}e^{-\frac{1}{2}\left(\beta\frac{N^2}{g}+\frac{g}{c_s^2}\right)z}\ 2A\sinh(mz)$$

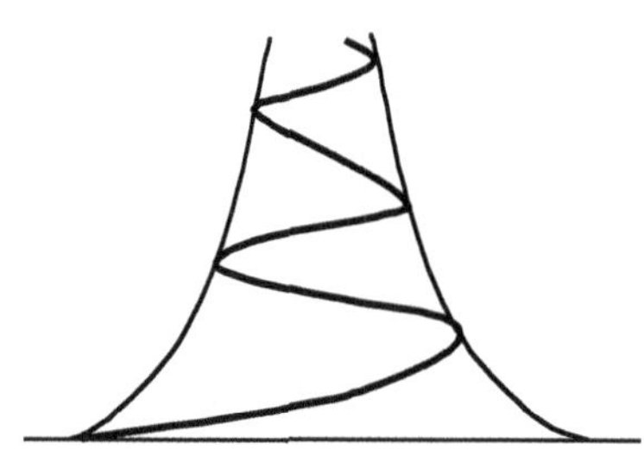

Figure 2.3.2: Schematic of density weighted internal (vertically propagating) waves.

which has an exponential behavior in z. Since $\sinh(mz)$ cannot be zero above the ground, an upper boundary condition of a rigid top can only be satisfied if $A = 0$. In other words, we cannot have external waves with rigid top and bottom boundary conditions: external waves require a free surface at the top (or at the bottom).

2.3.2.2 Internal waves

If $n^2 > 0$, the vertical wavenumber n is real:

$$w^*(x, z, t) = e^{i(kx-\nu t)}(Ae^{inz} + Be^{-inz})e^{-\frac{1}{2}\left(\beta\frac{N^2}{g}+\frac{g}{c_s^2}\right)z} \tag{2.3.24}$$

A, B are determined from the boundary conditions. Now there is both vertical and horizontal propagation. For example, if there is a rigid bottom, we have again $A + B = 0$, and the solution becomes

$$w^*(x, z, t) = A\left(e^{i(kx+nz-\nu t)} - e^{i(kx-nz-\nu t)}\right)e^{-\frac{1}{2}(\beta\frac{N^2}{g}+\frac{g}{c_s^2})z}$$

The shape of internal waves in the vertical is shown schematically in Fig. 2.3.2.

2.3.3 Analysis of the FDR of wave solutions in a resting, isothermal atmosphere

We will now plot the general FDR equation (2.3.22). We assume $T_{00} = 250$ K and $f = 2\Omega \sin 45^0 \approx 10^{-4}$ s^{-1}. Then, the speed of sound is $c_s^2 = \gamma RT \approx 10^5$ m^2/s^2, or $c_s \approx 320$ m/s, the scale height is $H = RT/g = 7.3$ km $= 7300$ m, and the Brunt–Vaïsälä frequency is $N^2 = gd(\ln\theta_0)/dz = g\kappa/H$ for the isothermal atmosphere, or about 4×10^{-4} s^{-2}. Note that the frequency associated with inertial oscillations is much lower than the frequency associated with gravitational oscillations.

$$f \sim 10^{-4}\,\mathrm{s}^{-1} << N \sim 10^{-2}\,\mathrm{s}^{-1} \tag{2.3.25}$$

We first plot in Fig. 2.3.3 the FDR (2.3.22), with $\alpha = \beta = 1$, i.e., without making either the hydrostatic or the quasi-Boussinesq approximations. Note that this equation contains four solutions for the frequency ν, plus an additional solution $\nu = 0$, the geostrophic mode that satisfies nontrivially (2.3.19).

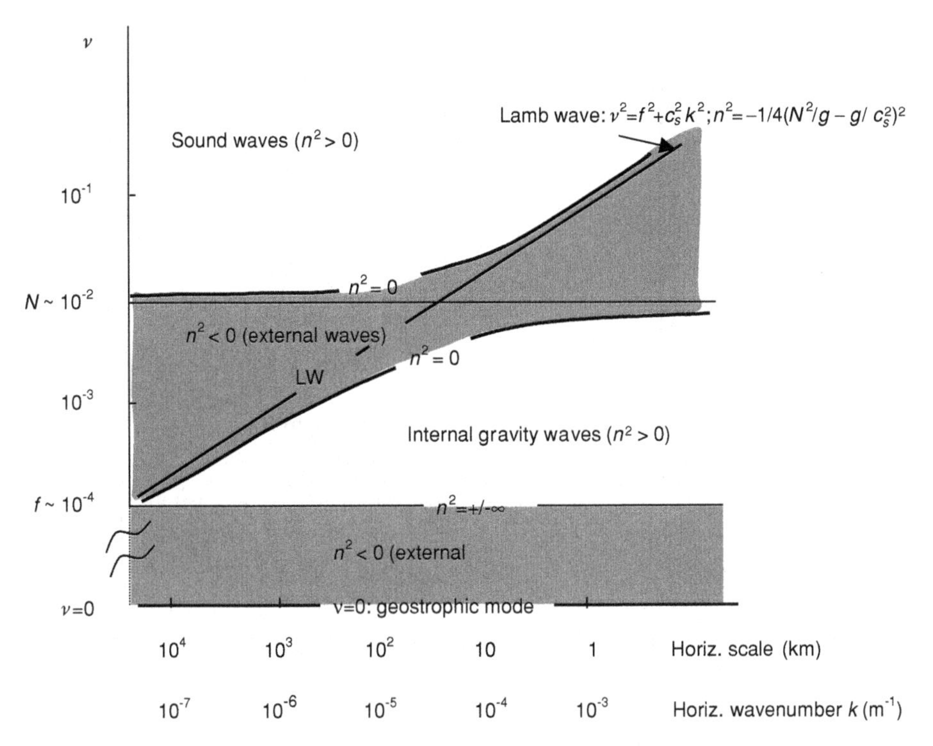

Figure 2.3.3: Schematic of the frequencies of small perturbations in an isothermal resting atmosphere as a function of k, the horizontal wavenumber (the horizontal scale is its inverse), and the vertical wavenumber n. Shaded regions represent $n^2 < 0$, external waves.

2.4 Filtering approximations

When we neglect the time derivative of one of the equations of motion, we convert it from a prognostic equation into a diagnostic equation, and eliminate with it one type of solution. Physically, we eliminate a restoring force that supports a certain type of wave. We call this a "filtering approximation". Use of the quasi-geostrophic filtering approximation that eliminates both sound and gravity waves made possible the successful forecast of Charney *et al.* (1950). Currently most global models and some regional models use the hydrostatic approximation, which filters sound waves. In this section we explore the effect of the filtering approximations.

2.4.1 Quasi-geostrophic approximation

As we have already seen, without rotation, if we assume a steady state, the solution of (2.3.19) would be a trivial solution: all perturbations would be equal to zero. However, with rotation, if we assume steady state solutions, and neglect all time derivatives $\nu = 0$, we obtain from the perturbed equations (2.3.17), the geostrophic mode, a nontrivial solution:

$$\left.\begin{aligned} &V = ikP/f \\ &U = 0 \\ &\frac{dP}{dz} = -Rg \\ &\frac{dW}{dz} = 0 \\ &P = c_s^2 R \end{aligned}\right\} \tag{2.4.1}$$

For the continuous perturbation equations (2.3.17), this means:

$$\left.\begin{aligned} &fv^* = -\frac{\partial p'}{\partial x} && \text{(geostrophically balanced flow)} \\ &\frac{\partial v^*}{\partial t} = 0 && \text{(steady state flow)} \\ &0 = -\frac{\partial p'}{\partial z} - \rho' g && \text{(hydrostatically balanced flow)} \\ &w^* = 0 && \\ &\frac{\partial w^*}{\partial z} = \frac{\partial u^*}{\partial x} = 0 && \text{(horizontal, nondivergent flow)} \\ &s^* = C_p\left(\frac{p'}{c_s^2} - \rho'\right) = 0 && \text{(pressure perturbations are proportional to density perturbations multiplied by the speed of sound squared, which is true whenever the hydrostatic equation is valid)} \end{aligned}\right\} \tag{2.4.2}$$

This is the "ultimate" filtering approximation: it filters out sound waves, inertia and gravity oscillations.

For large horizontal scales we have to include the effects of varying rotation, and the f-plane becomes a β-plane: $f = f_0 + \beta y$. When horizontal advection by the basic flow is included, the stationary geostrophic flow solution becomes quasi-stationary (slowly varying). The waves corresponding to the geostrophic mode are Rossby-type waves with a frequency small compared with the Coriolis or inertial frequency $\nu \approx Uk - \beta/k \sim 10^{-5} - 10^{-6}\ \mathrm{s}^{-1}$. Rossby waves are quasi-geostrophic ($\nu^2 << f^2$), hydrostatically balanced, and the flow is quasi-horizontal ($w^*/H << U^*/L$), and therefore quasi-nondivergent ($\nabla \cdot \mathbf{v}_h \approx 0$).

Note that this type of quasi-geostrophic solution, fundamental for NWP, is still present in the general equations of motion, and survives as a solution when we make either the anelastic or the hydrostatic approximation in order to filter out sound waves.

2.4.2 Quasi-Boussinesq or anelastic approximation (Ogura and Phillips, 1962)

We now substitute $\beta = 0$ in (2.3.19d). This means that we neglected the time derivative $\partial\rho'/\partial t$ compared with $\nabla_H \cdot \mathbf{v}^*$, $\partial w^*/\partial z$ in the continuity equation. With this approximation, the equations become "anelastic", i.e., they do not allow the presence of sound waves, which require three-dimensional divergence and convergence for their propagation. Consider the terms that are neglected in the FDR (2.3.22):

(1) $\nu^2 - f^2 << c_s^2 k^2$, i.e., the frequency of retained solutions is much smaller than that of sound waves, therefore this also filters out the Lamb waves, i.e., horizontally propagating sound waves.

(2) $N^2/g << g/c_s^2$. This approximation is justified if

$$\frac{N^2}{g} = \frac{1}{\theta_0}\frac{d\theta_0}{dz} << \frac{g}{\gamma R T_0}, \text{ i.e., } \frac{\gamma H}{\theta_0}\frac{d\theta_0}{dz} << 1$$

In other words, the deep anelastic approximation is justified for a model for which the potential temperature does not change too much within the depth $\gamma RT/g \sim 10$ km. This is a reasonable approximation for the standard troposphere (not for deep flow into the stratosphere), since for the troposphere: $\Delta\theta_0/\theta_0 \sim 30\ \text{K}/300\ \text{K} \sim 0.1$.

For models that are so shallow that not only $\Delta\theta_0/\theta_0 << 1$, but also $\Delta T_0/T_0 << 1$, we can also neglect $\partial\rho_0/\partial z$ in the continuity equation, and assume $\nabla_3 \cdot \mathbf{v}' = 0$, not just $\nabla_3 \cdot \mathbf{v}^* = 0$. In this case we treat the atmosphere as if it was an incompressible fluid. This approximation is only accurate for very shallow atmospheric models (less than 1 km depth), but is very appropriate for ocean models, since water is well approximated as an incompressible fluid.

Fig 2.4.1 schematically shows the FDR when we make the anelastic approximation. From (2.3.22), and letting $\beta = 0$ (with $\alpha = 1$), we can derive the frequency of inertia-gravity waves with the anelastic approximation:

$$\nu^2 = f^2 \frac{n^2 + p^2}{n^2 + p^2 + k^2} + N^2 \frac{k^2}{n^2 + p^2 + k^2} \tag{2.4.3}$$

where p is like the inverse of a vertical wavelength

$$p = \frac{1}{2}\frac{g}{c_s^2} \sim \frac{1}{20}\ \text{km}$$

From (2.4.3) we see that, for internal ($n^2 > 0$) inertia-gravity waves, $f^2 < \nu^2 < N^2$, the frequency ν is between the Coriolis and Brunt–Vaïsälä frequencies. Note from Fig. 2.4.1 that for these waves, $\partial\nu^2/\partial k^2 > 0$, *but* $\partial\nu^2/\partial n^2 < 0$. This implies (since we can assume without loss of generality that $k > 0$) that the horizontal group velocity for gravity waves $\partial\nu/\partial k$ has the same sign as the phase velocity (the energy of gravity

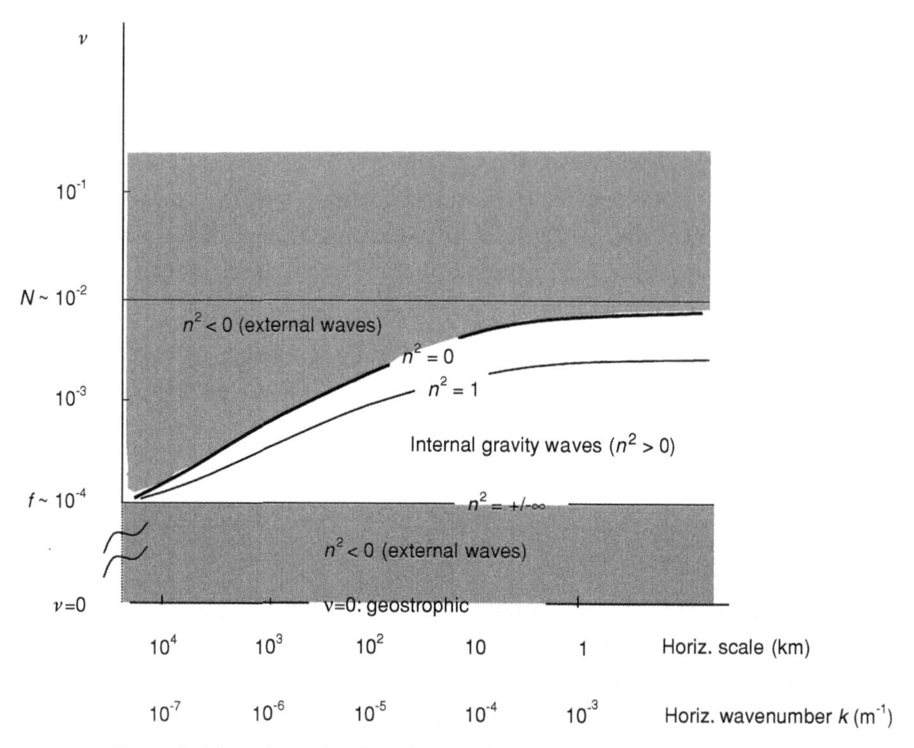

Figure 2.4.1: Schematic of the frequencies of small perturbations in an isothermal resting atmosphere when the quasi-Boussinesq or anelastic approximation is made ($\beta = 0$).

waves moves in the same direction as the phase speed in the horizontal). In the vertical the opposite is true: if the group velocity is upwards, which happens for example when gravity waves are generated by mountain forcing, the phase velocity is downwards.

Because the anelastic equation filters out acoustic internal waves (as well as the Lamb wave) it is widely used for problems in which the hydrostatic approximation cannot be made, as is the case for convection. For example, the ARPS model is based on deep anelastic equations. The FDR with the quasi-Boussinesq approximation is shown schematically in Fig. 2.4.1.

2.4.3 Hydrostatic approximation

If we neglect the vertical acceleration $\partial w^*/\partial t$ in the vertical momentum equation (2.3.17c), letting $\alpha = 0$ (with $\beta = 1$), we get the FDR

$$n^2 = \frac{-(\nu^2 - f^2 - c_s^2 k^2)N^2}{c_s^2(\nu^2 - f^2)} - \frac{1}{4}\left(\frac{N^2}{g} - \frac{g}{c_s^2}\right)^2 \tag{2.4.4}$$

This FDR has two solutions: the horizontally propagating external sound wave (Lamb wave) solution, which unfortunately is retained:

$$\nu^2 = f^2 + c_s^2 k^2, \quad n^2 = -1/4(N^2/g - g/c_s^2)^2 \tag{2.4.5}$$

and inertia-gravity waves. From (2.4.4) we can derive the following relationship for inertia-gravity waves: using $N^2 = \kappa g/H,\ H = RT/g,\ c_s^2 = \gamma RT$

$$n^2 = \frac{N^2 k^2}{\nu^2 - f^2} - \frac{1}{4H^2}, \text{ or } \nu^2 = f^2 + \frac{N^2 k^2}{n^2 + \dfrac{1}{4H^2}} \tag{2.4.6}$$

Fig. 2.4.2 shows the relationship between frequency and horizontal and vertical wavenumbers with the hydrostatic equation.

Exercise 2.4.1: Derive (2.4.6) from (2.4.5).

When are we justified in using the hydrostatic equation? By taking $\alpha = 0$, we neglected the time derivative of the vertical velocity compared to $\rho'/\rho_0 g$. Note that it is not enough to find $dw/dt << g$ to make the hydrostatic approximation: the vertical acceleration is small compared to gravity even for strong vertical motions, as in a

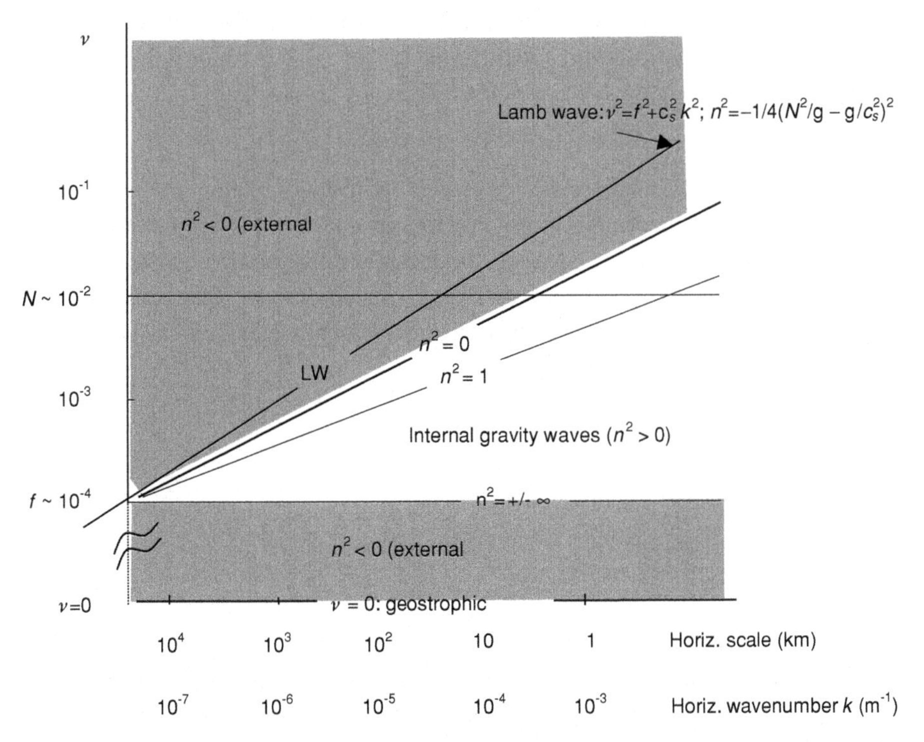

Figure 2.4.2: Schematic of the frequencies of small perturbations in an isothermal resting atmosphere when the hydrostatic approximation is made ($\alpha = 0$).

cumulus cloud. The hydrostatic approximation requires that the vertical acceleration be small compared with the buoyancy $(\rho'/\rho_0)g$, or gravitational acceleration within the fluid. It can be shown by scale analysis that the hydrostatic approximation is valid as long as we are dealing with shallow flow ($H/L << 1$). For quasi-geostrophic flow, the condition for hydrostatic balance is valid even if $H/L \sim 1$. This implies that the hydrostatic approximation is very accurate for models with grid sizes of the order of 100 km or larger, and still quite acceptable for quasi-geostrophic flow, even when the horizontal grid size of the model approaches 10 km. However, the hydrostatic equation is not valid for models with grid sizes of the order of 10 km that attempt to resolve explicitly cumulus convection. Fig. 2.4.2 shows that for high frequencies $\nu \sim N$ or larger, or small horizontal scales the hydrostatic approximation distorts the original FDR (compare with Fig. 2.3.3).

Exercise 2.4.2: Show by scale analysis that as long as we are dealing with shallow flow ($H/L << 1$) the hydrostatic approximation is valid.

Exercise 2.4.3: Show that the condition for quasi-geostrophic balance, the hydrostatic approximation is valid even if $H/L \sim 1$.

We now summarize in Table 2.4.1 the characteristics of the different types of waves and the approximations that can be used to filter them out. For more details about Rossby waves and the filtering of inertia gravity waves, see Section 2.5, where these topics are discussed in the context of the SWEs.

Notes

(1) In normal mode analysis of large-scale (hydrostatic) motion, or of atmospheric models, it is customary to find a horizontal structure equation and a vertical structure equation, associated by a separation constant h, where h is denoted as "equivalent depth" (e.g., Williamson and Temperton, 1981). In our simple f-plane case, the horizontal structure equation for the inertia gravity waves (2.4.6) is

$$\frac{\nu^2 - f^2}{k^2} = gh \tag{2.4.7}$$

and the vertical structure equation

$$n^2 = \frac{1}{H^2}\left(\frac{\kappa H}{h} - \frac{1}{4}\right) \tag{2.4.8}$$

where we have used $N^2 = \kappa g/H$.

The reason h is called the equivalent depth is that internal modes are governed by equations similar to the SWEs with depth h. However, h is not a constant but a function of vertical wavenumber, and therefore the analogy is only approximate.

Table 2.4.1. *Summary of wave characteristics and the filtering approximations (adapted from Zhang, personal communication, 1996).*

Type of wave (typical amplitude)	Phase speed	Restoring force	Filtering approximations
Acoustic (less than 0.1 hPa, noise level)	$\sqrt{\gamma RT}$ (320 m/s)	Compression	Hydrostatic anelastic quasi-geostrophic
External gravity (if initial conditions are not balanced, 10 hPa)	$\sqrt{gH}$ (320 m/s for $H = 10$ km)	Gravity	No free surface at the top or the bottom, or no net column mass convergence
Internal gravity (0.1–1 hPa)	$\sim \frac{1}{k}\sqrt{\frac{N^2k^2}{k^2+n^2}} \sim N/k$ (50 m/s for $L =$ 30 km)	Buoyancy (gravitational acceleration within fluid)	Neutral stratification ($N = 0$), or $\frac{\partial \nabla \cdot \mathbf{v}_H}{\partial t} = 0$
Inertia	f/k (15 m/s for $L = 1000$ km)	Coriolis force (f)	No rotation ($f = 0$)
Rossby (20 hPa)	$U - \beta/k^2$ (relative phase speed $\sim$ 20–50 m/s depending on L)	Variation of f with latitude (β effect) $d\varsigma/dt = -\beta v$	Constant f ($\beta = 0$)

(2) With the hydrostatic approximation, the geopotential energy gz and the internal energy $C_v T$ of an air column are related to each other, since$\int_{z_s}^{\infty} \rho gz dz = \int_{z_s}^{\infty} -(\partial p/\partial z)z\, dz = [-pz]_{z_s}^{\infty} + \int_{z_s}^{\infty} p dz =$ $p_s z_s + \int_{z_s}^{\infty} \rho RT dz$. Here the subscript s represents the earth's surface, and $\lim_{z\to\infty} pz = 0$ is assumed. So, when $z_s = 0$, the ratio of the potential to the internal energy of a column is equal to $R/C_v = 0.4$. When z_s is not constant, the total potential energy (Lorenz, 1955) is given by

$$\int_{z_s}^{\infty} \rho(gz + C_v T) dz = p_s z_s + \int_{z_s}^{\infty} \rho C_p T dz$$

2.5 Shallow water equations, quasi-geostrophic filtering, and filtering of inertia-gravity waves

Consider now the SWEs (Fig. 2.5.1), valid for an incompressible hydrostatic motion of a fluid with a free surface $h(x, y, t)$. "Shallow" means that the vertical depth is much smaller than the typical horizontal depth, which justifies the hydrostatic

Figure 2.5.1: Schematic of the shallow water model: a hydrostatic, incompressible fluid with a rigid bottom $h_s(x, y)$, a free surface $h(x, y, t)$, and horizontal scales L much larger than the mean vertical scale H.

approximation. These equations are not only appropriate for representing a shallow mass of water (e.g., river flow, storm surges), but they are prototypical of the primitive equations based on the hydrostatic approximation and are frequently used to test numerical schemes. The shallow water horizontal momentum equations are

$$\frac{d\mathbf{v}}{dt} = -f\mathbf{k} \times \mathbf{v} - \nabla\phi \tag{2.5.1}$$

where

$$\frac{d}{dt} = \frac{\partial}{\partial t} + \mathbf{v} \cdot \nabla \qquad \mathbf{v} = \mathbf{v}_H = u\mathbf{i} + v\mathbf{j} \qquad \phi = gh$$

The continuity equation is

$$\frac{d(\phi - \phi_s)}{dt} = -(\phi - \phi_s)\nabla \cdot \mathbf{v}$$

which can also be written as

$$\frac{\partial\phi}{\partial t} = -\nabla \cdot [(\phi - \phi_s)\mathbf{v}] \tag{2.5.2}$$

Here $\phi_s = gh_s(x, y)$ and h_s is the bottom topography.

Exercise 2.5.1: Derive the SWE from the primitive equations assuming hydrostatic, incompressible motion, and that the horizontal velocity is uniform in height. Is the vertical velocity uniform in height as well?

We now derive the equation of conservation of potential vorticity: expanding the total derivative of the momentum equation and making use of the relationship

$$\mathbf{v}_H \cdot \nabla\mathbf{v}_H = \nabla\left(\mathbf{v}_H^2/2\right) + \varsigma\,\mathbf{k} \times \mathbf{v}_H$$

where $\varsigma = \mathbf{k} \cdot \nabla \times \mathbf{v}_H$ we obtain

$$\frac{\partial\varsigma}{\partial t} + \mathbf{v} \cdot \nabla\varsigma + \varsigma\nabla \cdot \mathbf{v} = -f\nabla \cdot \mathbf{v} - \mathbf{v} \cdot \nabla f \tag{2.5.3}$$

or (since $df/dt = \mathbf{v} \cdot \nabla f$)

$$\frac{d(f + \varsigma)}{dt} = -(f + \varsigma)\nabla \cdot \mathbf{v} \tag{2.5.4}$$

which indicates that the absolute vorticity $(f + \varsigma)$ of a parcel of "water" increases with its convergence (or vertical stretching).

Eliminating the divergence, we obtain

$$\frac{d}{dt}\left(\frac{f+\varsigma}{\phi-\phi_s}\right)=0 \tag{2.5.5}$$

where

$$q=\left(\frac{f+\varsigma}{\phi-\phi_s}\right) \tag{2.5.6}$$

is the potential vorticity: the absolute vorticity divided by the depth of the fluid.

Exercise 2.5.2: Give a physical interpretation of the equation of conservation of potential vorticity.

The conservation of potential vorticity is an extremely powerful dynamical constraint. In a multilevel primitive equation model, the *isentropic potential vorticity* (the absolute vorticity divided by the distance between two surfaces of constant potential temperature) is also individually conserved. If the initial potential vorticity distribution is accurately represented in a numerical model, and the model is able to transport potential vorticity accurately, then the forecast will also be accurate.

We now consider small perturbations on a flat bottom and a mean height $\Phi = gH = const.$ on a constant f-plane.

$$\frac{\partial \mathbf{v}'}{\partial t}=-f\mathbf{k}\times\mathbf{v}'-\nabla\phi' \tag{2.5.7}$$

and

$$\frac{\partial \phi'}{\partial t}=-\Phi\nabla\cdot\mathbf{v}' \tag{2.5.8}$$

(note that (2.5.7) and (2.5.8) are the same equations as in Section 2.4.5 on horizontal sound (Lamb) waves, with $gH = c_s^2 = \gamma RT_0$, $g\phi = p'/\rho_0$).

Assume solutions of the form $(u', v', \phi')e^{i(kx-\nu t)}$. Then the FDR is

$$\nu(\nu^2-f^2-\Phi k^2)=0 \tag{2.5.9}$$

with three solutions for ν:

$$\nu^2=f^2+\Phi k^2 \tag{2.5.10}$$

the frequency of inertia-gravity waves, analogous to the inertia-Lamb wave, and $\nu = 0$, the geostrophic mode. As before, this is a geostrophic, nondivergent steady state solution $\frac{\partial}{\partial t}() = 0$, $\mathbf{v} = \frac{\mathbf{k}\times\nabla\phi}{f}$, $\nabla\cdot\mathbf{v} = 0$.

Following Arakawa (1997), we can now compare the FDR of inertia-gravity waves in the SWE with the FDR of a three-dimensional isothermal system using the hydrostatic approximation (2.4.4)–(2.4.6). We see that (2.5.10) is analogous to internal inertia-gravity waves for an isothermal hydrostatic atmosphere (2.4.6) if we define an equivalent depth such that $\Phi = gh_{eq}$:

$$h_{eq} = \frac{N^2/g}{n^2 + \dfrac{1}{4H_0^2}} = \frac{d\ln\theta_0/dz}{n^2 + \dfrac{1}{4H_0^2}} \tag{2.5.11}$$

and is analogous to the (external) inertia Lamb waves (2.4.5) if we define the equivalent depth as

$$h_{eq} = c_s^2/g = \gamma H_0 \tag{2.5.12}$$

2.5.1 Quasi-geostrophic scaling for the SWE

If we want to filter the inertia-gravity waves, as Charney did in the first successful numerical weather forecasting experiment (Chapter 1), we can develop a quasi-geostrophic version of the SWE. We can do it first for an f-plane ($f = f_0$).

Assume that the atmosphere is in *quasi-geostrophic* balance: $\mathbf{v} = \mathbf{v}_g + \mathbf{v}_{ag} = \mathbf{v}_g + \varepsilon\mathbf{v}'$ where we assume that the typical size of the ageostrophic wind is much smaller (order $\varepsilon = U/fL$, the Rossby number) than the geostrophic wind $\varepsilon\mathbf{v}' << \mathbf{v}_g$, and that the same is true for their time derivatives $\varepsilon\partial\mathbf{v}'/\partial t << \partial\mathbf{v}_g/\partial t$. The geostrophic wind is given by

$$\mathbf{v}_g = \frac{1}{f}\mathbf{k} \times \nabla\phi \tag{2.5.13}$$

Plugging these into the perturbation equations (2.5.7) and (2.5.8) we obtain

$$\frac{\partial \mathbf{v}_g}{\partial t} + \varepsilon\frac{\partial \mathbf{v}'}{\partial t} = -\nabla\phi - f\mathbf{k} \times \mathbf{v}_g - \varepsilon f\mathbf{k} \times \mathbf{v}' = -\varepsilon f\mathbf{k} \times \mathbf{v}' \tag{2.5.14}$$

In this equation, the dominant terms (pressure gradient and Coriolis force on the geostrophic flow) cancel each other (geostrophic balance), so that the smaller effect of the Coriolis force acting on the ageostrophic flow is left to balance the time derivative. From (2.5.8),

$$\frac{\partial\phi}{\partial t} = -\Phi\nabla \cdot \mathbf{v}_g - \varepsilon\Phi\nabla \cdot \mathbf{v}' = -\varepsilon\Phi\nabla \cdot \mathbf{v}' \tag{2.5.15}$$

Here the geostrophic wind is nondivergent, so that the time derivative of the pressure is given by the divergence of the smaller ageostrophic wind.

From (2.5.14) and (2.5.15) we can conclude that $\partial\mathbf{v}_g/\partial t$ and $\partial\phi/\partial t$ are of order ε, i.e., the geostrophic flow changes *slowly* (it is almost stationary compared with other types of motion), and that $\partial\mathbf{v}_{ag}/\partial t = \varepsilon\partial\mathbf{v}'/\partial t$, which is smaller than $\partial\mathbf{v}_g/\partial t$, is of order ε^2. With *quasi-geostrophic scaling* we neglect terms of $O(\varepsilon^2)$ and we obtain the linearized quasi-geostrophic SWE:

$$\frac{\partial \mathbf{v}_g}{\partial t} = -\nabla\phi - f\mathbf{k} \times \mathbf{v} = -f\mathbf{k} \times \mathbf{v}_{ag} \tag{2.5.16a}$$

$$\frac{\partial \phi}{\partial t} = -\Phi \nabla \cdot \mathbf{v} = -\Phi \nabla \cdot \mathbf{v}_{ag} \tag{2.5.16b}$$

$$\mathbf{v}_g = 1/f \, \mathbf{k} \times \nabla \phi \tag{2.5.16c}$$

Note that in (2.5.16) there is only one independent time derivative because of the geostrophic relationship (we lost the other two time derivatives when we neglected the term $\partial \mathbf{v}_{ag}/\partial t$). Physically, this means that we only allow divergent motion to exist as required to maintain the quasi-geostrophic balance, and eliminate the degrees of freedom necessary for the propagation of gravity waves.

We can rewrite (2.5.16) as

$$\frac{\partial u_g}{\partial t} = -\frac{\partial \phi}{\partial x} + f v = f v_{ag} \tag{2.5.17a}$$

$$\frac{\partial v_g}{\partial t} = -\frac{\partial \phi}{\partial y} - f u = -f u_{ag} \tag{2.5.17b}$$

$$\frac{\partial \phi}{\partial t} = -\Phi \left(\frac{\partial u}{\partial x} + \frac{\partial v}{\partial y} \right) = \Phi \left(\frac{\partial u_{ag}}{\partial x} + \frac{\partial v_{ag}}{\partial y} \right) \tag{2.5.17c}$$

$$u_g = -\frac{1}{f}\frac{\partial \phi}{\partial y}; v_g = \frac{1}{f}\frac{\partial \phi}{\partial x} \tag{2.5.17d}$$

We can compute the equation for the geostrophic vorticity evolution from (2.5.17) by taking the x-derivative of (2.5.17b) minus the y-derivative of (2.5.17a):

$$\frac{\partial \zeta}{\partial t} = -f_0 \left(\frac{\partial u}{\partial x} + \frac{\partial v}{\partial y} \right) - \beta v \tag{2.5.18}$$

where the last term in (2.5.18) appears if we are on a β-plane: $f = f_0 + \beta y$. Then we can eliminate the (ageostrophic) divergence between (2.5.18) and (2.5.17c) and obtain the linear quasi-geostrophic potential vorticity equation on a β-plane:

$$\frac{\partial}{\partial t} \left(\frac{\zeta}{f_0} - \frac{\phi}{\Phi} \right) = -\frac{\beta}{f_0^2}\frac{\partial \phi}{\partial x} \tag{2.5.19}$$

or, since $\zeta = \nabla^2 \phi / f_0$,

$$\frac{\partial}{\partial t} \left(\frac{\nabla^2 \phi}{f_0^2} - \frac{\phi}{\Phi} \right) = -\frac{\beta}{f_0^2}\frac{\partial \phi}{\partial x} \tag{2.5.20}$$

Note that there is a single independent variable (ϕ) so that there is a single solution for the frequency. If we neglect the β-term (i.e., assume an f-plane) and allow for plane-wave-type solutions $\phi = F e^{i(kx - \nu t)}$, the only solution of the FDR in (2.5.20) is $\nu = 0$, the geostrophic mode. This confirms that by eliminating *the time derivative* of the ageostrophic (divergent) wind $\mathbf{v}_{ag}$, we have eliminated the inertia-gravity wave solution. If we assume a β-plane, i.e., keep the β term in (2.5.20), the quasi-geostrophic FDR becomes

$$\nu = \frac{-\beta k}{k^2 + f_0^2/\Phi} \tag{2.5.21}$$

The Rossby waves are the essential "weather waves", and as shown in Table 2.4.1, have rather large amplitudes (up to 50 hPa). The ageostrophic flow associated with these waves is responsible for the upward motion that produces precipitation ahead of the troughs.

In a multilevel model, the FDR (2.5.21) can be used with the equivalent depths (2.5.11), (2.5.12) applied to the baroclinic (internal) and barotropic Rossby waves, respectively. With these definitions, we can say that the waves in the atmosphere are analogous to the SWE waves. However, because h_{eq} appears as a separation constant in the definition of the normal modes of the atmosphere, the equivalent depth depends on the vertical wavenumber, and on the type of wave considered (Lamb or inertia-gravity waves).

Exercise 2.5.3: Show that the quasi-geostrophic PVE (potential vorticity equation) for nonlinear SWE is

$$\left(\frac{\partial}{\partial t} + u_g \frac{\partial}{\partial x} + v_g \frac{\partial}{\partial y}\right)\left(\frac{\nabla^2 \phi}{f_0^2} - \frac{\phi}{\Phi}\right) = -\frac{\beta}{f_0^2}\frac{\partial \phi}{\partial x} \tag{2.5.22}$$

using similar scaling arguments.

Exercise 2.5.4: Allow for a basic flow $u_{g(total)} = U + u_g$; $v_{g(total)} = v_g$, in (2.5.22). How will this change the FDR (2.5.21)?

Exercise 2.5.5: Estimate the initial time derivative for typical values of the horizontal wavenumber $L = 2000$ km, 8000 km, that Richardson would have observed, i.e., compare the frequency v for the external (barotropic, $H \sim 10$ km) mode for inertia–gravity waves and for Rossby waves.

Exercise 2.5.6: Derive the formula for group velocity in the x-direction for Rossby waves.

Exercise 2.5.7: Using typical values of long and short synoptic waves (e.g., horizontal wavelengths of 8000 km and 2000 km respectively), calculate the phase speed and the group velocity of Rossby waves for the barotropic mode and the first baroclinic mode ($H \sim 10$ km and 1 km respectively).

2.5.2 Inertia-gravity waves in the presence of a basic flow

As we just saw, the SWEs are a simple version of the primitive equations, and are widely used to understand numerical and dynamical processes in primitive equations. As we noted in Chapter 1, filtered quasi-geostrophic models have been substituted by primitive equation models for NWP, because the quasi-geostrophic filtering is not an accurate approximation (it assumes that the Rossby number U/fL is much smaller than 1). Recall that quasi-geostrophic filtering was introduced by Charney *et al.* (1950) in order to eliminate the problem of gravity waves (which requires a small time step) whose high frequencies produced a huge time derivative in Richardson's computation, masking the time derivative of the actual weather signal.

An alternative way to deal with the presence of fast gravity waves without resorting to quasi-geostrophic filtering is the use of semi-implicit time schemes (to be discussed in Chapter 3). Consider small perturbations in the SWE including a basic flow U in the x-direction. Then the total linearized time derivative becomes

$$\frac{d}{dt} = \frac{\partial}{\partial t} + U\frac{\partial}{\partial x}$$

In that case, when we assume solutions of the form $Ae^{i(kx-\nu t)}$, $d/dt = i(-\nu + kU)$. Therefore the FDR remains the same except that ν is replaced by $\nu - kU$. The FDR for small perturbations in the SWE with a basic flow U is therefore

$$(\nu - kU)[(\nu - kU)^2 - f^2 - \Phi k^2] = 0 \tag{2.5.23}$$

As noted before, this has three solutions, quasi-geostrophic flow (which is steady state, except for the uniform translation with speed U) and two solutions for the inertia-gravity waves, modified by the basic flow translation:

$(\nu_G - kU) = 0$ (geostrophic mode)

$[(\nu_{IGW} - kU)^2 - f^2 - \Phi k^2] = 0$ (inertia gravity waves, modified by the basic flow U)

The phase speed of the inertia-gravity wave is given by

$$c_{IGW} = \frac{\nu_{IGW}}{k} = U \pm \sqrt{\frac{f^2}{k^2} + \Phi} \tag{2.5.24}$$

Finally, we note that for the Lamb wave (as well as for the external gravity wave), the phase speed of the inertia-gravity wave is dominated by the term $\sqrt{\Phi} \approx \sqrt{g \times 10\ \text{km}} \approx 300$ m/s. As we will see in Section 3.2.5, it is possible to avoid using costly small time steps by means of a *semi-implicit time scheme*. An implicit time scheme has no constraint on the time step. Therefore, in a semi-implicit scheme, the terms that give rise to the fast gravity waves, namely the horizontal divergence and the horizontal pressure gradient are written implicitly, while the rest of the SWE terms can be written explicitly. The terms generating the gravity wave are underlined in the following nonlinear SWE:

$$\left.\begin{aligned}
\frac{\partial u}{\partial t} + u\frac{\partial u}{\partial x} + v\frac{\partial u}{\partial y} &= \underline{-\frac{\partial \phi}{\partial x}} + fv \\
\frac{\partial v}{\partial t} + u\frac{\partial v}{\partial x} + v\frac{\partial v}{\partial y} &= \underline{-\frac{\partial \phi}{\partial y}} - fu \\
\frac{\partial \phi}{\partial t} + u\frac{\partial \phi}{\partial x} + v\frac{\partial \phi}{\partial y} &= \underline{-\Phi\left(\frac{\partial u}{\partial x} + \frac{\partial v}{\partial y}\right)} - (\phi - \Phi)\left(\frac{\partial u}{\partial x} + \frac{\partial v}{\partial y}\right)
\end{aligned}\right\} \tag{2.5.25}$$

2.6 Primitive equations and vertical coordinates

As Charney (1951) foresaw, most NWP modelers went back to using the primitive equations, with the hydrostatic approximation, but without quasi-geostrophic filtering. Quasi-geostrophic models are now reserved for simple problems where the main motivation is the understanding of atmospheric or ocean dynamics.

Exercise 2.6.1: Give two or more reasons why using the primitive equations, with the hydrostatic approximation but without quasi-geostrophic filtering was a desirable goal.

So far we have used z as the vertical coordinate. When we make the hydrostatic approximation, as in the primitive equations, the use of pressure vertical coordinates becomes very advantageous. We can also use any arbitrary variable $\zeta(x, y, z, t)$ as the vertical coordinate as long as it is a monotonic function of z (Kasahara, 1974). The most commonly used vertical coordinates are height z, pressure p, a normalized pressure σ (Phillips, 1957), potential temperature θ (Eliassen, 1949), and several kinds of *hybrid* coordinates (e.g., Simmons and Burridge, 1981, Johnson *et al.*, 1993, Purser, pers. comm., Bleck and Benjamin 1993).

2.6.1 General vertical coordinates

When we transform the vertical coordinate, a variable $A(x, y, z, t)$ becomes $A(x, y, \zeta(x, y, z, t), t)$. The horizontal coordinates and time remain the same. Let s represent x, y, or t. Then, from Fig. 2.6.1(a)

$$\frac{D-B}{\Delta s} = \frac{C-B}{\Delta s} + \frac{D-C}{\Delta z} \cdot \frac{\Delta z}{\Delta s}$$

so that

$$\left(\frac{\partial A}{\partial s}\right)_\zeta = \left(\frac{\partial A}{\partial s}\right)_z + \left(\frac{\partial A}{\partial z}\right)_s \left(\frac{\partial z}{\partial s}\right)_\zeta \tag{2.6.1}$$

and

$$\frac{\partial A}{\partial \zeta} = \frac{\partial A}{\partial z}\frac{\partial z}{\partial \zeta}$$

or

$$\frac{\partial A}{\partial z} = \frac{\partial A}{\partial \zeta}\frac{\partial \zeta}{\partial z} \tag{2.6.2}$$

Substituting (2.6.2) in (2.6.1), we get

$$\left(\frac{\partial A}{\partial s}\right)_\zeta = \left(\frac{\partial A}{\partial s}\right)_z + \left(\frac{\partial A}{\partial \zeta}\right)\left(\frac{\partial \zeta}{\partial z}\right)\left(\frac{\partial z}{\partial s}\right)_\zeta \tag{2.6.3}$$

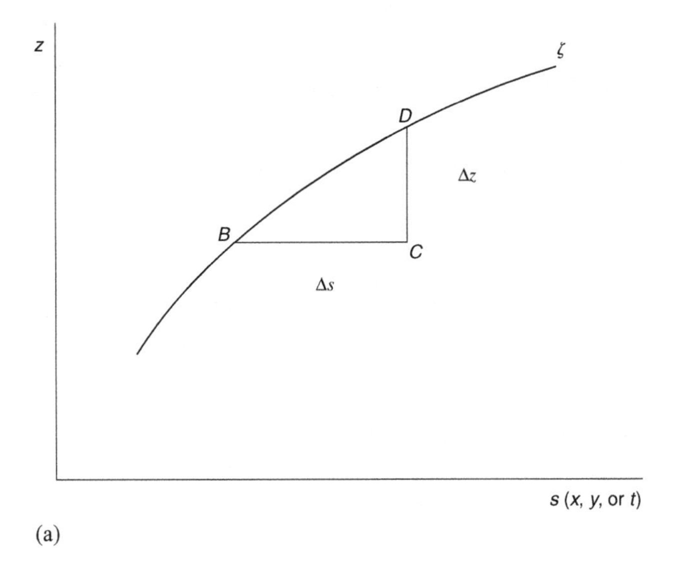

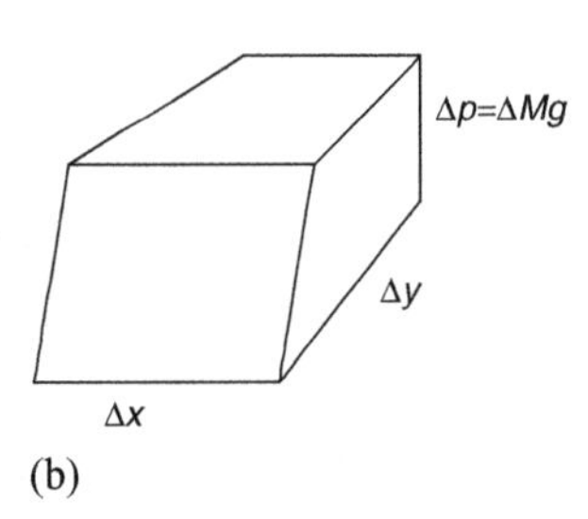

Figure 2.6.1: (a) Schematic showing the relationship between the derivatives of A at constant ζ and at constant Z. The points B and D represent values of A on a ζ-surface and B and C those on a constant z surface. (b) Schematic of a parcel of air in a hydrostatic system, where Δp is proportional to the change in mass per unit area ΔM.

From this relationship (for $s = x, y$) we can get an expression for the horizontal gradient of a scalar A in ζ coordinates:

$$\nabla_\zeta A = \nabla_z A + \left(\frac{\partial A}{\partial \zeta}\right)\left(\frac{\partial \zeta}{\partial z}\right)\nabla_\zeta z \tag{2.6.4}$$

and for the horizontal divergence of a vector $\mathbf{B}$:

$$\nabla_\zeta \cdot \mathbf{B} = \nabla_z \cdot \mathbf{B} + \left(\frac{\partial \mathbf{B}}{\partial \zeta}\right)\cdot\left(\frac{\partial \zeta}{\partial z}\right)\nabla_\zeta z \tag{2.6.5}$$

The total derivative of $A(x, y, \zeta, t)$ is given by

$$\frac{dA}{dt} = \left(\frac{\partial A}{\partial t}\right)_\varsigma + \mathbf{v}\cdot\nabla_\varsigma A + \dot{\varsigma}\frac{\partial A}{\partial \varsigma} \tag{2.6.6}$$

The horizontal pressure gradient is therefore

$$\frac{1}{\rho}\nabla_z p = \frac{1}{\rho}\left[\nabla_\varsigma p - \left(\frac{\partial p}{\partial \varsigma}\right)\left(\frac{\partial \varsigma}{\partial z}\right)\nabla_\varsigma z\right] \tag{2.6.7}$$

which becomes, using the hydrostatic equation $\partial p/\partial\phi = -\rho$,

$$\frac{1}{\rho}\nabla_z p = \frac{1}{\rho}\nabla_\varsigma p + \nabla_\varsigma \phi \tag{2.6.8}$$

In summary the horizontal momentum equations become

$$\frac{d\mathbf{v}}{dt} = -\alpha\nabla_\varsigma p - \nabla_\varsigma \phi - f\mathbf{k}\times\mathbf{v} + F \tag{2.6.9}$$

and the hydrostatic equation $\partial p/\partial z = -\rho g$ becomes

$$\frac{\partial p}{\partial \varsigma}\frac{\partial \varsigma}{\partial z} = -\rho g$$

or

$$\frac{\partial p}{\partial \varsigma} = -\rho\frac{\partial \phi}{\partial \varsigma} \tag{2.6.10}$$

The continuity equation can be derived from the conservation of mass for an infinitesimal parcel: the hydrostatic equation indicates that the mass of a parcel is proportional to the increase in pressure from the top to the bottom of the parcel (Fig. 2.6.1(b)):

$$g\Delta M = \Delta x\Delta y\Delta p \tag{2.6.11}$$

Now, $\Delta p = (\partial p/\partial\varsigma)\Delta\varsigma$, so that taking a logarithmic total derivative, and noting that

$$\frac{1}{\Delta x}\frac{d\Delta x}{dt} \to \frac{\partial u}{\partial x}$$

and the same with the other space variables, we obtain

$$\frac{d}{dt}\left(\ln\frac{\partial p}{\partial \varsigma}\right) + \nabla\cdot\mathbf{v}_H + \frac{\partial\dot{\varsigma}}{\partial\varsigma} = 0 \tag{2.6.12}$$

The thermodynamic equation is as before

$$C_p\frac{T}{\theta}\frac{d\theta}{dt} = C_p\frac{dT}{dt} - \alpha\frac{dp}{dt} = Q \tag{2.6.13}$$

The kinematic lower boundary condition is that the surface of the earth is a material surface: the flow can only be parallel to it, not normal. This means that once a parcel touches the surface it is "stuck" to it. This can be expressed as

$$\frac{d(\varsigma - \varsigma_s)}{dt} = 0 \qquad at \qquad \varsigma = \varsigma_s$$

or

$$\frac{d\varsigma}{dt} = \frac{\partial \varsigma_s}{\partial t} + \mathbf{v} \cdot \nabla \varsigma_s \quad \text{at} \quad \varsigma = \varsigma_s \tag{2.6.14}$$

This kinematic boundary condition is well defined although in practice it may not be accurate, for example, when there is subgrid-scale orography.

At the top, unfortunately, the boundary condition is not so well defined: As $z \to \infty$, $p \to 0$, but in general there is no satisfactory way to express this condition for a finite vertical resolution model. Most models assume a simple condition of a "rigid top" (i.e., making the top surface a material surface)

$$\frac{d\varsigma}{dt} = 0 \text{ at } \varsigma = \varsigma_T \tag{2.6.15}$$

but this is an artificial boundary condition that introduces spurious effects. For example, Kalnay and Toth (1996) showed that a rigid top introduces artificial "upside-down" baroclinic instabilities in the NCEP global model, and similar observations were made by Hartmann *et al.* (1997) with the ECMWF model. If the top of the model is sufficiently high, and there is enough vertical resolution, the upward moving perturbations are damped in the model (as they are in nature), and the spurious interaction with the artificial top may remain small. Alternatively, radiation conditions enforcing the condition that energy can only propagate upwards can be used, but they are not simple to implement.

2.6.2 Pressure coordinates

These coordinates are a natural choice for a hydrostatic atmosphere (Eliassen, 1949). They greatly simplify the equations of motion: the horizontal pressure gradient becomes irrotational, and the continuity equation becomes simply zero three-dimensional divergence, a diagnostic linear equation.

As a result the geostrophic wind relationship is also simpler: $\mathbf{v}_g = (1/f)\mathbf{k} \times \nabla\phi$. For this reason, rawinsonde measurements have been made in pressure coordinates since the early 1950s.

In pressure coordinates, $\partial p/\partial\varsigma \equiv 1$, the total derivative operator (2.6.6) is given by

$$\frac{d}{dt} = \frac{\partial}{\partial t} + \mathbf{v} \cdot \nabla + \omega \frac{\partial}{\partial p}$$

where the vertical velocity in pressure coordinates is $\omega = dp/dt$. The primitive equations become:

$$\frac{d\mathbf{v}}{dt} = -\nabla_p \phi - f\mathbf{k} \times \mathbf{v} + F \tag{2.6.16}$$

$$\frac{\partial \phi}{\partial p} = -\alpha \tag{2.6.17}$$

$$\nabla_p \cdot \mathbf{v} + \frac{\partial \omega}{\partial p} = 0 \tag{2.6.18}$$

and the thermodynamic equation (2.6.13) is unchanged.

The geostrophic and thermal wind relationships are especially simple in pressure coordinates:

$$\mathbf{v}_g = \frac{1}{f}\mathbf{k} \times \nabla\phi \qquad \text{and} \qquad \frac{\partial \mathbf{v}_g}{\partial p} = -\frac{R}{fp}\mathbf{k} \times \nabla T \tag{2.6.19}$$

On the other hand, the bottom boundary condition is not simple in pressure coordinates because the pressure surfaces intersect the surface:

$$\omega = \frac{\partial p_s}{\partial t} + \mathbf{v} \cdot \nabla p_s \qquad \text{at} \qquad p = p_s \tag{2.6.20}$$

This requires knowing the rate of change of p_s:

$$\frac{\partial p_s}{\partial t} + \mathbf{v} \cdot \nabla p_s = -\int_0^\infty \nabla_p \cdot \mathbf{v} dp \tag{2.6.21}$$

This complication of the surface boundary condition in pressure coordinates led Phillips (1957) to the invention of sigma coordinates (next subsection).

Instead of the horizontal momentum equations, we can use the prognostic equations for the vorticity ζ and divergence δ, obtained by applying the operators $\mathbf{k} \cdot \nabla x$ and $\nabla \cdot$ to the momentum equations. In pressure coordinates these equations are

$$\frac{\partial \varsigma}{\partial t} + \mathbf{v} \cdot \nabla(f + \varsigma) + \omega \frac{\partial \varsigma}{\partial p} + (f + \varsigma)\nabla \cdot \mathbf{v} + k \cdot \nabla\omega \times \frac{\partial \mathbf{v}}{\partial p} = 0 \tag{2.6.22}$$

$$\frac{\partial \delta}{\partial t} + \nabla \cdot (\mathbf{v} \cdot \nabla \mathbf{v}) + \omega \frac{\partial \delta}{\partial p} + \nabla\omega \cdot \frac{\partial \mathbf{v}}{\partial p} + \nabla \cdot (f\mathbf{k} \times \mathbf{v}) + \nabla^2 \phi = 0 \tag{2.6.23}$$

2.6.3 Sigma and eta coordinates

Because of the complication of the bottom boundary conditions, Phillips (1957) introduced "normalized pressure" or "sigma" coordinates, where $\sigma = p/p_s$ and $p_s(x, y, t)$ is the surface pressure. These are by far the most widely used vertical coordinates. At the surface, $\sigma = 1$, and at $p = 0$, $\sigma = 0$, so that the top and bottom boundary conditions are $\dot\sigma = 0$. More generally, allowing for a rigid top at a finite pressure $p_T = const.$,

$$\sigma = \frac{p - p_T}{p_S - p_T} = \frac{p - p_T}{\pi} \tag{2.6.24}$$

with $\dot\sigma = 0$ at $\sigma = 0, 1$.

The continuity equation is

$$\frac{\partial \pi}{\partial t} = -\nabla \cdot (\pi \mathbf{v}) - \frac{\partial \pi \dot\sigma}{\partial \sigma} \tag{2.6.25}$$

The surface pressure tendency equation is:

$$\frac{\partial \pi}{\partial t} = \frac{\partial p_S}{\partial_t} = -\nabla \cdot \int_0^1 (\pi v) d\sigma \tag{2.6.26}$$

Substituting back into the continuity equation, one can determine $\dot{\sigma}$ diagnostically from the horizontal wind field $\mathbf{v}$.

Exercise 2.6.2: Derive (2.6.25) and (2.6.26)

Despite their popularity, sigma coordinates have a serious disadvantage: the pressure gradient becomes the difference between two terms:

$$\frac{d\mathbf{v}}{dt} = -\alpha \nabla_\sigma p - \nabla_\sigma \phi - f\mathbf{k} \times \mathbf{v} + \mathbf{F} \tag{2.6.27}$$

where the first term, if sigma surfaces are steep, may not have the information that went into the finite difference calculation of the second. To avoid the resulting errors, Mesinger (1984) introduced a step-mountain coordinate denoted "eta" (used in the Eta model at NCEP, e.g., Mesinger *et al.* (1988), Janjic (1990), Black (1994)):

$$\eta = \frac{p}{p_S} \frac{p_o(z)}{1000 \text{ mb}} \tag{2.6.28}$$

The first factor is the standard sigma coordinate, the second is a scaling factor, with $p_o(z)$ the pressure in the standard atmosphere. Mountains are defined as boxes, whose tops have to coincide with a model eta level (Fig. 2.6.2). As a result of the scaling, the eta surfaces are almost horizontal, and the pressure gradient is computed accurately. At NCEP, the Eta model has proven to be very skillful especially in predicting storms.

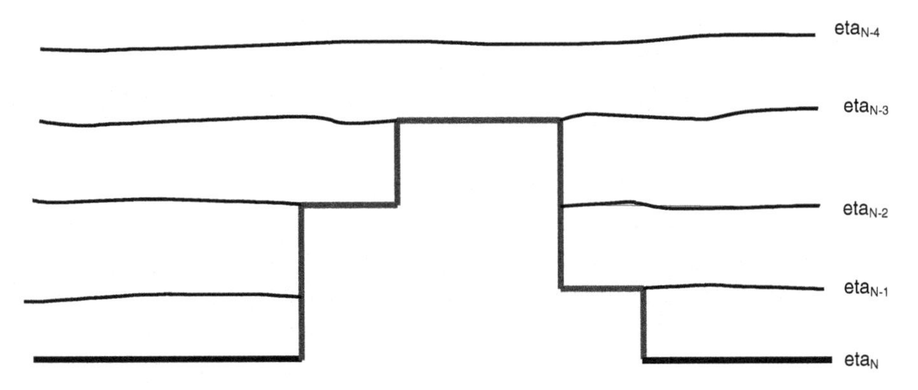

Figure 2.6.2: Schematic of the eta coordinate.

2.6.4 Isentropic coordinates

The fact that under adiabatic motion, potential temperature is individually conserved suggested long ago that it could be used as a vertical coordinate. The main advantage, which makes it an almost ideal coordinate, is that "vertical" motion $\dot{\theta}$ is approximately zero in these coordinates (except for diabatic heating). This reduces finite difference errors in areas such as fronts, where pressure or z-coordinates tend to have large errors associated with poorly resolved vertical motion.

Hydrostatic equation: from the definition of potential temperature, and using the hydrostatic and state equations, we get

$$\frac{d\theta}{\theta} = \frac{dT}{T} - \frac{R}{C_p}\frac{dp}{p} = \frac{dT}{T} - \frac{1}{C_p}\frac{d\phi}{T} \tag{2.6.29}$$

If we define the Exner function $\Pi = C_p T/\theta = C_p(p/p_0)^{R/C_p}$, and the Montgomery potential $M = C_p T + \phi$, we see from the previous equation that

$$\frac{\partial M}{\partial \theta} = \Pi \tag{2.6.30}$$

The horizontal pressure gradient becomes very simple, so that for $\zeta = \theta$ the momentum equation is

$$\frac{d\mathbf{v}}{dt} = -\nabla_\theta M - f\mathbf{k}\times\mathbf{v} + \mathbf{F} \tag{2.6.31}$$

The continuity equation is

$$\frac{d}{dt}\ln\frac{\partial p}{\partial \theta} + \nabla_\theta \cdot \mathbf{v} + \frac{\partial \dot{\theta}}{\partial \theta} = 0 \tag{2.6.32}$$

The potential vorticity is conserved for adiabatic, frictionless flow (Ertel's theorem). This general property can be posed in its simplest formulation in isentropic coordinates:

$$\frac{dq}{dt} = 0 \tag{2.6.33}$$

where $q = (f + \mathbf{k}\cdot\nabla_\theta \times \mathbf{v})\,\partial\theta/\partial p$, and integrating between two isentropic surfaces, the potential vorticity is

$$q = \frac{(f + \mathbf{k}\cdot\nabla_\theta \times \mathbf{v})}{\Delta p} \tag{2.6.34}$$

which is similar to the SWE potential vorticity.

Although the isentropic coordinates have many advantages, they have also two main disadvantages: The first is that isentropic surfaces intersect the ground (as do other vertical coordinates except for sigma-type coordinates). In practice this implies that it is difficult to enforce strict conservation of mass, and this is important for long (climate) integrations. For this reason, hybrid sigma–theta coordinates have been used (e.g., Johnson *et al.*, 1993). Other approaches have been those of Bleck and Benjamin (1993) for the operational RUC/MAPS model, and that of Arakawa and Konor (1996). The second disadvantage is that only statically stable solutions are allowed, since the vertical coordinate has to vary monotonically with height. There are situations, e.g., over hot surfaces, where this is not true even at a grid scale. Moreover, in regions of low static stability, the vertical resolution of isentropic coordinates can be poor.

Exercise 2.6.3: Derive (2.6.31) from (2.6.9) for $\zeta = \theta$, and (2.6.32) from the logarithmic derivative of

$$\Delta M = \frac{\partial p}{\partial \theta} \cdot \Delta\theta \cdot \Delta x \cdot \Delta y \tag{2.6.35}$$

where ΔM is proportional to the mass of a parcel in isentropic coordinates

3

Numerical discretization of the equations of motion

3.1 Classification of partial differential equations (PDEs)

3.1.1 Reminder about PDEs

Second order linear PDE

$$\alpha\frac{\partial^2 u}{\partial x^2} + 2\beta\frac{\partial^2 u}{\partial x \partial y} + \gamma\frac{\partial^2 u}{\partial y^2} + \delta\frac{\partial u}{\partial x} + \varepsilon\frac{\partial u}{\partial x} + \eta = 0$$

Second order linear partial differential equations are classified into three types depending on the sign of $\beta^2 - \alpha\gamma$ (e.g., Courant and Hilbert, 1962). Equations are hyperbolic, parabolic or elliptic if the sign is positive, zero, or negative, respectively. The simplest (canonical) examples of these equations are

(a) $\dfrac{\partial^2 u}{\partial t^2} = c^2\dfrac{\partial^2 u}{\partial x^2}$ Wave equation (hyperbolic).
Example: vibrating string.

(b) $\dfrac{\partial u}{\partial t} = \sigma\dfrac{\partial^2 u}{\partial x^2}$ Diffusion equation (parabolic).
Example: heated rod.

(c) $\dfrac{\partial^2 u}{\partial x^2} + \dfrac{\partial^2 u}{\partial y^2} = 0\ (or\ f(x, y))$ Laplace's or Poisson's equations (elliptic).
Examples: steady state temperature of a plate, streamfunction/vorticity relationship.

The behavior of the solutions, the proper initial and/or boundary conditions, and the numerical methods that can be used to find the solutions *depend essentially on the type of PDE* that we are dealing with. Although nonlinear multidimensional PDEs cannot in general be reduced to these canonical forms, we need to study these prototypes of the PDEs to develop an understanding of their properties, and then apply similar methods to the more complicated NWP equations.

(d) $\dfrac{\partial u}{\partial t} = -c\dfrac{\partial u}{\partial x}$ Advection equation, with solution $u(x, t) = u(x - ct, 0)$.

The advection equation is a first order PDE, but it can also be classified as a hyperbolic, since its solutions satisfy the wave equation (a), and the latter is usually written as the system

$$\frac{\partial \mathbf{u}}{\partial t} = \mathbf{A}\frac{\partial \mathbf{u}}{\partial x}$$

where

$$\mathbf{u} = \begin{pmatrix} \dfrac{\partial u}{\partial t} \\ c\dfrac{\partial u}{\partial x} \end{pmatrix}$$

and

$$\mathbf{A} = \begin{bmatrix} 0 & c \\ c & 0 \end{bmatrix} \text{ or an equivalent transformation}$$

3.1.2 Well-posedness, initial and boundary conditions

A *well-posed* initial/boundary condition problem has a unique solution that depends continuously on the initial/boundary conditions. Clearly, the specification of proper initial conditions and boundary conditions for a PDE is essential in order to have a well-posed problem. If too many initial/boundary conditions are specified, there will be no solution. If too few are specified, the solution will not be unique. If the number of initial/boundary conditions is right, but they are specified at the wrong place or time, the solution will be unique, but it will not depend smoothly on initial/boundary conditions, i.e., small errors in the initial/boundary conditions will produce huge errors in the solution. In any of these cases we have an *ill-posed problem.* And we can *never* find a numerical solution of a problem that is ill posed: the computer will show its disgust by "blowing up".

We briefly discuss well-posed initial/boundary conditions:

- Second order elliptic equations require one boundary condition on each point of the spatial boundary. These are "boundary value", time-independent problems, and the methods used to solve them are introduced in Section 3.4. The boundary conditions may be on the value of the function (Dirichlet problem), as when we specify the temperature in the borders of a plate, or on its normal derivative (Neumann problem), as when we specify the heat flux. We could also have a mixed "Robin" boundary condition, involving a linear combination of the function and its derivative.

- Linear parabolic equations require one initial condition at the initial time and one boundary condition at each point of the spatial boundaries (if they exist).
- Linear hyperbolic equations require as many initial conditions as the number of characteristics that come out of every point in the surface $t = 0$, and as many boundary conditions as the number of characteristics that cross a point in the (space) boundary pointing *inwards* (into the spatial domain). For example: to solve $\partial u/\partial t = -c\partial u/\partial x$ for $x > 0, t > 0$; characteristics: solutions of $dx/dt = c$; space boundary: $x = 0$ (see Fig. 3.1.1(a),(b)). If $c > 0$, we need the initial condition $u(x, 0) = f(x)$ and the boundary condition $u(0, t) = g(t)$. If $c < 0$, we need the initial condition $u(x, 0) = f(x)$ but *no boundary conditions*.

For nonlinear equations, no general statements can be made, but physical insight and local linearization can help to determine proper initial/boundary conditions. For example, in the nonlinear advection equation $\partial u/\partial t = -u\partial u/\partial x$, the characteristics are $dx/dt = u$, and since we don't know *a priori* the sign of u at the boundary, and whether the characteristics will point inwards or outwards, we have to estimate the value of u from the nearby solution, and define the boundary condition accordingly.

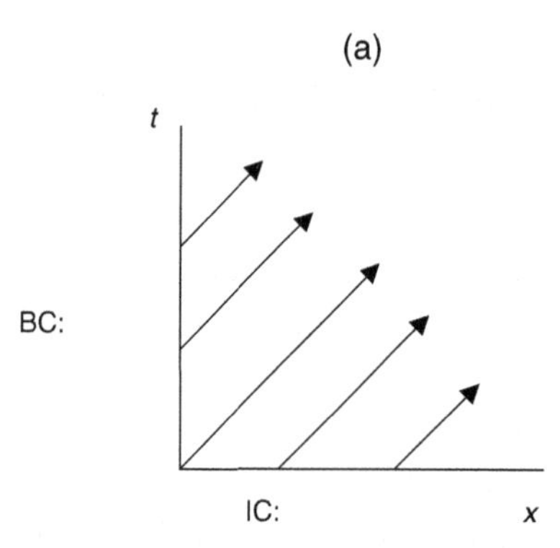

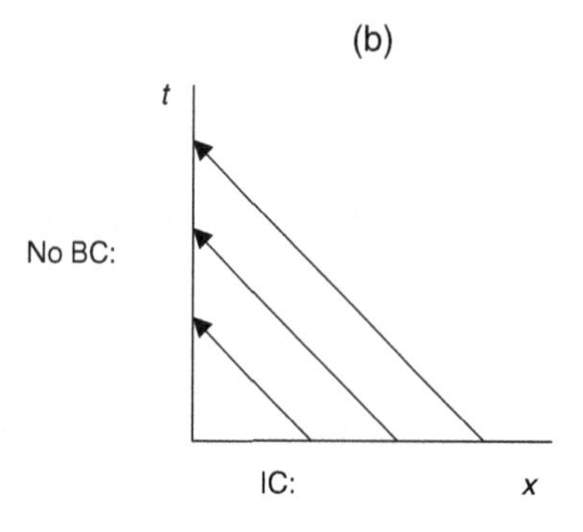

Figure 3.1.1: Schematic of the characteristics of the advection equation $\partial u/\partial t = -c\partial u/\partial x$ for (a) positive and (b) negative velocity c and the corresponding well-posed initial/boundary conditions (IC/BC).

One method of solving simple PDEs is the method of separation of variables, but unfortunately in most cases it is not possible to use it (hence the need for numerical models!). Nevertheless, it is useful to try to solve some simple PDE's analytically.

Exercise 3.1.1: Solve by the method of separation of variables these prototype PDEs:

(1) $$\frac{\partial^2 u}{\partial x^2} + \frac{\partial^2 u}{\partial y^2} = 0 \quad 0 \le x \le 1 \quad 0 \le y \le 1$$

Boundary condition: $u(x, 0) = f(x)$, $u(x, 1) = u(0, y) = u(1, y) = 0$

Assume

$$f(x) = \sum_{k=1}^{\infty} a_k \sin k\pi x \text{ with } \sum_{k=1}^{\infty} k^2 |a_k| < \infty$$

Find the solution

$$u(x, y) = \sum_{k=1}^{\infty} a_k \frac{\sinh k\pi(1-y)}{\sinh k\pi} \sin k\pi x$$

(2) $$\frac{\partial u}{\partial t} = \sigma \frac{\partial^2 u}{\partial x^2} \quad 0 \le x \le 1 \quad t \ge 0$$

Boundary condition: $u(0, t) = u(1, t) = 0$
Initial condition: $u(x, 0) = f(x) = \sum_{k=1}^{\infty} a_k \sin k\pi x$

Find

$$u(x, t) = \sum_{k=1}^{\infty} a_k e^{-\sigma k^2 \pi^2 t} \sin k\pi x$$

Note that the higher the wavenumber, the faster it goes to zero, i.e., the solution is smoothed as time goes on.

(3) $$\frac{\partial^2 u}{\partial t^2} = c^2 \frac{\partial^2 u}{\partial x^2} \quad 0 \le x \le 1 \quad 0 \le t \le 1$$

Boundary condition: $u(0, t) = u(1, t) = 0$

Initial condition: $u(x, 0) = f(x) = \sum_{k=1}^{\infty} a_k \sin k\pi x$;

$$\frac{\partial u}{\partial t}(x, 0) = g(x) = \sum_{k=1}^{\infty} b_k \sin k\pi x$$

(4) Same as (3), but now, instead of two initial conditions, we give an initial and a "final" condition:

Boundary condition: $u(0, t) = u(1, t) = 0$
Initial condition: $u(x, 0) = f(x)$; "final condition" $u(x, 1) = g(x)$

In other words, we try to solve a hyperbolic (wave) equation as if it was a boundary value problem. Show that the solution is unique but it does not depend continuously on the boundary conditions, and therefore it is not a well-posed problem.

Conclusion: Before trying to solve a problem numerically, make sure that it is well posed: it has a unique solution that depends continuously on the data that define the problem.

Exercise 3.1.2: Lorenz showed that the atmosphere has a finite limit of predictability: even if the models and the observations were perfect, "the flapping of a butterfly in Brazil (not taken into account in the model) will result in a completely different forecast over the US after a couple of weeks". Does this mean that the problem of NWP is not well posed?

3.2 Initial value problems: numerical solution

Hyperbolic and parabolic PDEs are *initial value* or *marching* problems: The solution is obtained by using the known initial values and marching or advancing in time. If boundary values are necessary, they are called "mixed initial–boundary value problems". Again, the simplest prototypes of these initial value problems are:

$$\frac{\partial u}{\partial t} = -c\frac{\partial u}{\partial x} \tag{3.2.1}$$

the wave or advection equation, with solution $u(x, t) = u(x - ct, 0)$, a hyperbolic equation, and

$$\frac{\partial u}{\partial t} = \sigma\frac{\partial^2 u}{\partial x^2} \tag{3.2.2}$$

the diffusion equation, a parabolic equation.

3.2.1 Finite difference method

We take discrete values for x and t: $x_j = j\Delta x$, $t_n = n\Delta t$. The solution of the finite difference equation is also defined at the discrete points $(j\Delta x, n\Delta t)$: $U_j^n = U(j\Delta x, n\Delta t)$. We will use a small u to denote the solution of the PDE (continuous) and capital U to denote the solution of the finite difference equation (FDE), a discrete solution.

Consider again the advection equation (3.2.1). Suppose that we choose to approximate this PDE with the following FDE (called an "upstream scheme"):

$$\frac{U_j^{n+1} - U_j^n}{\Delta t} + c\frac{U_j^n - U_{j-1}^n}{\Delta x} = 0 \tag{3.2.3}$$

Note that both differences are noncentered with respect to the point $(j\Delta x, n\Delta t)$. We should now ask two fundamental questions:

(1) Is the FDE *consistent* with the PDE?

(2) For any given time $t > 0$, will the solution U of the FDE *converge* to u as $\Delta x \to 0$, $\Delta t \to 0$?

Let us now clarify these questions.

3.2.2 Truncation errors and consistency

We say that the FDE is *consistent* with the PDE if, in the limit $\Delta x \to 0$, $\Delta t \to 0$ the FDE coincides with the PDE. Obviously, this is a first requirement that the FDE should fulfill if its solutions are going to be good approximations of the solutions of the PDE. The difference between the PDE and the FDE is the discretization error or *local* (in space and time) *truncation error*. Consistency is rather simple to verify: Substitute U by u in the FDE, and evaluate all terms using a Taylor series expansion centered on the point (j, n), and then subtract the PDE from the FDE. If the difference (or local truncation error τ) goes to zero as $\Delta x \to 0$, $\Delta t \to 0$, then the FDE is consistent with the PDE.

Example 3.2.1: We verify the consistency of (3.2.3) with (3.2.1) by a Taylor series expansion:

$$\left.\begin{aligned} u_j^{n+1} &= \left(u + u_t \Delta t + u_{tt}\tfrac{\Delta t^2}{2} + \cdots\right)_j^n \\ u_{j-1}^n &= \left(u - u_x \Delta x + u_{xx}\tfrac{\Delta x^2}{2} + \cdots\right)_j^n \end{aligned}\right\} \tag{3.2.4}$$

Substitute the series (3.2.4) in the FDE (3.2.3)

$$\left(u_t + u_{tt}\frac{\Delta t}{2} + \cdots + cu_x - cu_{xx}\frac{\Delta x}{2} + \cdots\right)_j^n \simeq 0 \tag{3.2.5}$$

and when we subtract the PDE (3.2.1) we get the (local) *truncation error*

$$\tau = u_{tt}\frac{\Delta t}{2} - cu_{xx}\frac{\Delta x}{2} + \text{higher order terms} = 0(\Delta t) + 0(\Delta x) \tag{3.2.6}$$

so that $\lim_{\Delta t \to 0, \Delta x \to 0} \tau \to 0$. Therefore *the FDE is consistent.* Note that both the *time and the space truncation errors are of first order*, because the finite differences are uncentered in both space and time. Truncation errors for *centered differences are second order*, and therefore centered differences are more accurate than uncentered differences (see Fig. 3.2.1(a) and the leapfrog scheme, based on centered differences in space and in time, later in this section).

3.2.3 Convergence and stability criteria

The second question posed in Section 3.2.1 was whether the solution of the FDE converges to the PDE solution, i.e., whether $U(j\Delta x, n\Delta t) \to u(x, t)$ when $j\Delta x \to x, n\Delta t \to t$, $\Delta x \to 0$, $\Delta t \to 0$. This is of evident practical importance, but can only be answered after considering another problem, that of *computational stability*. Consider again the PDE (3.2.1), which has the solution $u(x, t) = u(x - ct, 0)$, shown schematically in Fig. 3.2.1(b) (the initial shape of u translates with velocity c).

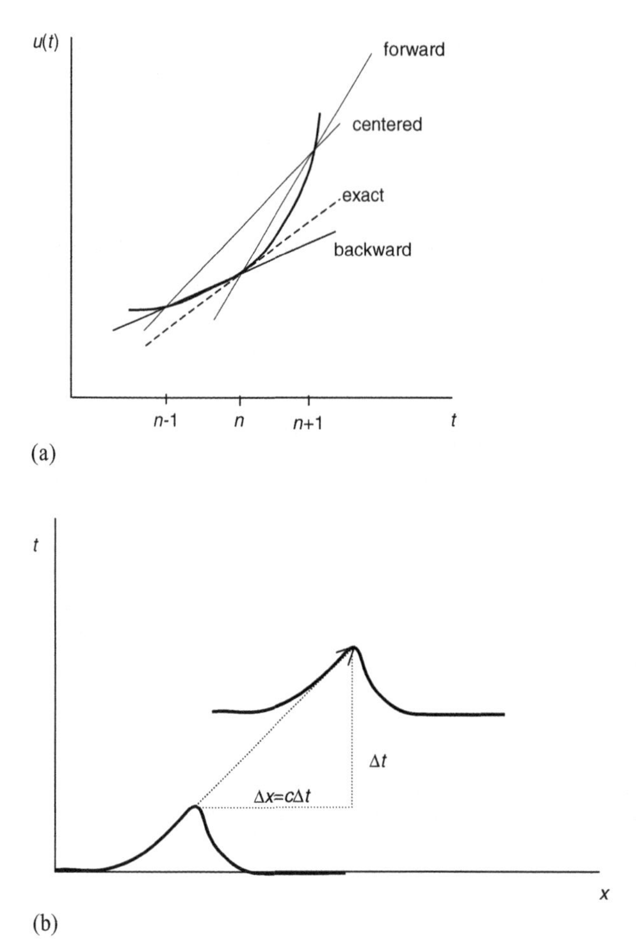

Figure 3.2.1: (a) Schematic of centered $(\partial u/\partial t)_n \approx (u_{n+1} - u_{n-1})/(2\Delta t)$, forward $(\partial u/\partial t)_n \approx (u_{n+1} - u_n)/\Delta t$, and backward $(\partial u/\partial t)_n \approx (u_n - u_{n-1})/\Delta t$ finite differences estimating the time derivative $\partial u/\partial t$ at time $t_n = n\Delta t$. The three estimates are *consistent* with $\partial u/\partial t$ since they all converge to $\partial u/\partial t$ as $\Delta t \to 0$. However, the slope calculated from centered differences is much closer to the exact derivative because its truncation errors are second order. (b) Schematic of the solution of the wave equation.

The FDE (3.2.3) can be written as

$$U_j^{n+1} = (1-\mu)U_j^n + \mu U_{j-1}^n \tag{3.2.7}$$

where $\mu = c\Delta t/\Delta x$ is the *Courant number*. Assume that $0 \leq \mu = c\Delta t/\Delta x \leq 1$, as in Fig. 3.2.2(a). Then the FDE solution at the new time level U_j^{n+1} is *interpolated* between the values U_j^n and U_{j-1}^n. In this case the advection scheme works the way it should, because we know the true solution is in between those values. However, if this condition is not satisfied, and $\mu = c\Delta t/\Delta x > 1$ (as in Fig. 3.2.2(b)) or $\mu = c\Delta t/\Delta x < 0$ (as in Fig. 3.2.2(c)), then the value of U_j^{n+1} is *extrapolated* from the values U_j^n and U_{j-1}^n. The problem with extrapolation is that the maximum absolute value of the solution U_j^n increases with each time step. Taking absolute values of (3.2.7) and letting $U^n = \max_j |U_j^n|$, we get

$$\left|U_j^{n+1}\right| \leq \left|U_j^n\right| |1-\mu| + \left|U_{j-1}^n\right| |\mu|$$

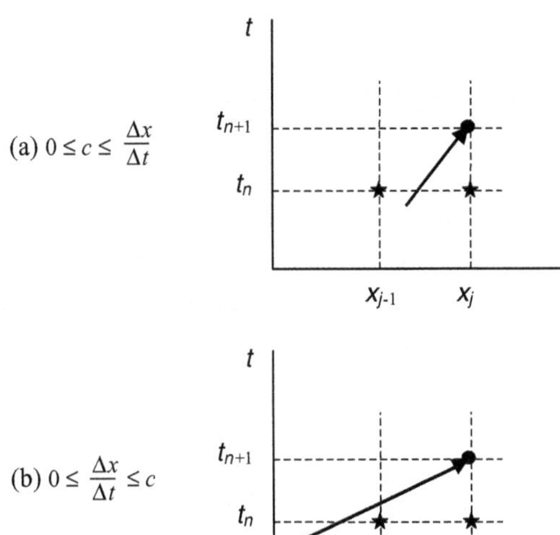

Figure 3.2.2: Schematic of the relationship between Δx, Δt, and c leading to interpolation of the solution at time level $n+1$ (case (a)), or to extrapolation (cases (b) and (c)) depending on the value of the Courant number $\mu = c\Delta t/\Delta x$.

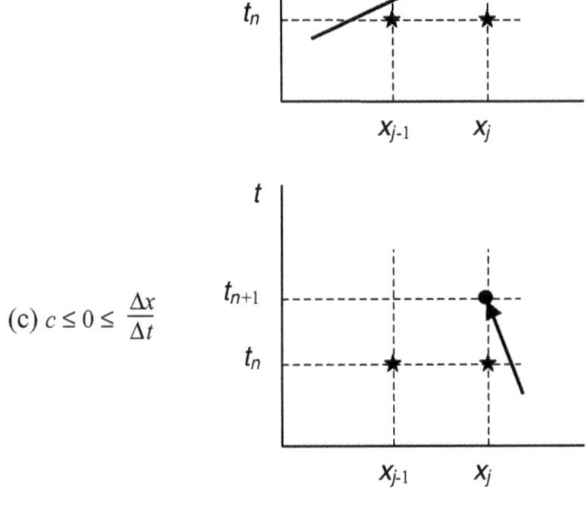

(c) $c \le 0 \le \frac{\Delta x}{\Delta t}$
t
t_{n+1}
t_n
x_{j-1}
x_j

so that

$$U^{n+1} \le \{|1-\mu| + |\mu|\}U^n$$

Then $U^{n+1} \le U^n$ if and only if $0 \le \mu \le 1$.

If the condition $0 \le \mu \le 1$ is *not* satisfied, then *the solution is not bounded* and it grows with n. If we let Δt, $\Delta x \to 0$ with $\mu = const.$, it only makes things worse, because then $n \to \infty$. In practice, if the condition $0 \le \mu \le 1$ is not satisfied, the FDE "blows up" in a few time steps, faster for nonlinear problems. We define now computational stability: we say that an FDE is *computationally stable* if the solution of the FDE at a fixed time $t = n\Delta t$ remains bounded as $\Delta t \to 0$. The condition on the Courant number of being less than 1 in absolute value is usually known as the Courant–Friedrichs–Lewy or CFL condition.

We can now state the fundamental Lax–Richtmyer theorem: Given a *properly posed linear* initial value problem, and a finite difference scheme that satisfies the *consistency* condition, then the *stability* of the FDE is the necessary and sufficient condition for *convergence*.

The theorem is useful because it allows us to establish convergence by examining separately the easier questions of consistency and stability. We are interested in convergence not because we want to let Δt, $\Delta x \to 0$, but because we want to make sure that if Δt, Δx are small, then the errors $u(j\Delta x, n\Delta t) - U_j^n$ (accumulated or *global truncation errors* at a finite time) are acceptably small.

To determine the necessary condition for stability of the FDE (3.2.3) we used the "criterion of the maximum" method. We can also use it to study the stability condition of the following FDE, which approximates the parabolic diffusion equation $\partial u/\partial t = \sigma \partial^2 u/\partial x^2$:

$$\frac{U_j^{n+1} - U_j^n}{\Delta t} = \sigma \frac{U_{j+1}^n - 2U_j^n + U_{j-1}^n}{\Delta x^2} \tag{3.2.8}$$

The verification of consistency is immediate. Note that, because the differences are centered in space but forward in time, the truncation error is first order in time and second order in space $O(\Delta t) + O(\Delta x)^2$.

We can write (3.2.8) as

$$U_j^{n+1} = \mu U_{j+1}^n + (1 - 2\mu)U_j^n + \mu U_{j-1}^n$$

where $\mu = \sigma \Delta t/\Delta x^2$. If we take absolute values, and let $U^n = \max_j |U_j^n|$, we get

$$U^{n+1} \leq \{|\mu| + |1 - 2\mu| + |\mu|\}U^n \tag{3.2.9}$$

So we obtain a condition $0 \leq \mu \leq 1/2$ to insure that the solution remains bounded as $n \to \infty$, i.e., as the necessary condition for stability of the FDE.

Exercise 3.2.1: The condition on the wave equation $0 \leq \mu \leq 1$ for the upstream FDE is interpreted as "the time step should be chosen so that a signal cannot travel more than one grid size in one time step." Give a physical interpretation of the stability condition and the equivalent "Courant number" $\mu = \sigma\,(\Delta t/\Delta x^2) \leq 1/2$ for the diffusion equation.

Unfortunately, the criterion of the maximum, which is intuitively very clear, can only be applied in very few cases. In most FDEs some coefficients of the equations analogous to (3.2.9) are negative, and the criterion cannot be applied.

Another stability criterion that has much wider application is *the von Neumann stability criterion*: Assume that the boundary conditions allow expansion of the solution of the FDE in an appropriate set of eigenfunctions. For simplicity we will assume an expansion into Fourier series (e.g., periodic boundary conditions):

$$U(x, t) = \sum_{\mathbf{k}} Z_{\mathbf{k}} e^{i\mathbf{k}\cdot\mathbf{x}} \tag{3.2.10}$$

The space variable, $\mathbf{x}$, and the wavenumber $\mathbf{k}$ can be multidimensional, e.g., $\mathbf{x} = (x_1, x_2, x_3)$, $\mathbf{k} = (k_1, k_2, k_3)$. The dependent variable U can also be a vector for a system of equations.

Let $x_j = j\Delta x$ (or $\mathbf{x}_j = (j_1\Delta x_1, j_2\Delta x_2, j_3\Delta x_3)$). We define p as the wavenumber for the finite Fourier series: $p = k\Delta x$ or $\mathbf{p} = (k_1\Delta x_1, k_2\Delta x_2, k_3\Delta x_3)$. Let $t_n = n\Delta t$. Then the Fourier expansion is

$$U_j^n = \sum_p Z_p^n e^{ipj} \tag{3.2.11}$$

(where for multiple dimensions $\mathbf{p}\cdot\mathbf{j} = p_1 j_1 + p_2 j_2 + p_3 j_3$).

When we substitute this Fourier expansion into a linear FDE, we obtain a system of equations

$$Z_p^{n+1} = G_p Z_p^n$$

G is an "amplification" matrix that, when applied to the pth Fourier component of the solution at time $n\Delta t$ "advances" it to the time $(n+1)\Delta t$; G depends on p, Δt and Δx. If we know the initial conditions

$$U_j^0 = \sum_p Z_p^0 e^{ipj} \tag{3.2.12}$$

then the solution of the FDE in (3.2.11) is

$$Z_p^n = G_p^n Z_p^0 \tag{3.2.13}$$

Therefore, stability, i.e., boundedness of the solution for any permissible initial condition at any fixed time, is guaranteed if the matrix G^n *is bounded for all* p when $\Delta t \to 0$ and $n \to \infty$. So, we must have $||G^n|| < M$ for all p, as $n \to \infty$. Here $||\mathbf{A}||$ is a norm or measure of the "size" of a matrix $\mathbf{A}$. If $\sigma(G)$ is the *spectral radius* of G, i.e., $\sigma(G) = \max_i |\lambda_i|$, where λ_i are the eigenvalues of G, then it can be shown that for any norm,

$$[\sigma(G)]^n \le ||G^n|| \le ||G||^n \tag{3.2.14}$$

The equal sign is valid if G is normal, i.e., if $GG^\star = G^\star G$, where $G^\star$ is the transpose-conjugate of G, but in general the amplification matrices arising from FDEs are not normal.

Thus a necessary condition for stability of an FDE, and therefore a necessary condition for convergence, is that

$$\lim_{\Delta t \to 0, n\Delta t \to t} [\sigma(G)]^n = \text{finite} = e^{const.} \tag{3.2.15}$$

Then

$$\sigma(G) \le \left[\sigma(G)^n\right]^{1/n} \le e^{const./n} = e^{const.\Delta t/t} \approx 1 + \frac{const.\Delta t}{t}$$

or

$$\sigma(G) \leq 1 + O(\Delta t) \tag{3.2.16}$$

the von Neumann necessary condition for computational stability.

The term $O(\Delta t)$ allows bounded growth with time if this growth is "legitimate", i.e., if it arises from a physical instability present in the PDE. If the exact solution grows with time, then the FDE cannot both satisfy $\sigma(G) \leq 1$ and be consistent with the PDE.

Sufficient conditions are very complicated, and are known only for special cases. In practice it is generally observed that eliminating the equal sign in (3.2.16) is enough to ensure computational stability.

In principle this method can also be used to study the stability of the boundary conditions, if they are appropriately included in the amplification matrix. In practice this is complicated, and computational stability of the boundary conditions is usually obtained by ensuring well-posedness, and testing the stability experimentally. For simple equations, and without considering the effect of boundary conditions, the von Neumann criterion can be simplified by assuming solutions with an amplification factor ρ rather than a matrix. The solution for the amplification factor ρ then coincides with the eigenvalues of the amplification matrix, and the von Neumann stability criterion is $\rho \leq 1 + O(\Delta t)$.

Example 3.2.2:

$$\text{PDE: } \frac{\partial u}{\partial t} + c\frac{\partial u}{\partial x} = 0$$

$$\text{FDE: } \frac{U_j^{n+1} - U_j^n}{\Delta t} + c\frac{U_j^n - U_{j-1}^n}{\Delta x} = 0 \quad \text{(upstream scheme)} \tag{3.2.17}$$

We have already studied consistency, and used the criterion of the maximum to get a sufficient condition for stability. Let us now apply the von Neumann criterion: Assume

$$U_j^n = \sum_p Z_p^n e^{ipj} = \sum_p A\rho_p^n e^{ipj}$$

We substitute in (3.2.17) and eliminate Ae^{ipj} and obtain

$$\frac{\rho_p^{n+1} - \rho_p^n}{\Delta t} + c\frac{\rho_p^n(1 - e^{-ip})}{\Delta x} = 0 \quad \text{for all } p \tag{3.2.18}$$

The amplification factor ρ is the same as the 1×1 amplification matrix $\mathbf{G}$, and therefore the same as its spectral radius $\sigma(G)$, and the stability condition is $|\rho| \leq 1$ for all wavenumbers p. We need to estimate the maximum value of the spectral radius (or amplification factor in this case):

$$\rho = 1 - \mu(1 - e^{-ip}) = 1 - \mu(1 - \cos p + i \sin p) \tag{3.2.19}$$

$$|\rho|^2 = (1 - \mu(1 - \cos p))^2 + \mu^2 \sin^2 p \tag{3.2.20}$$

We make use of the trigonometrical relationships

$$\cos p = \cos^2 \frac{p}{2} - \sin^2 \frac{p}{2} \quad \sin p = 2 \sin \frac{p}{2} \cos \frac{p}{2}$$

and obtain

$$|\rho|^2 = 1 - 4\mu(1 - \mu) \sin^2 \frac{p}{2} \tag{3.2.21}$$

Now consider the $\sin^2 p/2$ term: *The shortest wave that can be present in the finite difference solution is* $L = 2\Delta x$, therefore the maximum value that $p = k\Delta x = 2\pi \Delta x / L$ can take is $p = \pi$, and the maximum value of $\sin^2 p/2$ is therefore 1. The other factor, $\mu(1 - \mu)$, is a parabola whose maximum value is 0.25 when $\mu = 0.5$. So the amplification factor squared will remain less than or equal to 1 as long as $0 \leq \mu \leq 1$. This coincides with the condition we obtain from the criterion of the maximum (and also with the notion that we should not extrapolate but interpolate the new values at time level $t = (n + 1)\Delta t$, cf. Fig. 3.2.2).

It is important to note that the amplification factor ρ indicates how much the amplitude of each wavenumber will decrease or increase with each time step. The upstream scheme decreases the amplitude of all Fourier wave components of the solution, since, if $0 < \mu < 1$, $\rho < 1$. This is therefore a very dissipative FDE: it has strong "numerical diffusion." Fig. 3.2.3 shows the decrease in amplitude when using the upstream scheme after one time step and after 100 time steps for each

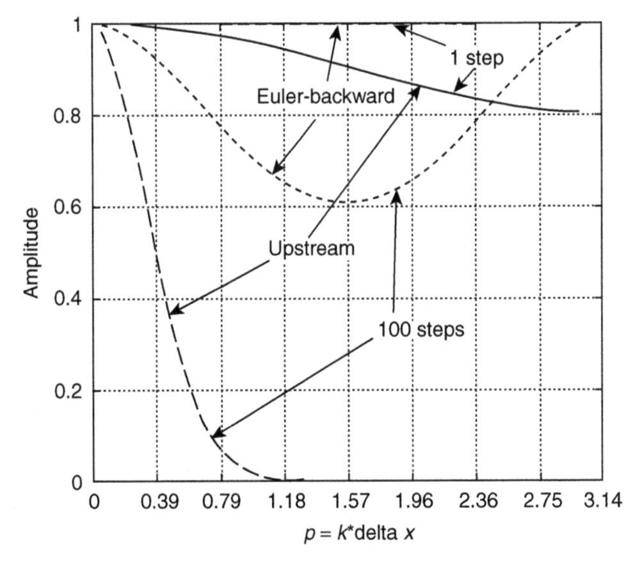

Figure 3.2.3: Amplification factor of wave components of the wave equation using either the "upstream" FDE, and the Matsuno or Euler-backward schemes with $\mu = 0.1$; L is the wavelength in units of $\Delta x \rho_{EB} = [1 - \mu^2 \sin^2 p + \mu^4 \sin^4 p]^{1/2}$ and $\rho_{Upxy} = [1 - 4\mu(1 - \mu) \sin^2 p]^{1/2}$.

wavenumber p, using a Courant number $\mu = 0.1$, a typical value for advection given the presence of fast gravity waves. Since its truncation errors are large (of first order), the upstream scheme is in general not recommended except for special situations (e.g., for outflow boundary conditions, or when modified in such a way that the dissipation rate becomes lower). An alternative, less damping scheme known as the Matsuno or Euler-backward scheme, frequently used in combination with the leapfrog scheme is also shown. Note that a "downstream" scheme (Fig. 3.2.2(c)) is unstable.

Example 3.2.3: Leapfrog scheme for the wave equation

$$\text{PDE}: \frac{\partial u}{\partial t} + c\frac{\partial u}{\partial x} = 0$$

$$\text{FDE}: \frac{U_j^{n+1} - U_j^{n-1}}{2\Delta t} + c\frac{U_{j+1}^n - U_{j-1}^n}{2\Delta x} = 0 \tag{3.2.22}$$

This is the most popular of all schemes used for hyperbolic equations.

Exercise 3.2.2: Find that the leapfrog scheme is consistent, and the local truncation error is of second order in space and time.

Stability: Assume $U_j^n = \sum_p Z_p^n e^{ipj} = \sum_p A\rho_p^n e^{ipj}$

Substitute in the FDE

$$\frac{\rho^{n+1} - \rho^{n-1}}{2\Delta t} + c\frac{\rho^n(e^{ip} - e^{-ip})}{2\Delta x} = 0 \tag{3.2.23}$$

Therefore

$$\rho^2 + 2i\mu \sin p\rho - 1 = 0 \tag{3.2.24}$$

Because we have three, not two, time levels ρ^{n+1}, ρ^n, and ρ^{n-1}, we have a quadratic equation and two solutions for the amplification factor ρ:

$$\rho = (-i\mu \sin p) \pm \sqrt{(-\mu^2 \sin^2 p + 1)} \tag{3.2.25}$$

Since the last term in the quadratic equation (3.2.24) is –1, and this is the product of the roots, the term inside the root $(-\mu^2 \sin^2 p + 1)$ must be positive, since otherwise the roots would be purely imaginary, and one of them would be larger than 1, which violates the stability criterion. In order for $\sqrt{(-\mu^2 \sin^2 p + 1)}$ to be real for all p, we must have $\mu^2 \leq 1$. The stability condition for the leapfrog scheme therefore becomes

$$-1 \leq c\Delta t/\Delta x \leq 1 \tag{3.2.26}$$

Exercise 3.2.3: Draw a schematic like Fig. 3.2.2, and explain why the sign of the Courant number does not matter for its stability criterion, unlike the sign of the upstream scheme. Why did the term $O(\Delta t)$ not appear in the stability condition?

We can actually find the exact solution of the leapfrog FDE (3.2.22), as well as of the PDE. Recall that the PDE $\partial u/\partial t + c\partial u/\partial x = 0$ has plane wave solutions of the form $Ae^{ik(x-ct)} = Ae^{i(kx-\omega t)}$, since the exact solution is of the form $u(x,t) = u(x-ct,0)$. The FDR $\omega = kc$ gives the exact frequency of the PDE.

By analogy we try to find solutions of the FDE of the form $A_p e^{i(pj-\theta n)}$, where $\theta = \nu\Delta t$ represents the computational frequency ν multiplied by Δt (the computational frequency ν is in general different than the exact frequency ω). Substituting in the FDE and dividing by $e^{i(pj-\theta n)}$, we get

$$(e^{-i\theta} - e^{i\theta}) + \mu(e^{ip} - e^{-ip}) = 0$$

or

$$\sin\theta = \mu \sin p \tag{3.2.27}$$

the FDR for the leapfrog scheme. Because $\sin\theta = \sin(\pi - \theta)$, the two solutions for the finite difference FDR are

$$\left.\begin{aligned} \theta_1 &= \arcsin(\mu \sin p) \\ \theta_2 &= \pi - \arcsin(\mu \sin p) \end{aligned}\right\} \tag{3.2.28}$$

Substituting into the FDR, and assuming that the initial amplitude for the wavenumber p is 1, we obtain that the solution of the FDE is a sum of two terms corresponding to θ_1 and θ_2 respectively:

$$U_j^n = A_p e^{i(pj-\theta n)} + (1 - A_p)e^{i(pj+\theta n)}(-1)^n \tag{3.2.29}$$

where $\theta = \arcsin(\mu \sin p)$, and $e^{i\pi} = -1$. (This can also be obtained by noting that when we assume solutions of the form $e^{i(pj-\theta n)}$, they imply an amplification factor $\rho = e^{-i\theta} = \cos\theta - i\sin\theta = -i\mu\sin p \pm \sqrt{1-\mu^2\sin^2 p}$, i.e., $\sin\theta = \mu\sin p$, with two solutions as indicated above.)

Of the two terms in the solution, the first one is the "legitimate" solution, which approximates the PDE solution. Note that the second term changes sign every time step, and it moves in the wrong direction: for this reason this unphysical term is called "computational mode". It arises because the leapfrog scheme has three time levels, rather than two, giving rise to an additional spurious solution. Although the leapfrog scheme is simple and accurate, its three-time level character gives rise to two problems that need to be dealt with.

The first problem is that the leapfrog scheme needs a special initial step to get to the first time level U^1 from the initial conditions U^0, before it can be started (Fig. 3.2.4). This can be done in several simple ways:

(a) Simply set $U^1 = U^0$. Since $u^1 = u^0 + u_t\Delta t + \cdots$, this introduces errors of order $O(\Delta t)$, and is not recommended.

(b) Use for the first time step a forward time scheme. The forward scheme has truncation errors of order $O(\Delta t)$, but since the time step is only used once, its

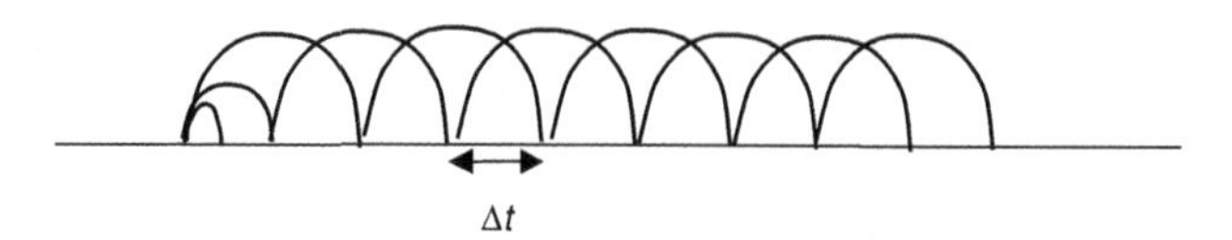

Figure 3.2.4: Schematic of the leapfrog scheme with a half time step starting step.

contribution to the global error is multiplied by Δt, so that the total error is still of $O(\Delta t)^2$. For the same reason, the computational instability is not a significant problem. Alternatively, use an Euler-backwards (Matsuno) scheme for the first time step (see Table 3.2.1).

(c) Use half (or a quarter, eighth, etc.) of the initial time step for the forward time step (Fig. 3.2.4), followed by leapfrog time steps. This will halve (or reduce by a quarter, eighth, etc.) the error introduced in the unstable first step.

The second problem is that for nonlinear examples, the leapfrog scheme has a tendency to increase the amplitude of the computational mode with time, separating the space dependence in a checkerboard fashion between the even and odd time steps. This can be solved by restarting every 50 steps or so, or by applying a Robert–Asselin time filter.

Exercise 3.2.4: Show that a forward time scheme is unstable for hyperbolic equations.

Robert–Asselin time filter (Robert, 1969, Asselin, 1972)

After the leapfrog scheme is used to obtain the solution at $t = (n + 1)\Delta t$, a slight time smoothing is applied to the solution at time $n\Delta t$:

$$\overline{U^n} = U^n + \alpha(U^{n+1} - 2U^n + \overline{U^{n-1}}) \tag{3.2.30}$$

replacing the solution at time n. Note that the added term is like smoothing in time, an approximation of an ideally time-centered smoother:

$$\overline{U^n} = U^n + \alpha(U^{n+1} - 2U^n + U^{n-1}) \tag{3.2.31}$$

which cannot be carried out without knowing the complete unsmoothed series.

The smoother (3.2.31) is centered in time, and reduces the amplitude of different frequencies ν by a factor $(1 - 4\alpha \sin^2(\nu\Delta t/2))$. The computational mode, whose period is $2\Delta t$, is reduced by $(1 - 4\alpha)$ every time step. Because the field at $t = (n - 1)\Delta t$ is replaced by the already filtered value, the filter (3.2.30) introduces a slight distortion of the centered filter (Asselin, 1972). This filter is widely used with the leapfrog scheme, with α of the order of 1%.

Exercise 3.2.5: Integrate the linear wave equation using values typical of large scale models. You can write your own FORTRAN or MATLAB program.

$$\frac{\partial u}{\partial t} = -c\frac{\partial u}{\partial x}$$

Table 3.2.1. *Time schemes for initial value problems dU/dt = F(U) (schemes (a)–(i)); dU/dt = $F_1(U)$ +$F_2(U)$ (schemes (j)–(k))*

(a) $\dfrac{U^{n+1}-U^{n-1}}{2\Delta t}=F(U^n)$	Leapfrog (good for hyperbolic equations, unstable for parabolic equations)
(a′) $\dfrac{U^{n+1}-\overline{U}^{n-1}}{2\Delta t}=F(U^n)$; $\overline{U}^n=U^n+\alpha(U^{n+1}-2U^n+\overline{U}^{n-1})$	Leapfrog smoothed with the Robert–Asselin time filter; $\alpha\sim 1\%$
(b) $\dfrac{U^{n+1}-U^n}{\Delta t}=F(U^n)$	Euler (forward, good for diffusive terms, unstable for hyperbolic equations)
(c) $\dfrac{U^{n+1}-U^n}{\Delta t}=F\left(\dfrac{U^n+U^{n+1}}{2}\right)$	Crank–Nicholson or centered implicit
(c′) $\dfrac{U^{n+1}-U^n}{\Delta t}=F\left(\dfrac{\beta U^n+(1-\beta)U^{n+1}}{2}\right)$; $\beta<0.5$	Implicit, slightly damping
(d) $\dfrac{U^{n+1}-U^n}{\Delta t}=F(U^{n+1})$	Fully implicit or backward
(e) $\dfrac{U^*-U^n}{\Delta t}=F(U^n)$; $\dfrac{U^{n+1}-U^n}{\Delta t}=F(U^*)$	Euler-backward or Matsuno: good for damping high frequency waves
(f) $\dfrac{U^*-U^n}{\Delta t}=F(U^n)$; $\dfrac{U^{n+1}-U^n}{\Delta t}=F\left(\dfrac{U^n+U^*}{2}\right)$	Another predictor–corrector scheme (Heun)
(g) $\dfrac{U^{n+1}-U^n}{\Delta t}=F\left(\dfrac{3}{2}U^n-\dfrac{1}{2}U^{n-1}\right)$	Adams–Bashford (second order in time).
(h) $\dfrac{U^{n+1/2^*}-U^n}{\Delta t/2}=F(U^n)$; $\dfrac{U^{n+1/2^{**}}-U^n}{\Delta t/2}=F(U^{n+1/2^*})$; $\dfrac{U^{n+1^*}-U^n}{\Delta t}=F(U^{n+1/2^{**}})$ $\dfrac{U^{n+1}-U^n}{\Delta t}=\dfrac{1}{6}[F(U^n)+2F(U^{n+1/2^*})+2F(U^{n+1/2^{**}})+F(U^{n+1^*})]$	Runge–Kutta (fourth order)

Table 3.2.1. *(cont.)*

(i) $a = 0;\ b = 1/\Delta t$ $U^* \leftarrow (aU^* + F(U^n))/b$ $U^n \leftarrow U^n + U^*$ $a \leftarrow a - 1/(N\Delta t);\ b \leftarrow b - 1/(N\Delta t)$	N-times	Lorenz's N-cycle, N = multiple of 4; Nth order
(j) $\dfrac{U^{n+1} - U^{n-1}}{2\Delta t} = F_1(U^n) + F_2\left(\dfrac{U^{n+1} + U^{n-1}}{2}\right)$		Semi-implicit
(k) $\dfrac{U^* - U^n}{\Delta t} = F_1(U^n);\ \dfrac{U^{n+1} - U^*}{\Delta t} = F_2(U^*)$		Fractional steps

Boundary conditions: periodic
Initial conditions: $u(x, 0) = c + A\sin(kx)$
$c = 20\,\text{m/s}$, $A = 10\,\text{m/s}$, $\Delta x = 200\,\text{km}$, $k = 2\pi/L$ with $L = 10\Delta x$

(a) Choose two time steps, one of which satisfies the CFL condition and one which violates it. How long does it take to "blow up"?

(b) Compare with the exact solution, compute the rms error R and the relative error *RE*.
Repeat with $A = 25$ m/s.
Repeat with $L = 4\Delta x$.

(c) Prepare a table that summarizes R and *RE*.

Exercise 3.2.6: Modify the equation and the program used before to integrate a nonlinear wave equation using values typical of large scale models:

$$\frac{\partial u}{\partial t} = -(u + c)\frac{\partial u}{\partial x}$$

Boundary conditions: periodic
Initial conditions: $u(x, 0) = A\sin(kx)$
$c = 20\,\text{m/s}$, $A = 10\,\text{m/s}$, $\Delta x = 200\,\text{km}$, $k = 2\pi/L$ with $L = 10\Delta x$

Choose two time steps, one of which satisfies the CFL condition and one which violates it. How long does it take to "blow up"? Compare with the linear equation results.

Repeat with $A = 25\,\text{m/s}$.
Repeat with $L = 4\Delta x$.

Compute a nonlinear solution with high resolution, taking it as "truth," and then prepare a table summarizing R and *RE*.

Example 3.2.4:

$$\frac{\partial u}{\partial t} = \sigma \frac{\partial^2 u}{\partial x^2} + bu \tag{3.2.32}$$

This is the heat or diffusion equation with a "source of growth" bu.

$$\text{FDE}: \frac{U_j^{n+1} - U_j^n}{\Delta t} = \sigma \frac{U_{j+1}^n - 2U_j^n + U_{j-1}^n}{\Delta x^2} + bU_j^n \tag{3.2.33}$$

Exercise 3.2.7: Show that the amplification factor is

$$\rho = 1 - \frac{4\sigma \Delta t}{(\Delta x)^2} \sin^2 \frac{p}{2} + b\Delta t \leq 1 + O(\Delta t) \tag{3.2.34}$$

Therefore the stability criterion is still $\sigma \Delta t / \Delta x^2 \leq 1/2$, as we obtained with the criterion of the maximum.

Exercise 3.2.8: Explain physically why the term $b\Delta t$ does not influence the stability criterion.

3.2.4 Implicit time schemes

In these schemes the advection or diffusion terms are written in terms of the new time level variables.

Example 3.2.5:

$$\text{PDE:} \frac{\partial u}{\partial t} + c\frac{\partial u}{\partial x} = 0$$

$$\begin{aligned} \text{FDE:} & \frac{U_j^{n+1} - U_j^n + U_{j+1}^{n+1} - U_{j+1}^n}{2\Delta t} \\ & + c\frac{\alpha\left(U_{j+1}^n - U_j^n\right) + (1-\alpha)\left(U_{j+1}^{n+1} - U_j^{n+1}\right)}{\Delta x} = 0 \end{aligned} \tag{3.2.35}$$

The factor α determines the weight of the "old" time values compared with the "new" time values in the right-hand side of the FDE. Using the von Neumann method, we substitute $U_j^n = A\rho^n e^{ipj} = Ae^{i(pj-\theta n)}$ into (3.2.35).

Note that the scheme is centered in time (if $\alpha = 1/2$) at the point $U_{j+1/2}^{n+1/2}$. For this reason, we multiply by $e^{-ip/2}$, and obtain the amplification factor:

$$\rho = \frac{\cos\frac{p}{2} - i2\mu\alpha \sin\frac{p}{2}}{\cos\frac{p}{2} + i2\mu(1-\alpha)\sin\frac{p}{2}} \tag{3.2.36}$$

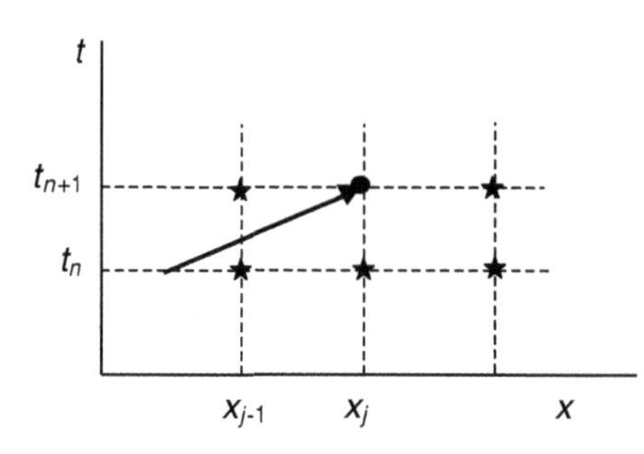

Figure 3.2.5: Schematic of an implicit scheme. The dot represents the value being updated and the stars the values that influence it. Note that with the implicit scheme there is no extrapolation, and allows no limit to the size of Δt.

or

$$|\rho|^2 = \frac{1 + 4\mu^2\alpha^2 \tan^2 \frac{p}{2}}{1 + 4\mu^2(1-\alpha)^2 \tan^2 \frac{p}{2}} \tag{3.2.37}$$

This implies that $\rho \leq 1$ if $\alpha \leq 0.5$, i.e., if the new values are given at least as much weight as the old values in computing the RHS. In this case *there is no restriction on the size that Δt can take!* This result (absolute stability, independent of the Courant number) is typical of implicit time schemes. In Fig. 3.2.5 we show that in an implicit scheme, a point at the new time level is influenced by all the values at the new level, which avoids extrapolation, and therefore is *absolutely stable*. Note also that if $\alpha < 0.5$ the implicit time scheme reduces the amplitude of the solution: it is an example of a *damping scheme*. This property is useful for solving some problems such as spuriously growing mountain waves in semi-Lagrangian schemes.

In summary, if we consider a marching equation

$$\frac{dU}{dt} = F(U) \tag{3.2.38}$$

explicit methods such as the forward scheme

$$\frac{U^{n+1} - U^n}{\Delta t} = F(U^n) \tag{3.2.39}$$

or the leapfrog scheme

$$\frac{U^{n+1} - U^{n-1}}{2\Delta t} = F(U^n) \tag{3.2.40}$$

are either *conditionally stable* (when there is a condition on the Courant number or the equivalent stability number for parabolic equations) or *absolutely unstable*.

A fully implicit scheme

$$\frac{U^{n+1} - U^n}{\Delta t} = F(U^{n+1}) \tag{3.2.41}$$

and a centered implicit scheme (Crank–Nicholson)

$$\frac{U^{n+1} - U^n}{\Delta t} = F\left(\frac{U^n + U^{n+1}}{2}\right) \tag{3.2.42}$$

are *absolutely stable.* The latter scheme is attractive because it is centered in time (around $t_{n+1/2}$), and it can be written with centered space differences, which makes it second order in space and in time. Also, it only has two time levels so it does not have a computational mode. But, like all implicit schemes, it also has a great disadvantage. Since U^{n+1} appears on the left- and on the right-hand sides, the solution for U^{n+1}, unlike explicit schemes, in general requires the solution of a system of equations. If it involves only tridiagonal systems, this is not an obstacle, because there are fast methods to solve them. There are also methods, such as fractional steps (with each spatial direction solved successively), where one space dimension is considered at a time, that allow taking advantage of the large time steps permitted by implicit schemes without paying a large additional computational cost.

Moreover, we will see in the next section that the possibility of using a time step with a Courant number much larger than 1 in an implicit scheme does not imply that we will obtain accurate results economically. The implicit scheme maintains stability by *slowing down* the solutions, so that the slower waves do satisfy the CFL condition. For this reason implicit schemes are only useful for those modes (such as the Lamb wave or vertical sound waves) that are very fast but of little meteorological importance (semi-implicit schemes, see next section).

Notes

(1) It is easy to check the properties of the time schemes in Table 3.2.1 when applied to hyperbolic equations by testing them with a simple harmonic equation:

$$\frac{\partial U}{\partial t} = -i\nu U \tag{3.2.43}$$

with solution $U(t) = U(0)e^{-i\nu t}$. After one time step, the exact solution is

$$U((n+1)\Delta t) = U(n\Delta t)e^{-i\nu\Delta t} \tag{3.2.44}$$

which indicates that the exact magnification factor is $e^{-i\nu\Delta t}$.

In (3.2.43), ν is the computational frequency for a wave equation for a given space discretization. For example, if we were using second order centered differences in space, $\nu = (\sin k\Delta x/\Delta x)\, c$, for a spectral scheme, $\nu = kc$. For the fully implicit time scheme (d), the amplification factor is

$$\frac{1}{1+i\nu\Delta t} = \frac{1-i\nu\Delta t}{1+(\nu\Delta t)^2}$$

Since the exact amplification factor has an amplitude equal to 1, this shows that the implicit scheme is dissipative; similarly, comparing the imaginary components of the exact and approximate amplification factors, it is clear that the implicit solution is slowed down by a factor of about $1/[1+(\nu\Delta t)^2]$.

Exercise 3.2.9: Show that the Crank–Nicholson scheme significantly slows down the angular speed of the solution by deriving the magnification factor for this scheme, and comparing it with the exact magnification factor $e^{-i\nu\Delta t}$. Determine the limit of the Crank–Nicholson amplification factor for the Courant number $\nu\Delta t \to \infty$.

$$U^{n+1} = U^n \rho_{CN} \tag{3.2.45}$$

(2) Equations with damping terms (such as the parabolic equation) can also be simply represented by the equation:

$$\frac{\partial U}{\partial t} = -\mu U \tag{3.2.46}$$

In (2.46), μ can be considered as the computational rate of damping. For example, for the diffusion equation, using centered differences in space,

$$\mu = \frac{4\sigma}{(\Delta x)^2}\sin^2\frac{k\Delta x}{2}$$

Exercise 3.2.10: Show that the leapfrog scheme is unstable for a damping term.

Exercise 3.2.11: Write a numerically stable scheme for the equation with both wave-like and damping terms $\partial U/\partial t = -(i\nu + \mu)U$ using a three-time level scheme.

Exercise 3.2.12: Show that for a wave equation the forward time scheme with centered differences in space is absolutely unstable. Note that this scheme shows that the "no extrapolation" rule is a necessary but not a sufficient condition for stability of wave equations.

3.2.5 Semi-implicit schemes

Consider the SWEs that we discussed in Section 2.4.1:

$$\left.\begin{aligned}
\frac{\partial u}{\partial t} + u\frac{\partial u}{\partial x} + v\frac{\partial u}{\partial y} &= \underline{-\frac{\partial \phi}{\partial x}} + fv \\
\frac{\partial v}{\partial t} + u\frac{\partial v}{\partial x} + v\frac{\partial v}{\partial y} &= \underline{-\frac{\partial \phi}{\partial y}} - fu \\
\frac{\partial \phi}{\partial t} + u\frac{\partial \phi}{\partial x} + v\frac{\partial \phi}{\partial y} &= \underline{-\Phi\left(\frac{\partial u}{\partial x} + \frac{\partial v}{\partial y}\right)} - (\phi - \Phi)\left(\frac{\partial u}{\partial x} + \frac{\partial v}{\partial y}\right)
\end{aligned}\right\} \tag{3.2.47}$$

As indicated in that section, the phase speed of the inertia-gravity wave is given by

$$c_{IGW} = \frac{\nu_{IGW}}{k} = U \pm \sqrt{\frac{f^2}{k^2} + \Phi} \approx U \pm 300\,\text{m/s}$$

and the terms that give rise to the fast gravity waves are underlined. This means that the Courant number $\mu = c_{IGW}\Delta t/\Delta x$ is dominated by the speed of external inertia-gravity waves (equivalent to the Lamb waves, horizontal sound waves), and an explicit scheme would therefore require a time step an order of magnitude smaller than that required for advection. For this reason, Robert (1969) introduced the use of semi-implicit schemes to slow down the gravity waves. We write such a scheme using the compact finite difference notation for differences and averages:

$$\left.\begin{aligned} \delta_x f &= \frac{f_{j+1/2} - f_{j-1/2}}{\Delta x} \\ \overline{f}^x &= (f_{j+1/2} + f_{j-1/2})/2 \end{aligned}\right\} \tag{3.2.48}$$

and similarly for differences in y or t. With this notation, assuming uniform resolution,

$$\left.\begin{aligned} \delta_{2x} f &= \delta_x \overline{f}^x = \frac{f_{i+1} - f_{i-1}}{2\Delta x} \\ \overline{f}^{2x} &= (f_{i+1} + f_{i-1})/2 \end{aligned}\right\} \tag{3.2.49}$$

Using this compact finite difference notation we can write the leapfrog semi-implicit SWE as

$$\left.\begin{aligned} \delta_{2t} u + u\delta_{2x} u + v\delta_{2y} u &= -\delta_{2x}\overline{\phi}^{2t} + fv \\ \delta_{2t} v + u\delta_{2x} v + v\delta_{2y} v &= -\delta_{2y}\overline{\phi}^{2t} - fu \\ \delta_{2t}\phi + u\delta_{2x}\phi + v\delta_{2y}\phi &= -\Phi\overline{(\delta_{2x} u + \delta_{2y} v)}^{2t} - (\phi - \Phi)(\delta_{2x} u + \delta_{2y} v) \end{aligned}\right\} \tag{3.2.50}$$

Everything that does not have a time average involves only terms evaluated explicitly at the nth time step. We can rewrite the FDEs (3.2.50) as

$$\left.\begin{aligned} \frac{u^{n+1} - u^{n-1}}{2\Delta t} &= -\delta_{2x}(\phi^{n+1} + \phi^{n-1})/2 + R_u \\ \frac{v^{n+1} - v^{n-1}}{2\Delta t} &= -\delta_{2y}(\phi^{n+1} + \phi^{n-1})/2 + R_v \\ \frac{\phi^{n+1} - \phi^{n-1}}{2\Delta t} &= -\Phi[\delta_{2x}(u^{n+1} + u^{n-1})/2 + \delta_{2y}(v^{n+1} + v^{n-1})/2] + R_\phi \end{aligned}\right\} \tag{3.2.51}$$

where the "R" terms are the "rest" of the terms evaluated at the center time $n\Delta t$. For example, $R_u = fv - u\delta_{2x} u - v\delta_{2y} u$, and similarly for R_v and R_ϕ.

From these three equations we can eliminate u^{n+1}, v^{n+1} and obtain an elliptic equation for ϕ^{n+1}:

$$\begin{aligned} \left(\delta_{2x}^2 + \delta_{2y}^2 - \frac{1}{\Phi\Delta t^2}\right)\phi^{n+1} &= -\left(\delta_{2x}^2 + \delta_{2y}^2 + \frac{1}{\Phi\Delta t^2}\right)\phi^{n-1} \\ &+ 2(\delta_{2x} R_u + \delta_{2y} R_v) + \frac{2}{\Delta t}(\delta_{2x} u^{n-1} + \delta_{2y} v^{n-1}) - \frac{2}{\Phi\Delta t} R_\phi = F_{i,j}^n \end{aligned} \tag{3.2.52}$$

Note that the right-hand side of this elliptic equation is evaluated at $t = n\Delta t$ or $(n-1)\Delta t$, so that it is known. Solving this elliptic equation provides ϕ^{n+1}, and once this is known, it can be plugged back into the first two equations of (3.2.51), and thus (u^{n+1}, v^{n+1}) can be obtained.

The elliptic operator in brackets in the left-hand side of (3.2.52), is a finite difference equivalent to $(\nabla^2 - \lambda^2)$,

$$\left(\delta_{2x}^2 + \delta_{2y}^2 - \frac{1}{\Phi\Delta t^2}\right)\phi = \frac{\phi_{i+2,j} + \phi_{i-2,j} + \phi_{i,j+2} + \phi_{i,j-2} - \left(4 + \frac{1}{\mu^2}\right)\phi_{i,j}}{4\Delta^2} \tag{3.2.53}$$

where we have assumed for simplicity that $\Delta x = \Delta y = \Delta$, and $\mu^2 = \Phi\Delta t^2/\Delta^2$ is the square of the Courant number for gravity waves. Since $\mu^2 = \Phi\Delta t^2/\Delta^2 >> 1$, the semi-implicit scheme distorts the gravity wave solution, slowing the gravity wave down until they satisfy the von Neumann criterion. This is an acceptable distortion since we are interested in the slower "weather-like" processes, and since the slower modes satisfy the CFL (von Neumann) stability criterion, and they are written explicitly, they are not slowed down or distorted in a significant way.

In the same way that the terms giving rise to gravity waves can be written semi-implicitly, the terms giving rise to sound waves can also be written semi-implicitly (Robert, 1982). They are the three-dimensional divergence in the continuity equation (Sections 2.3.2, 2.3.3). This has allowed the use of nonhydrostatic models without the use of the anelastic approximation or the hydrostatic approximation. André Robert (1982) created a model that can be considered the "ultimate" atmospheric model. It treats the terms generating sound waves (anelastic terms, i.e., three-dimensional divergence), and the terms generating gravity waves (pressure gradient and horizontal divergence) semi-implicitly, and it uses a three-dimensional semi-Lagrangian scheme for all advection terms. This model, denoted the "Mesoscale Compressible Community" (MCC) model, is a "universal" model designed so that it can tackle accurately atmospheric problems from the planetary scale through mesoscale, convective and smaller (Laprise *et al.*, 1997).

There is another approach followed by several major nonhydrostatic models (e.g., MM5 and ARPS): the use of fractional steps (see Table 3.2.1, scheme (k)), with the sound-wave terms integrated with small time steps. In addition, the ARPS model uses a semi-implicit scheme for vertically propagating sound waves (Xue *et al.*, 1995).

Exercise 3.2.13: Consider the diffusion equation $\partial u/\partial t = \sigma\partial^2 u/\partial x^2$ with initial conditions $u = x$ for $x \le 0.5$ and $u = 1 - x$ for $x \ge 0.5$. Compute the first two time steps using an explicit scheme (forward in time, centered in space) with five points between $x = 0$ and $x = 1$, and a time step such that $r = \sigma\Delta t/(\Delta x)^2$ is equal to $r = 0.1,\ 0.5,\ 1.0$. Repeat using Crank–Nicholson's scheme.

3.3 Space discretization methods

3.3.1 Space truncation errors. Computational phase speed. Second and fourth order schemes.

It is convenient to separate the truncation errors in a discretized model into space truncation errors and time truncation errors. For explicit finite difference models, the errors introduced by space truncation tend to dominate the total forecast errors because for "weather waves" the time step and the Courant number used are much smaller than would be required to physically resolve the frequency. Let's neglect for the moment time truncation errors and consider the wave equation $\partial U/\partial t = -c\partial U/\partial x$ discretized only in space.

If we approximate $\partial U/\partial x$ using space centered differences, we get

$$\begin{aligned}\delta_{2x} U_j = \frac{U_{j+1} - U_{j-1}}{2\Delta x} &= U_x + \frac{\Delta x^2}{6} U_{xxx} + \frac{\Delta x^4}{120} U_{xxxxx} + HOT \\ &= U_x + A\Delta x^2 + B\Delta x^4 + \cdots \end{aligned} \tag{3.3.1}$$

If instead of the closest neighboring points $j+1$, $j-1$, we use the points $j+2$, $j-2$, we get

$$\delta_{4x} U_j = \frac{U_{j+2} - U_{j-2}}{4\Delta x} = U_x + 4A\Delta x^2 + 16B\Delta x^4 + \cdots \tag{3.3.2}$$

This is also a second order scheme, but the truncation errors are four times as large. We can now eliminate from (3.3.1) and (3.3.2) the term $A\Delta x^2$, and obtain

$$\frac{4}{3}\delta_{2x} U_j - \frac{1}{3}\delta_{4x} U_j = U_x - 4B\Delta x^4 + \cdots \tag{3.3.3}$$

Now (3.3.3) is a *fourth order* approximation of the space derivative. So

$$\frac{dU_j}{dt} = -c\delta_{2x} U_j \tag{3.3.4}$$

is a second order FDE and

$$\frac{dU_j}{dt} = -c\left(\frac{4}{3}\delta_{2x} U_j - \frac{1}{3}\delta_{4x} U_j\right) \tag{3.3.5}$$

is a fourth order FDE.

Assume solutions of the form

$$U_j(t) = Ae^{ik(x_j - c't)} = Ae^{i(kx_j - \nu' t)} \tag{3.3.6}$$

where c' is the computational phase speed, and ν' the computational frequency, so that $dU_j/dt = -i\nu' U_j$. Making use of $\delta_{2x} U_j = i\,(\sin k\Delta x/\Delta x)\,U_j$, and replacing in (3.3.4) and (3.3.5) we find that for second order differences,

$$c_2' = \frac{\sin k\Delta x}{k\Delta x} c \tag{3.3.7}$$

and for fourth order differences,

$$c_4' = \left(\frac{4}{3}\frac{\sin k\Delta x}{k\Delta x} - \frac{1}{3}\frac{\sin 2k\Delta x}{2k\Delta x}\right) c \tag{3.3.8}$$

Note that (3.3.7) and (3.3.8) imply that the phase speed is always underestimated by space finite differences. For the smallest possible wavelength, $L = 2\Delta x$, $k\Delta x = \pi$, the computational phase speed is zero for both second and fourth order differences: the shortest waves don't move at all (Fig. 3.3.1)! For $L = 4\Delta x$, $k\Delta x = \pi/2$, a much more accurate approximation is obtained with fourth order than with second order differences: $c_2' = 0.64c$, $c_4' = 0.85c$, and the fourth order advantage becomes even better for longer waves: for $L = 8\Delta x$, $c_2' = 0.90c$, $c_4' = 0.99c$.

We can also compute the *computational group velocity* $\partial \nu'/\partial k$, where

$$\nu' = c'k = c\frac{\sin k\Delta x}{\Delta x} \tag{3.3.9}$$

for second order differences. Then,

$$\frac{\partial \nu'}{\partial k} = c \cos k\Delta x \tag{3.3.10}$$

for second order differences. Therefore, for the shortest waves, $L = 2\Delta x$, $k\Delta x = \pi$, with both second and fourth order differences the energy moves in the opposite direction to the real group velocity (equal to the phase speed c): $c_g' = -c_g$. Figure. 3.3.1 shows the computational phase speed and group velocity for second and fourth order differences. As a result of the negative group velocity, space centered FDEs of the wave equation tend to leave a trail of short-wave computational noise upstream of where the real perturbation should be. This problem is greatly reduced using more advanced recent schemes such as those of Takacs (1985) and Smolarkiewicz and Grawoski (1990).

A second type of fourth order finite difference scheme, known as the *compact* or *implicit* fourth order scheme, can be obtained by again making use of (3.3.1) but replacing the third derivative in the truncation error for the centered differences by its finite difference approximation $U_{xxxj} \approx (U_{xj+1} - 2U_{xj} + U_{xj-1})/(\Delta x)^2 + O(\Delta x)^2$. The new fourth order scheme then becomes

$$U_{xj+1} + 4U_{xj} + U_{xj-1} = 6\frac{U_{j+1} - U_{j-1}}{2\Delta x} \tag{3.3.11}$$

It is called "compact" because it involves only the point j and its closest neighbors, and "implicit" because (3.3.11) results in a system of (tridiagonal) equations for the x-derivative, rather than an explicit estimate such as (3.3.4) or (3.3.5).

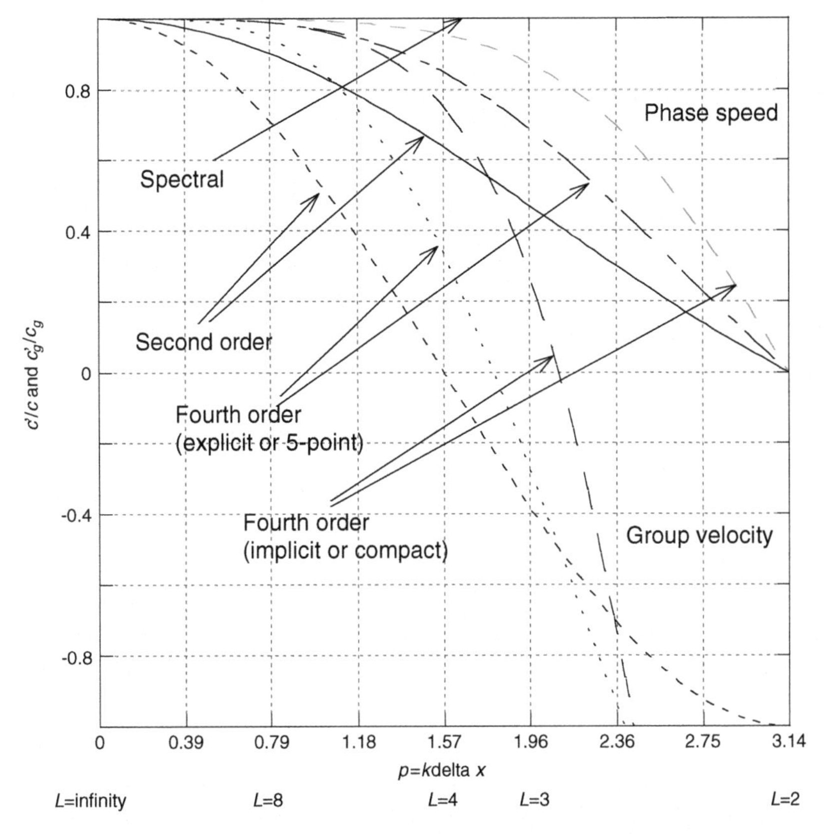

Figure 3.3.1: Ratio of the computational to the physical phase speed and group velocity for a simple wave equation, neglecting time truncation errors, for second order, fourth order explicit and implicit and spectral schemes.

With this scheme, the finite difference space derivative for a given wavenumber is given by

$$U_x \simeq U \frac{i \sin k\Delta x}{\Delta x} \frac{6}{4 + 2\cos k\Delta x}$$

so that

$$\frac{dU_j}{dt} = -cU_j \frac{i \sin k\Delta x}{\Delta x} \frac{6}{4 + 2\cos k\Delta x} = -i\upsilon_{4I}{}'U_j$$

and the computational phase speed becomes

$$c'_{4I} = \frac{\sin k\Delta x}{k\Delta x} \frac{6}{4 + 2\cos k\Delta x} c \tag{3.3.12}$$

and for $L = 4\Delta x$, $k\Delta x = \pi/2$, the phase speed is $c'_{14} = 0.955c$, which is considerably better even than the regular fourth order differences phase speed.

The group velocity for this scheme,

$$\frac{\partial v'_{4I}}{\partial k} = \left[\frac{6\cos k\Delta x}{4 + 2\cos k\Delta x} + \frac{2\sin^2 k\Delta x}{(4 + 2\cos k\Delta x)^2}\right]c \tag{3.3.13}$$

is already positive for $L = 3\Delta x$ (Fig. 3.3.1). For implicit schemes where one is already solving a tri-diagonal equation (see Section 3.4.2), this compact fourth order scheme, which has an accuracy equivalent to linear finite elements, is very accurate and involves little additional computational cost. The compact scheme is similar to Galerkin finite element approximation to space derivatives (Durran, 1999).

3.3.2 Galerkin and spectral space representation

The use of spatial finite differences, as we saw in the previous section, introduces errors in the space derivatives, resulting in a computational phase speed slower than the true phase speed, especially for short waves.

The Galerkin approach to ameliorate this problem is to perform the space discretization using a sum of basis functions $U(x,t) = \sum_{k=1}^{K} A_k(t)\varphi_k(x)$. Then, the residual (error) $R(U) = \partial U/\partial t + F(U)$ of the original PDE $\partial u/\partial t + F(u) = 0$ is required to be orthogonal to the basis functions $\varphi(x)$. The space derivatives are computed directly from the known $d\varphi(x)/dx$. This procedure leads to a set of ordinary differential equations for the coefficients $A_k(t)$. If the basis functions chosen for the discretization are orthogonal and satisfy the boundary conditions, the derivation becomes simpler. The use of *local* basis functions (e.g., $\varphi_i(x)$ a piecewise linear function equal to 1 at a grid point i and zero at the neighboring points) gives rise to the *finite element method*, with accuracy similar to that of the compact fourth order scheme. Another popular type of Galerkin approach is the use of a global spectral expansion for the space discretization, which allows the space derivatives to be computed analytically rather than numerically. In one dimension, periodic boundary conditions suggest the use of complex Fourier series as a basis.

Consider a periodic domain of length L, with a number of grid points $J_{max} = JM$, and scale x by $2\pi/L$. If we use discrete complex Fourier series truncated to include wavenumbers up to K, the spectral representation is:

$$U(x_j,t) = \sum_{k=-K}^{K} A_k(t)e^{ikx_j} \tag{3.3.14}$$

where $A_{-k}(t) = A_k^*(t)$, and the star represents the complex conjugate.

Alternatively, (3.3.14) can be written using real Fourier series as

$$U(x_j,t) = a_0 + \sum_{k=1}^{K} a_k\cos(kx) + \sum_{k=1}^{K} b_k\sin(kx) \tag{3.3.15}$$

where

$$A_k(t) = \frac{a_k}{2} - i\frac{b_k}{2} \quad k > 0 \quad A_0 = a_0$$

There are $2K + 1$ distinct real coefficients that are determined by

$$A_k(t) = \frac{1}{JM} \sum_{j=0}^{JM-1} U(x_j, t) e^{-ikx_j} \tag{3.3.16}$$

Here we have used the orthogonality property

$$\frac{1}{JM} \sum_{j=0}^{JM-1} e^{-ikx_j} e^{ilx_j} = \delta_{kl} = \begin{cases} 1 & \text{if } k = l \\ 0 & \text{otherwise} \end{cases} \tag{3.3.17}$$

If $JM = 2K + 1$, the grid representation (left-hand side of (3.3.14)) and the spectral representation (right-hand side of (3.3.14)) contain the same number of degrees of freedom, and the same information.

Then, in the wave equation $\partial U/\partial t = -c\partial U/\partial x$, we can discretize U in space as in (3.3.14) and compute the space derivative analytically:

$$\frac{\partial U(x, t)}{\partial x} = \sum_{k=-K}^{K} ikA_k(t)e^{ikx} \tag{3.3.18}$$

If we neglect the time discretization errors, as before, and assume solutions of the form $U(x, t) = Ae^{ik(x-c't)}$, we find that $c' = c$, i.e., the computational phase speed is *equal* to the true speed (Fig. 3.3.1). *The space discretization based on a spectral representation is extremely accurate* (the space truncation errors are of "infinite" order). This is because the space derivatives are computed analytically, not numerically, as done in finite differences.

If the PDE is nonlinear, for example $\partial U/\partial t = -U\partial U/\partial x$, then both the grid-point ("physical space") representation and the spectral representation are very useful: derivatives are computed efficiently and accurately in spectral space, whereas nonlinear products are computed efficiently in physical space. This leads to the so-called *transform method* used for spectral models: the space derivative is computed in spectral space, then U is transformed back into grid space, and the product $U_j\,(\partial U/\partial x)_j$ is computed locally in grid space. We will see later that in order to avoid nonlinear instability introduced by aliasing of wavenumbers beyond K that appear in quadratic terms, the grid representation requires about 3/2 as many points as the minimum number of points required for a linear transform ($JM = 2K + 1$). For this reason the new values of U at time $(n + 1)\Delta t$ are usually stored in their spectral representation, which is more compact.

We can use von Neumann's criterion to determine the maximum time step allowed for stability using, for example, the leapfrog time scheme. The FDE is

$$\frac{U^{n+1} - U^{n-1}}{2\Delta t} = -ikcU^n \tag{3.3.19}$$

Assuming solutions for the wave equation of the form $U^n = \rho^n e^{ikx}$, we obtain that the amplification factor is $\rho = -ikc\Delta t \pm \sqrt{1 - k^2c^2\Delta t^2}$, and in order to have $|\rho| \leq 1$ we need to satisfy the stability condition

$$(kc\Delta t)^2 \leq 1 \tag{3.3.20}$$

Since the highest wavenumber present corresponds to $L = 2\Delta x$, the stability criterion for spectral models is therefore $c\Delta t/\Delta x \leq 1/\pi$. So, the stability criterion is more restrictive for spectral models than for finite difference models, but this is compensated by the fact that the accuracy, especially for shorter waves, is much higher, and therefore fewer short waves need to be included (Fig. 3.3.1).

The basis functions used in spectral methods are usually the eigensolutions of the Laplace equation. In a rectangular domain, they are sines and cosines (e.g., the Regional Spectral Model (RSM), Juang *et al.*, 1997). On a circular plate, one would instead use Bessel functions.

Global atmospheric models use as basis functions spherical harmonics, which are the eigenfunctions of the Laplace equation on the sphere:

$$\begin{aligned}\nabla^2 Y_n^m &= \frac{1}{a^2}\left[\frac{1}{\cos^2\varphi}\frac{\partial^2 Y_n^m}{\partial\lambda^2} + \frac{1}{\cos\varphi}\frac{\partial}{\partial\varphi}\left(\cos\varphi\frac{\partial Y_n^m}{\partial\varphi}\right)\right] \\ &= \frac{-n(n+1)}{a^2}Y_n^m\end{aligned} \tag{3.3.21}$$

The spherical harmonics are products of Fourier series in longitude and associated Legendre polynomials in latitude:

$$Y_n^m(\lambda, \varphi) = P_n^m(\mu)e^{im\lambda} \tag{3.3.22}$$

where $\mu = \sin\varphi$, m is the zonal wavenumber and n is the "total" wavenumber in spherical coordinates (as suggested by the Laplace equation (3.3.21)). P_n^m are the associated Legendre polynomials in $\mu = \sin\varphi = \cos\theta$, where $\theta = \pi - \varphi$ is the colatitude. For example, the $P_0^0 = 1$; $P_1^0 = \cos\theta$; $P_1^1 = \sin\theta$; $P_2^0 = 1/2\,(3\cos^2\theta - 1)$; $P_2^1 = 3\sin\theta\cos\theta$; $P_2^1 = 3\sin^2\theta; \ldots$
Using "triangular" truncation

$$U(\lambda, \varphi, t) = \sum_{n=0}^{N}\sum_{m=-n}^{n} U_n^m(t)Y_n^m(\lambda, \varphi) \tag{3.3.23}$$

the spatial resolution is uniform throughout the sphere. This is a major advantage over finite differences based on a latitude–longitude grid, where the convergence of the meridians at the poles requires very small time steps. Although there are solutions for this "pole problem" for finite differences, the natural approach to solve the pole problem for global models is the use of spherical harmonics. Williamson and Laprise (1998) provide a comprehensive description of numerical methods for global models.

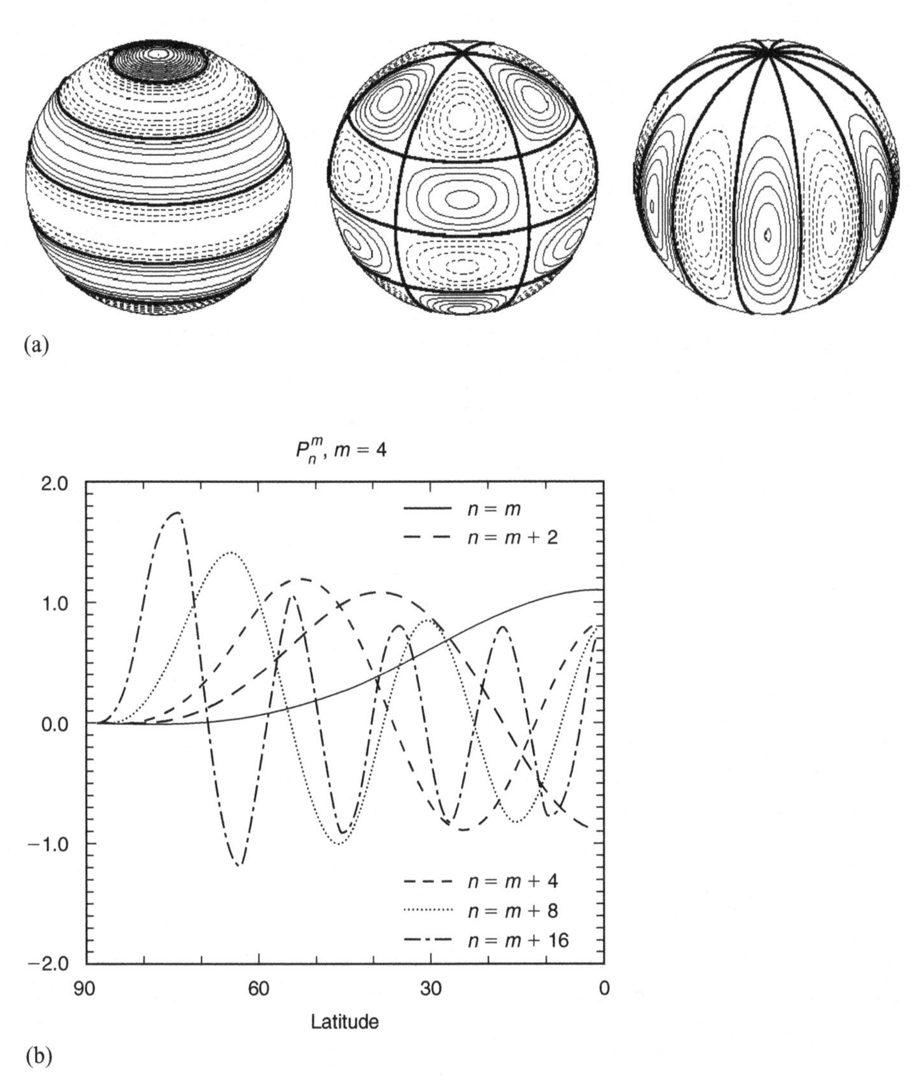

Figure 3.3.2: Illustration of the characteristics of spherical harmonics, adapted from Williamson and Laprise (1998). (a) Depiction of three spherical harmonics with total wavenumber $n = 6$. Left, zonal wavenumber $m = 0$; center, $m = 3$; right, $m = 6$. Note that n is associated with the total wavelength (twice the distance between a maximum and a minimum), which is the same for the three figures. (b) and (c) Amplitude of Legendre polynomials for different combinations of m and n showing how high zonal wavenumbers are suppressed near the poles, so that the horizontal resolution is uniform when using a spectral representation with triangular truncation.

Fig. 3.3.2(a) shows the shape of three spherical harmonics with total wavenumber $n = 6$, and zonal wavenumber $m = 0$, 3 and 6. Note that the distance between neighboring maxima and minima is similar for the three harmonics, and is associated with the "total" (two-dimensional) wavenumber n. Figures 3.3.2(b) and (c) show that the amplitude of the Legendre polynomials for high zonal wavenumbers are indeed

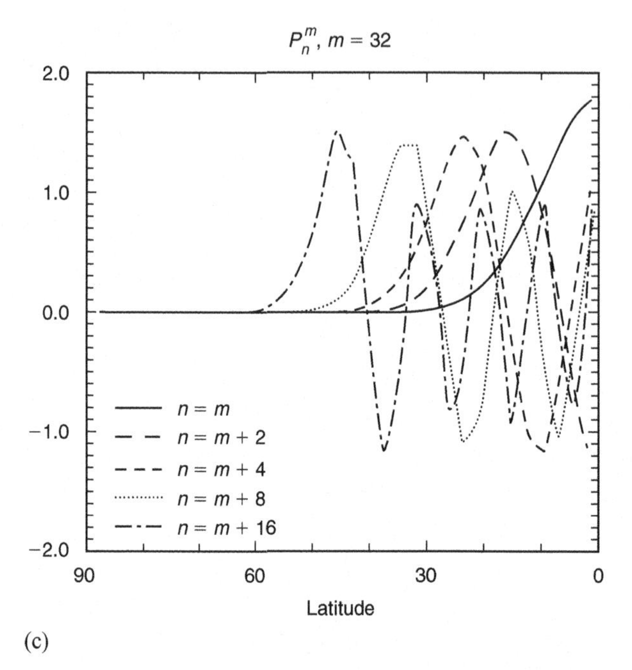

(c)

Figure 3.3.2: (*cont.*)

suppressed near the poles. This suppression eliminates the need for small time steps due to the convergence of the meridians in the poles, which are not singular points spectral models.

3.3.3 Semi-Lagrangian schemes

Another numerical method that has become very popular in NWP models is the *semi-Lagrangian* scheme. The equations of motion, as we have seen, can in general be written as conservation equations

$$\frac{du}{dt} = S(u) \tag{3.3.24}$$

where the left-hand side of the equation represents a *total time derivative* (following an individual parcel) of the vector of dependent variables u. The total time derivative (also known as individual, substantial or Lagrangian time derivative) is conserved for a parcel, except for the changes introduced by the source or sink S.

In a truly Lagrangian scheme, one would follow individual parcels (transporting them with the three-dimensional fluid velocity), and then add the source term at the right time. This is not practical in general because one has to keep track of many individual parcels, and with time they may "bunch up" in certain areas of the fluid, and leave others without parcels to track.

The semi-Lagrangian scheme avoids this problem by using a regular grid as in the previous schemes discussed (which are denoted Eulerian, because the partial

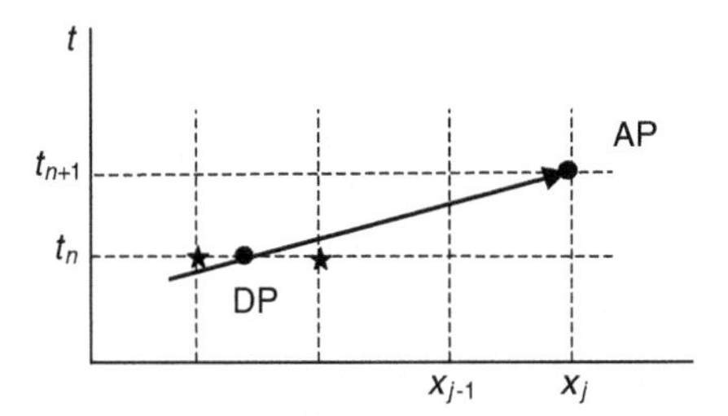

Figure 3.3.3: Schematic of the semi-Lagrangian scheme. The circles represent the arrival point AP at the new time level (a point) and the departure point DP at the previous time level. The thick arrow represents the advection from DP to AP. The value of the variables at AP is equal to their value at DP, which is obtained by interpolation between neighboring points. Because there is no extrapolation, the semi-Lagrangian schemes are absolutely stable.

derivative $\partial u/\partial t$ is estimated instead of the total derivative). At every new time step we find out where the parcel arriving at a grid point (denoted arrival point or AP) *came from* in the previous time step (denoted departure point or DP). The value of u at the DP is obtained by interpolating the values of the grid points surrounding the departure point. Figure 3.3.3 suggests that, because there is no extrapolation, the semi-Lagrangian scheme is absolutely stable with respect to advection, which can be confirmed by doing a von Neumann criterion check (Bates and McDonald, 1982).

The semi-Lagrangian scheme can then be written using two or three time levels. In a three-level time scheme, for example, if MP is the middle point between the DP and AP, the scheme can be written as

$$\left(U_j^{n+1}\right)_{AP} = (U^{n-1})_{DP} + 2\Delta t S(U^n)_{MP} \tag{3.3.25}$$

In a two-time level scheme it could be written as

$$\left(U_j^{n+1}\right)_{AP} = (U^n)_{DP} + \frac{\Delta t}{2}\left[S(U^n)_{DP} + S\left(U_j^{n+1}\right)_{AP}\right]$$

In general, for nonlinear equations $\partial q/\partial t = -u\, \partial q/\partial x + S(q)$, so the semi-Lagrangian scheme for the quantity q can be written as

$$q_{AP} = q_{DP} + S(q_{MP}) \tag{3.3.26}$$

However, the DP has to be determined from the trajectory $dx/dt = u$ integrated between the DP and AP, for example as

$$x_{DP} = x_{AP} - \frac{\Delta t}{2}(U_{DP} + U_{AP}) \tag{3.3.27}$$

Since u evolves with time, U_{AP} and U_{DP} are not known until the DP has been determined, this is an implicit equation that needs to be solved iteratively. For three-level semi-Lagrangian schemes, the approximation

$$x_{DP} = x_{AP} - 2\Delta t U_{MP} \tag{3.3.28}$$

also has to be solved iteratively for U_{MP}, but this is simpler than for the two-level time scheme.

The accuracy of the semi-Lagrangian scheme depends on the accuracy of the determination of the DP, and on the determination of the value of U_{DP} and the other conserved quantities q by interpolation from the neighboring points. A linear interpolation between neighboring points results in excessive smoothing, especially for the shortest waves. For this reason cubic interpolation is preferred (Williamson and Laprise, 1998). This is a costly overhead of semi-Lagrangian schemes. Despite the additional costs, in practice this scheme has been found to be accurate and efficient (see the general review of semi-Lagrangian methods by Staniforth and Côté (1991)). A "cascade" method has been proposed that results in a very efficient high order interpolation between the distorted Lagrangian grid and the regular Eulerian grid (Purser and Leslie, 1991, Leslie and Purser, 1995). This allowed Purser and Leslie to suggest a forward trajectory semi-Lagrangian approach instead of the conventional backward trajectory that we have so far described, which has additional advantages. (See Staniforth and Côté (1991), Bates *et al.* (1995), Purser and Leslie (1996), Williamson and Laprise (1998) for further details.) Combining the semi-Lagrangian approach with a semi-implicit treatment of gravity waves (Section 3.2.5), as first suggested by Robert (1982) and Robert *et al.* (1985), increases its efficiency. Laprise *et al.* (1997) have documented a "mesoscale compressible community" model, which is nonhydrostatic, three-level semi-Lagrangian, and uses the semi-implicit approach for both the elastic terms (three-dimensional divergence) and the gravity wave terms. As such, it is a flexible and accurate model that can be used for a wide range of scales.

3.3.4 Nonlinear computational instability. Quadratically conservative schemes. The Arakawa Jacobian

In 1957 Phillips published the first "climate" or "general circulation" simulation ever made with a numerical model of the atmosphere. He started with a baroclinically unstable zonal flow using a two-level quasi-geostrophic model, added small random perturbations, and was able to follow the baroclinic growth of the perturbations, and their nonlinear evolution. He obtained very realistic solutions that contributed significantly to the understanding of the atmospheric circulation in mid-latitudes.

However, his climate simulation only lasted for about 16 days: the model "blew up" despite the fact that care had been taken to satisfy the von Neumann criterion for linear computational instability. In 1959, Phillips pointed out that this instability, which he named nonlinear computational instability (NCI), was associated with nonlinear terms in the quasi-geostrophic equations, in which products of short waves create new waves shorter than $2\Delta x$. Since these waves cannot be represented in the grid, they are "aliased" into longer waves. The shortest wave that can be represented with a grid (with a wavelength $2\Delta x$) corresponds to the maximum computational wavenumber $p_{max} = 2\pi / L_{min} \Delta x = \pi$. However, quadratic terms with Fourier

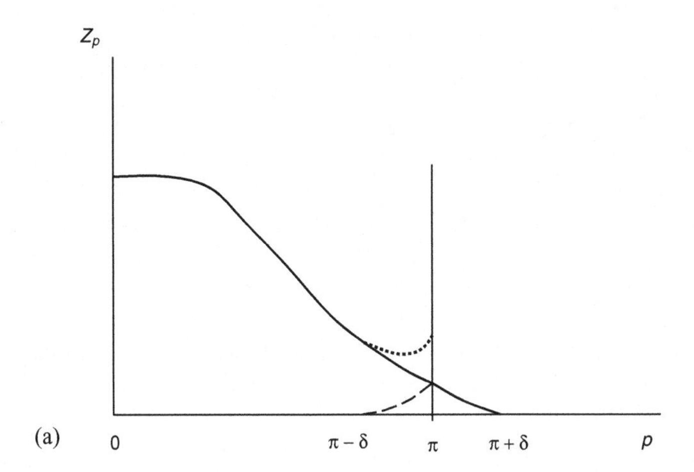

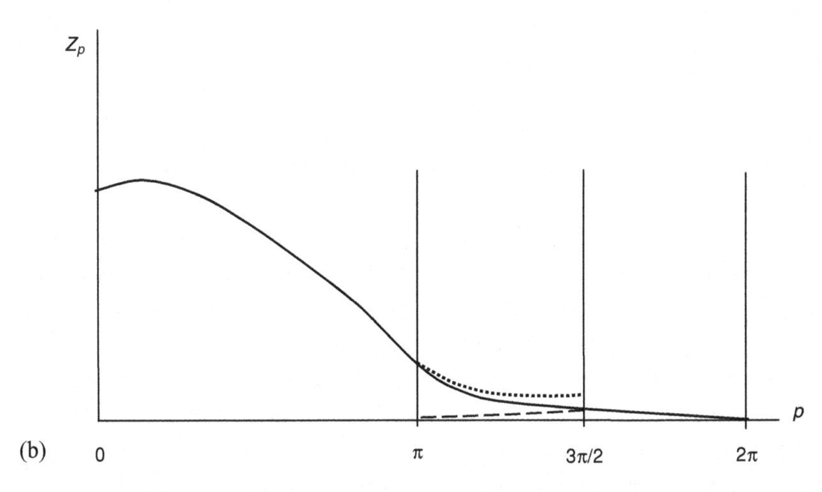

Figure 3.3.4: (a) Schematic of the effect of aliasing: the waves above $p = \pi$ (solid line) become folded back (dashed line) and are added to the original spectrum, producing a spurious maximum in the energy spectrum at the cut-off wavelength (dotted line). (b) Schematic showing that if we use a grid with 3/2 as many grid points as the original grid, the total spectrum in the Fourier transform of a quadratic product is increased by 3/2 (i.e., $p_{max} = 3\pi/2$). Then aliasing of wavenumbers between $3\pi/2$ and 2π occurs outside the original spectrum and it is avoided within the range 0 to π.

components will generate higher wavenumbers: $e^{\pm ip_1}e^{\pm ip_2} = e^{\pm i(p_1 \pm p_2)}$, doubling the maximum wavenumber. The new shorter waves, with wavenumbers $p = \pi + \delta$, cannot be represented in the grid, and become folded back (aliased) into $p' = \pi - \delta$, leading to a spurious accumulation of energy at the shortest range (Fig. 3.3.4).

The effect of NCI can be seen clearly in the following simple example: consider the nonlinear (quasi-linear) PDE $\partial u/\partial t = -u\partial u/\partial x$ and the corresponding FDE $\partial U_j/\partial t = -U_j(U_{j+1} - U_{j-1})/2\Delta x$. Suppose that at a given time t we have $U_1 = 0,\ U_2 > 0,\ U_3 < 0,\ U_4 = 0$. Then $\partial U_1/\partial t = 0$, $\partial U_2/\partial t > 0$, $\partial U_3/\partial t < 0$, $\partial U_4/\partial t = 0$, i.e., U_2 and U_3 will grow without bound and the FDE will blow up. In fact this will happen even for a linear model $\partial u/\partial t = -a(x)\partial u/\partial x$ if $a_1 = 0, a_2 > 0, a_3 < 0$, $a_4 = 0$. On the other hand, if $a(x)$ is always of the same sign, and we use the same FDE

$$\frac{\partial U_j}{\partial t} = -a_j \frac{U_{j+1} - U_{j-1}}{2\Delta x} \tag{3.3.29}$$

we can show that

$$\frac{\partial}{\partial t} \sum_j \frac{U_j^2}{a_j} = 0$$

i.e., that the solution will remain bounded. Numerical experiments show that nonlinear computational instability arises only when there are changes in sign in the velocity.

Exercise 3.3.1: Prove that the above solution will remain bounded.

There are basically two approaches for dealing with the problem of nonlinear computational instability.

(a) *Filtering out high wavenumbers.*
Phillips (1959) proposed transforming the grid-space solution into Fourier series (with sine and cosine wavenumbers from 0 to π), and chopping the upper half of the spectrum (wavenumbers above $\pi/2$). Since the maximum wavenumber generated in a quadratic term is twice the original wavenumber, this avoids spurious aliasing, and, indeed, Phillips found that the model could then be run indefinitely. However, the procedure is rather inefficient, since half of the spectrum is not used.

For grid-point models, complete Fourier filtering of the high wavenumbers has been found to be an unnecessarily strong measure to avoid nonlinear computational instability. Some models filter high wavenumbers but only enough to maintain computational stability. Experience shows that as long as the amplitude of the highest wavenumbers is not allowed to acquire finite amplitude, nonlinear computational stability can be avoided. For example, Kalnay-Rivas *et al.* (1977) combined the use of an energy-conserving fourth order model with a sixteenth order filter (similar to the eighth power of the horizontal Laplacian (Shapiro, 1970)). This efficiently filtered out the shortest waves (mostly between $2\Delta x$ and $3\Delta x$) without affecting waves of wavelength $4\Delta x$ or longer, and resulted in an accurate and economic model.[1]

1 The Shapiro filter of order n of a field U_i is a simple and efficient operator given by $\overline{U_j}^{2n} = [1 - (-D)^n]U_j$, where the "diffusion" operator $DU_j = (U_{j+1} - 2U_j + U_{j-1})/4$ is applied to the original field n times. For a Fourier component e^{ip} with wavenumber $p = 2\pi\,\Delta x/L$, the response of the operator is $De^{ipj} = -(\sin^2 p/2)e^{ipj}$ so that the second order Shapiro filter $\overline{U_j}^2 = [1 - (-D)^2]U_j = \frac{1}{4}(U_{j+1} + 2U_j + U_{j-1})$ has a response $\overline{U_j}^2 = (1 - \sin^2 p/2)U_j$. This is a strong filter that zeroes out the highest wavenumber ($L = 2\Delta x$), and reduces the amplitude of even the longer waves. A higher order filter, for example $2n = 16$, however, has the following desirable response: $\overline{U_j}^{16} = (1 - \sin^{16} p/2)U_j$, which still filters out $2\Delta x$ waves, dampens waves shorter than $4\Delta x$, and essentially leaves longer waves unaffected.

Spectral models with a wavenumber cut-off of M (i.e., with $2M + 1$ degrees of freedom) require at least $2M + 1$ grid points to be transformed into equivalent solutions in grid space. Orszag (1971) showed that if they are transformed into $3M + 1$ grid points before a quadratic term is computed in physical space, then aliasing is avoided. In other words, it is not necessary to perform the space transform into $4M$ points. The reason for this is shown schematically in Fig. 3.3.4(b): even if there is aliasing, it only occurs on the part of the spectrum (above $p = \pi$) that is eliminated anyway on the back transformation into spectral space. For this reason, in two horizontal dimensions, spectral models use "quadratic grids" with about $(3/2)^2$ as many grid points as spectral degrees of freedom, and therefore spectral models are "alias-free" for quadratic computations. Triple products in spectral models still suffer from aliasing, but this is generally not a serious problem.

(b) *Using quadratically conserving schemes*

Lilly (1965) showed that it is possible to create a spatial finite difference scheme that conserves both the mean value and its mean square value when integrated over a closed domain. Quadratic conservation will generally ensure that catastrophic NCI does not take place. Arakawa (1966) created a numerical scheme for the vorticity equation that conserves the mean vorticity, the mean square vorticity (enstrophy), and the kinetic energy. This ensures that the mean wavenumber is also conserved (as it is in the continuous equation), and therefore that even in the absence of diffusion the solution remains realistic. Arakawa and Lamb (1977) showed how an equivalent "Arakawa Jacobian" can be written for primitive equation models.

Consider first a conservation equation for the SWE written in advective form (as an example relevant to primitive equations):

$$\left.\begin{aligned} &\frac{\partial \alpha}{\partial t} + \mathbf{v} \cdot \nabla \alpha = 0 \\ &\frac{\partial h}{\partial t} + \mathbf{v} \cdot \nabla h + h \nabla \cdot \mathbf{v} = \frac{\partial h}{\partial t} + \nabla \cdot h\mathbf{v} = 0 \end{aligned}\right\} \tag{3.3.30}$$

If we multiply the first equation by h and the second by α, we can write the conservation equations in *flux* form:

$$\left.\begin{aligned} &\frac{\partial h\alpha}{\partial t} + \nabla \cdot h\mathbf{v}\alpha = 0 \\ &\frac{\partial h}{\partial t} + \nabla \cdot h\mathbf{v} = 0 \end{aligned}\right\} \tag{3.3.31}$$

Note that from the continuity equation written in flux form (second form), the total mass is conserved in time (Exercise 3.3.2).

Exercise 3.3.2: Show that from the continuity equation written in flux form, the total mass is conserved in time:

$$\frac{\partial}{\partial t}\iint h dx dy = 0$$

Now, consider any function of $G(\alpha)$. Multiply the conservation equation by $g(\alpha) = dG/d\alpha$ and integrate over a closed domain (i.e., a domain bounded by walls with zero normal velocities or by periodic boundary conditions). It is easy to show that the mean value of $G(\alpha)$ will be conserved in time:

$$\frac{\partial}{\partial t}\iint hG(\alpha) dx dy = 0 \tag{3.3.32}$$

Therefore the mass weighted mean and the mean squared value of α (as well as all its higher moments) will be conserved. With finite differences, we can only enforce two independent conservation properties (Arakawa and Lamb, 1977). We discuss now how to enforce mean and quadratic conservation, as suggested by Lilly (1965). The simplest approach is to write first the FDE continuity equation in flux form. This constitutes the backbone of a quadratically conservative scheme, and it is also similar to the simplest finite volume schemes (Section 3.3.6).

Exercise 3.3.3: Consider any function $G(\alpha)$ and multiply the conservation equation by $g(\alpha) = dG/d\alpha$ and integrate over a closed domain. Show that the mean value of $G(\alpha)$ will be conserved as in (3.3.32).

Consider Fig. 3.3.5, which shows a typical grid element with the value of α defined in the center, and estimates of the normal mass fluxes at its boundaries (e.g., $(hu)_{i+1/2,j}$ at the right wall). These estimates are used for casting the continuity FDE and for constructing a quadratically conservative FDE for α. The continuity FDE in flux form is

$$\frac{\partial h_{ij}}{\partial t} + \frac{(hu)_{i+1/2,j} - (hu)_{i-1/2,j}}{\Delta x} + \frac{(hv)_{i,j+1/2} - (hv)_{i,j-1/2}}{\Delta y} = 0 \tag{3.3.33}$$

It is easy to check that this FDE will conserve total mass $(\partial/\partial t)\sum_{i,j} h_{i,j}\Delta x_{i,j}\Delta y_{i,j} = 0$ since the mass flux into one grid box will cancel the mass flux out of the neighboring element. We now write the FDE for α using *any* consistent estimate of α at the normal walls of the grid box:

$$\frac{\partial h_{ij}\alpha_{i,j}}{\partial t} + \frac{(hu)_{i+1/2,j}(\alpha)_{i+1/2,j} - (hu)_{i-1/2,j}(\alpha)_{i-1/2,j}}{\Delta x} + \frac{(hv)_{i,j+1/2}(\alpha)_{i,j+1/2} - (hv)_{i,j-1/2}(\alpha)_{i,j-1/2}}{\Delta y} = 0 \tag{3.3.34}$$

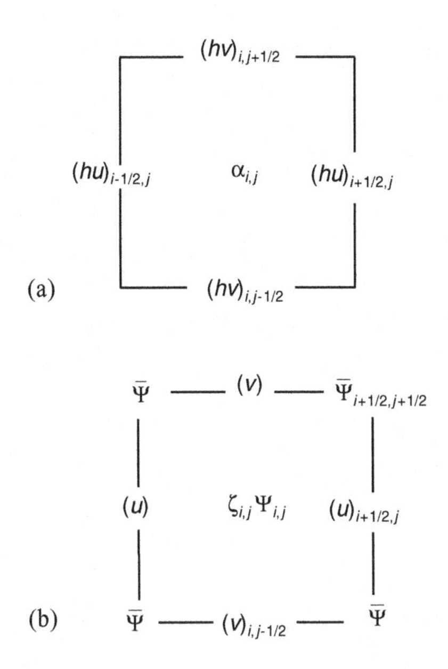

Figure 3.3.5: (a) A typical grid element with the value of α defined in the center, and estimates of the normal fluxes at its boundaries. These estimates are used for casting the continuity FDE and for constructing a quadratically conservative FDE for α (primitive equation model). (b) Grid for a simple enstrophy conserving FDE (quasi-geostrophic model).

Again it is easy to check that this FDE will conserve the total (mass weighted) value of α: $(\partial/\partial t)\,\Sigma_{i,j} h_{i,j}\alpha_{i,j}\Delta x_{i,j}\Delta y_{i,j} = 0$. *This is a general property of FDEs written in flux form.*

Finally, we choose to estimate the value of α at the walls of the grid-cells as an average between the two contiguous cells $(\alpha)_{i+1/2,j} = (\alpha_{i,j} + \alpha_{i+1,j})/2$, and similarly for the other walls. With this particular choice, we obtain:

$$\frac{\partial h_{i,j}\alpha_{i,j}}{\partial t} + \frac{(hu)_{i+1/2,j}(\alpha_{i,j} + \alpha_{i+1,j}) - (hu)_{i-1/2,j}(\alpha_{i,j} + \alpha_{i-1,j})}{2\Delta x}$$
$$+ \frac{(hv)_{i,j+1/2}(\alpha_{i,j} + \alpha_{i,j+1}) - (hv)_{i,j-1/2}(\alpha_{i,j} + \alpha_{i,j-1})}{2\Delta y} = 0 \qquad (3.3.35)$$

We can show that this scheme is quadratically conservative. First note that we can construct a mass weighted quadratic conservation equation for α from either the advection of the flux form prognostic equation for α and the continuity equation (prognostic equation for h):

$$\frac{\partial\left(h\frac{\alpha^2}{2}\right)}{\partial t} = \frac{\alpha^2}{2}\frac{\partial h}{\partial t} + h\alpha\frac{\partial \alpha}{\partial t} = -\frac{\alpha^2}{2}\frac{\partial h}{\partial t} + \alpha\frac{\partial h\alpha}{\partial t} \qquad (3.3.36)$$

The second equality suggests how to test quadratic conservation of α. Multiply the FDE continuity equation (3.3.33) by ${\alpha_{i,j}}^2/2$ and subtract it from the flux form prognostic equation (3.3.35) multiplied by $\alpha_{i,j}$. If we do this, we find that (because of cancellations of mass weighted fluxes of $\alpha_{i,j}$ on the grid-box walls), there is indeed quadratic conservation:

$$\frac{\partial}{\partial t}\sum_{i,j} h_{i,j}\frac{1}{2}\alpha_{i,j}{}^2\Delta x_{i,j}\Delta y_{i,j} = 0 \tag{3.3.37}$$

Note that this is true no matter how the FDE for the continuity equation is written. We could choose several finite difference formulations, and as long as the flux form of the FDE for $h\alpha$ is consistent with the continuity equation, and as long as we estimate α at the walls by a simple average, we have quadratic conservation and the danger of NCI is small.

Exercise 3.3.4: Show that the FDE (3.3.33) will conserve total mass.

Exercise 3.3.5: Show that the FDE (3.3.34) will conserve the total mass weighted value of α.

Exercise 3.3.6: Prove from (3.3.36) that there is quadratic conservation.

Exercise 3.3.7: Write two different FDEs for the continuity equation, i.e., two different estimates of the normal mass fluxes at the walls, $(hu)_{i+1/2,j}$, etc.

Finally we consider the vorticity equation, which is representative of much of the dynamics of the real atmosphere:

$$\frac{\partial \zeta}{\partial t} = -\mathbf{v}\cdot\nabla\zeta = -\nabla\cdot(\mathbf{v}\zeta) = -J(\Psi,\zeta) \tag{3.3.38}$$

where $\zeta = \mathbf{k}\cdot\nabla\times\mathbf{v} = \nabla^2\Psi$, $\mathbf{v} = \mathbf{k}\times\nabla\Psi$. The flow is nondivergent, so that the continuity equation is simply $\nabla\cdot\mathbf{v} = 0$.

In this case a simple scheme that conserves the mean vorticity and its mean square (i.e., an enstrophy conserving scheme) can be written following the recipe given above (Kalnay-Rivas and Merkine, 1981). The continuity equation is (cf. Fig. 3.3.5(b))

$$0 = -\frac{(u)_{i+1/2,j}-(u)_{i-1/2,j}}{\Delta x} - \frac{(v)_{i,j+1/2}-(v)_{i,j-1/2}}{\Delta y} \tag{3.3.39}$$

where the normal velocity estimates are obtained from

$$u_{i+1/2,j} = -\frac{\partial\overline{\Psi}_{i+1/2,j}}{\partial y} \approx -\frac{\overline{\Psi}_{i+1/2,j+1/2}-\overline{\Psi}_{i+1/2,j-1/2}}{\Delta y} \tag{3.3.40}$$

and similarly for the other velocities. Note that this satisfies the continuity equation automatically. Then we write the forecast equation for the vorticity in a way consistent with the continuity equation, thus ensuring conservation of the mean vorticity and enstrophy (mean square vorticity).

$$\frac{\partial \zeta_{i,j}}{\partial t} = -\frac{(u)_{i+1/2,j}(\zeta_{i,j}+\zeta_{i+1,j})-(u)_{i-1/2,j}(\zeta_{i,j}+\zeta_{i-1,j})}{2\Delta x} - \frac{(v)_{i,j+1/2}(\zeta_{i,j}+\zeta_{i,j+1})-(v)_{i,j-1/2}(\zeta_{i,j}+\zeta_{i,j-1})}{2\Delta y} \tag{3.3.41}$$

After a new vorticity field is obtained at $t = (n+1)\Delta t$ using, for example, leapfrog, we have to determine the new streamfunction ψ. This is done by solving the elliptic equation $\zeta = \nabla^2 \Psi$, which in finite differences can be written as

$$\frac{\Psi_{i+1,j}-2\Psi_{i,j}+\Psi_{i-1,j}}{\Delta x^2} + \frac{\Psi_{i,j+1}-2\Psi_{i,j}+\Psi_{i,j-1}}{\Delta y^2} = \zeta_{i,j} \tag{3.3.42}$$

In Section 3.4 we will discuss how to solve this *boundary value problem.*

Once we obtain $\psi_{i,j}$ we can obtain $\overline{\Psi}_{i+1/2,j+1/2}$ by averaging the corresponding four surrounding values of $\psi_{i,j}$. This is probably the simplest FDE model of the barotropic atmosphere devoid of nonlinear computational instability that we can construct.

Before we discuss the Arakawa Jacobian, let's note that the continuous vorticity equation conserves total (kinetic) energy as well as enstrophy. Multiply the vorticity equation by the streamfunction:

$$\Psi\frac{\partial \zeta}{\partial t} = -\Psi\nabla\cdot(\mathbf{v}\zeta) = -\nabla\cdot(\mathbf{v}\zeta\Psi) + \zeta\mathbf{v}\cdot\nabla\Psi \tag{3.3.43}$$

The last term on the right-hand side vanishes because $\mathbf{v}$ is perpendicular to $\nabla\Psi$. The left-hand side can be shown to be the time derivative of the kinetic energy:

$$\begin{aligned}\Psi\frac{\partial \zeta}{\partial t} &= \Psi\frac{\partial \nabla\cdot\nabla\Psi}{\partial t} = \frac{\partial \nabla\Psi\cdot\nabla\Psi}{\partial t} - \nabla\Psi\cdot\frac{\partial\nabla\Psi}{\partial t} \\ &= \frac{\partial \mathbf{v}\cdot\mathbf{v}}{\partial t} - \frac{\partial \mathbf{v}\cdot\mathbf{v}/2}{\partial t} = \frac{\partial\,|\mathbf{v}|^2/2}{\partial t} = \frac{\partial KE}{\partial t}\end{aligned} \tag{3.3.44}$$

Therefore, integrating (3.3.43) over the domain, the mean kinetic energy is conserved. The simple scheme described above conserves vorticity and squared vorticity but not kinetic energy.

Arakawa (1966) introduced a Jacobian that conserves all three properties: it is based on the FDE corresponding to these three equivalent formulations of the Jacobian:

$$J(\Psi,\zeta) = J_1 = J_2 = J_3 \tag{3.3.45}$$

where

$$\left.\begin{aligned} J_1 &= \frac{\partial \Psi}{\partial x}\frac{\partial \zeta}{\partial y} - \frac{\partial \Psi}{\partial y}\frac{\partial \zeta}{\partial x} \\ J_2 &= \frac{\partial}{\partial x}\left(\Psi \frac{\partial \zeta}{\partial y}\right) - \frac{\partial}{\partial y}\left(\Psi \frac{\partial \zeta}{\partial x}\right) \\ J_3 &= \frac{\partial}{\partial y}\left(\zeta \frac{\partial \Psi}{\partial x}\right) - \frac{\partial}{\partial x}\left(\zeta \frac{\partial \Psi}{\partial y}\right) \end{aligned}\right\} \tag{3.3.46}$$

The Arakawa Jacobian is the finite difference Jacobian corresponding to $J_A = (J_1 + J_2 + J_3)/3$ and it conserves kinetic energy and enstrophy. Arakawa and Lamb (1977) showed how the Arakawa Jacobian could also be approximately constructed for primitive equation models.

Exercise 3.3.8: Derive the finite difference equivalent of J_1, J_2 and J_3

The ratio of enstrophy to kinetic energy is proportional to the mean square of the wavenumber, and this quantity is conserved by the continuous frictionless vorticity equation. Therefore, the Arakawa conservation ensures that a long model run will conserve the mean square of the wavenumber, and generally look realistic (i.e., not become dominated by small scale noise) even without horizontal diffusion. In the real atmosphere, however, turbulent dissipation acts as a control on the amplitude of the smallest waves, and leaks their energy out of the system, so that strict conservation is not truly relevant. For this reason, there is no consensus among the community of modelers on whether the use of strictly conserving FDEs is an essential requirement. On the one hand, some models are based on schemes that are as conservative as possible (remember that the continuous equations conserve all moments of the quantities being advected, whereas the FDEs can only conserve one or two moments). Other modelers prefer to use less conservative but more accurate and simpler schemes. They include dissipation acting at the highest wavenumbers that mimics the leakage of energy that takes place in reality. Experience shows that an energy-conserving scheme, for example, combined with a small amount of high-order horizontal diffusion, in practice also behaves very realistically, approximately conserving enstrophy. This is because a catastrophic loss of enstrophy occurs only when energy is allowed to accumulate in the shortest waves and they acquire large amplitudes. The dispute as to whether it is more important to have conservative FDEs or accurate (higher order or semi-Lagrangian) FDEs that are not conservative but avoid NCI has thus not been resolved.

3.3.5 Staggered grids

So far all the variables we have used (e.g. h, u, v) have been defined at the same location in a grid cell. This means that in order to compute centered space

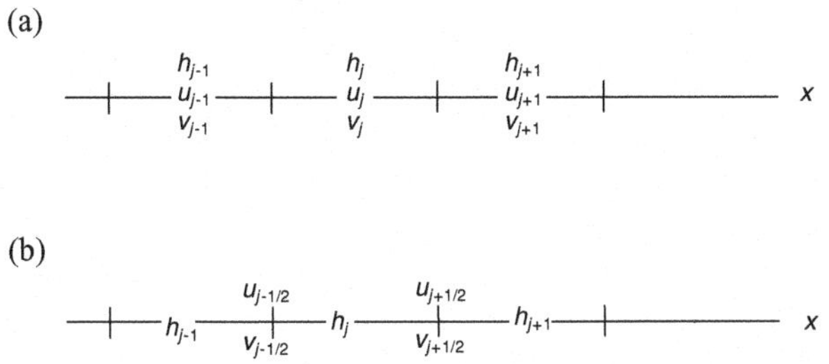

Figure 3.3.6: Staggered grids: (a) example of unstaggered grid in one dimension; (b) example of staggered grid in one dimension.

differences at a point j, for example, we need to go to $j+1$, and $j-1$, and the differences are computed over a distance of $2\Delta x$ (Fig. 3.3.6(a)). If we use instead a staggered grid, certain differences (such as the pressure gradient for the u equation and the horizontal convergence term for the h equation) can be computed over just $1\Delta x$, and, for those terms, it is equivalent to doubling the horizontal resolution. (Fig. 3.3.6(b)). However, the advection terms still have to be computed over $2\Delta x$ (or $2d$, where d is the distance between closest grid points of the same class).

Let's consider again the SWE in two dimensions:

$$\left.\begin{aligned}
\frac{\partial h}{\partial t} &= -\left[h\left(\frac{\partial u}{\partial x}+\frac{\partial v}{\partial y}\right)\right] - u\frac{\partial h}{\partial x} - v\frac{\partial h}{\partial y}\\
\frac{\partial u}{\partial t} &= -\left[g\frac{\partial h}{\partial x}+fv\right] - u\frac{\partial u}{\partial x} - v\frac{\partial u}{\partial y}\\
\frac{\partial v}{\partial t} &= -\left[g\frac{\partial h}{\partial y}-fu\right] - u\frac{\partial v}{\partial x} - v\frac{\partial v}{\partial y}
\end{aligned}\right\} \tag{3.3.47}$$

The terms in square brackets in (3.3.47) are the dominant terms for the geostrophic and the inertia-gravity wave dynamics. These terms are computed in different ways depending on the type of grid we use. The advective terms are less affected by the choice of alternative (staggered) grids.

In two dimensions there are several possibilities for staggered grids (Arakawa and Lamb, 1977), which are shown schematically in Fig. 3.3.7. Grid A (unstaggered) has several advantages and disadvantages. The advantages are its simplicity, and, because all variables are available at all the grid points, it is easy to construct a higher order accuracy scheme. Grid A tends to be favored by proponents of the philosophy "accuracy is more important than conservation". Its main disadvantage is that all differences occur on distances $2d$, and that neighboring points are not coupled for the pressure and convergence terms. This can give rise in time to a horizontal uncoupling (checkerboard pattern), which needs to be controlled by using a high order diffusion (e.g., Janjic, 1974, Kalnay-Rivas *et al.*, 1977).

Grid C has the advantage that the convergence and pressure terms in square brackets in (3.3.47) are computed over a distance of only $1d$, which is equivalent *to doubling the resolution of grid A*. For this reason geostrophic adjustment (the dispersion of gravity waves generated when the fields are not in geostrophic balance, see Chapter 5)

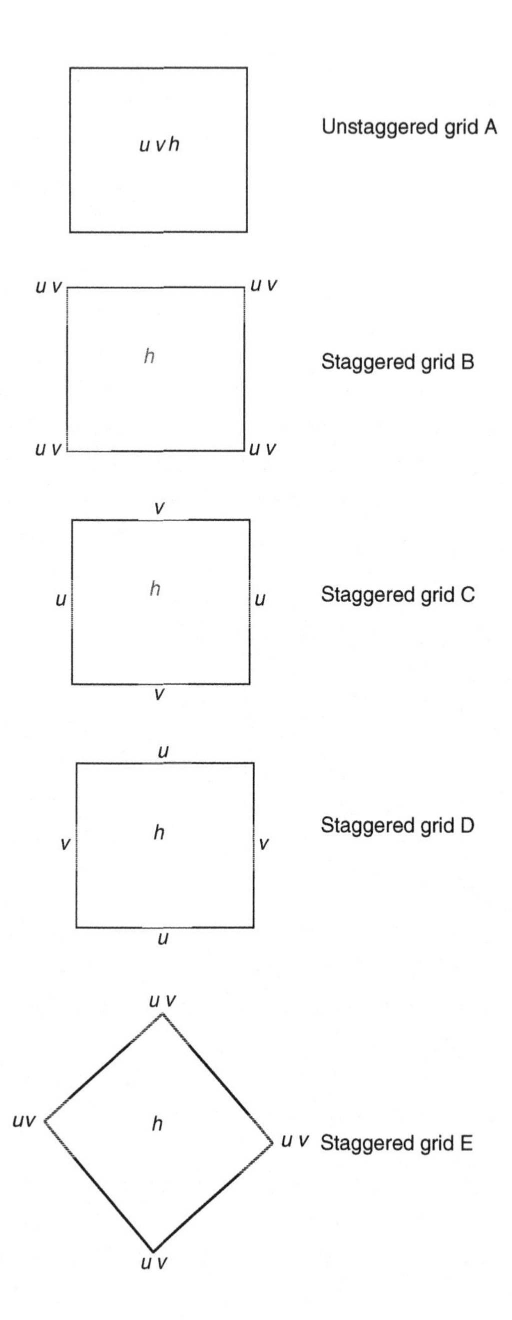

Figure 3.3.7: Staggered grids in two horizontal dimensions: Arakawa and Lamb (1977) classification.

is computed much more accurately (Arakawa, 1997). The Coriolis acceleration terms, on the other hand, require horizontal averaging, making the inertia-gravity waves less accurate. This makes grid C less attractive for situations in which the length of the Rossby radius of deformation $R_d = \sqrt{gH}/f$ is not large compared to the grid size d. The equivalent depth, H, is about 10 km for the external mode, so that R_d is

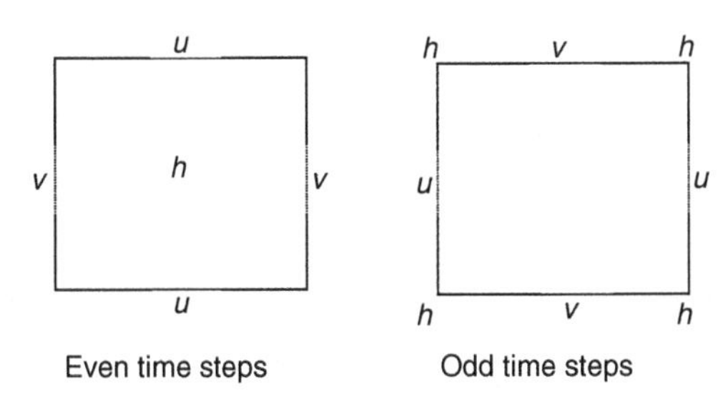

Figure 3.3.8: The Eliassen grid, staggered in both space and time.

about 3000 km (Chapter 6), but H is an order of magnitude smaller for the second vertical mode, and it becomes much smaller for higher vertical modes. Therefore, some atmospheric models use grid B, where the minimum distance for horizontal differences is $\sqrt{2}d$, rather than $1d$ as in grid C, but where u and v are available at the same locations. The NCEP Eta model is defined on a grid B rotated by 45°, denoted grid E by Arakawa and Lamb (1977), see Fig. 3.3.7.

The disadvantages of staggered grids are: (a) the terms in square brackets are hard to implement in higher order schemes, and (b) the staggering introduces considerable complexity in, for example, diagnostic studies and graphical output.

Grid D has no particular merit, but, if also staggered in time (as suggested by Eliassen), it becomes ideal for atmospheric flow using the leapfrog scheme (see Fig. 3.3.8). In the Eliassen grid *all* differences are computed on a distance d (the advection also requires a horizontal average over one grid length, but this is a small drawback). Despite its apparent optimality, this grid has not been adopted in any major model, probably because of the complications of the additional staggering, and because it would require special procedures for starting the leapfrog scheme. Lin and Rood (1997) have adopted a similar idea for a global atmospheric model on the sphere.

In the vertical direction, most models have adopted a staggered grid, with the vertical velocity defined at the boundary of layers and the prognostic variables in the center of the layer (Fig. 3.3.9). This type of grid, introduced by Lorenz in 1960, allows simple quadratic conservation, and the boundary conditions of no flux at the top and the bottom are easily fulfilled. However, as pointed out by Arakawa and Moorthi (1988), the Lorenz grid allows the development of a spurious computational mode, since the geopotential in the hydrostatic equation (and therefore the acceleration of the wind components) is insensitive to temperature oscillations of $2\Delta\sigma$ wavelength. The Lorenz grid is being replaced in some newer models by a vertical grid similar to the one introduced by Charney and Phillips (1953) for a two-level model. In the Charney–Phillips grid, the vertical staggering is more consistent with the hydrostatic equation and therefore it does not have the additional computational mode (Arakawa, 1997). A nonstaggered vertical grid, allowing a simple implementation of higher order differences in the vertical, would

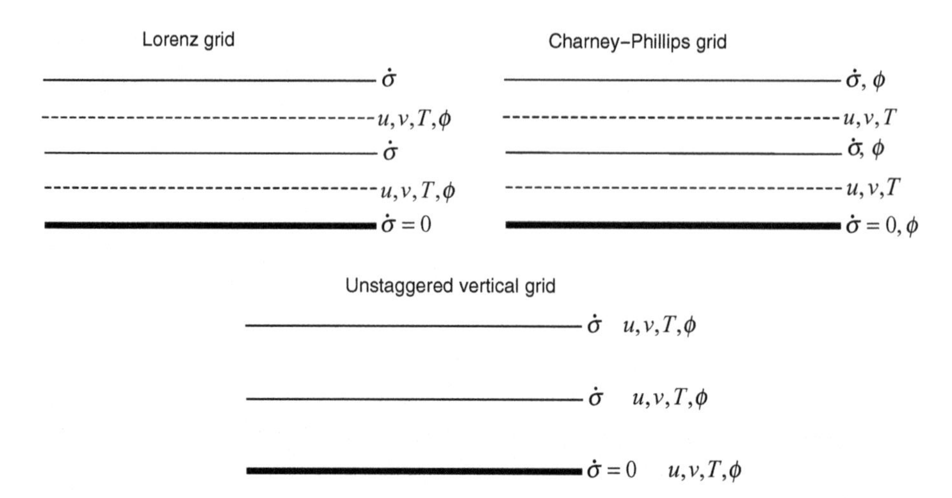

Figure 3.3.9: Staggering in vertical grids. (After Arakawa and Konor, 1996.)

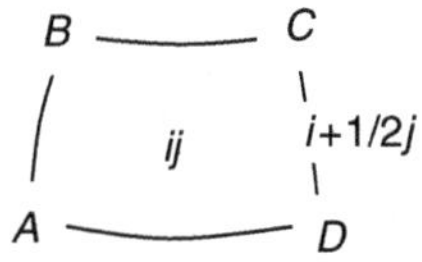

B C
ij i+1/2j
A D

Figure 3.3.10: Schematic of the two-dimensional volume centered at the point i, j and with walls at which fluxes are computed in a finite volume method.

also be possible, but it would also have more computational modes present in the solution.

3.3.6 Finite volume methods

We present here a brief introduction to the finite volume approach, which is discussed in more detailed in texts such as Durran (1999), Fletcher (1988) and Gustaffson *et al.* (1996). The basic idea of this method is that the governing equations are first written in an integral form for a finite volume, and only then are they discretized. This is in contrast to the methods we have seen so far, in which the equations in differential form are discretized using finite differences or spectral methods. The two approaches may or may not lead to similar discretized schemes.

Consider for example the continuity equation and a conservation equation for a shallow water model written in flux form, as in (3.3.31), and integrate them over a volume limited by walls $\overline{AB}$, $\overline{BC}$, $\overline{CD}$, and $\overline{DA}$ (in this two-dimensional case, the volume of integration is the horizontal area, Fig. 3.3.10).

If we integrate (3.3.31) within the volume $\overline{ABCD}$, and apply Green's theorem, we obtain

$$\left.\begin{aligned}\frac{d}{dt}\int hdxdy + \oint \mathbf{H}\cdot\mathbf{n}ds = 0\\ \frac{d}{dt}\int h\alpha dxdy + \oint (\mathbf{H}\alpha)\cdot\mathbf{n}ds = 0\end{aligned}\right\} \tag{3.3.48}$$

where $\mathbf{H}$ is the normal flux of h across the walls, and $\mathbf{n}$ is the normal vector to the wall. These equations can be discretized, for example, as

$$\left.\begin{aligned}\frac{d}{dt}\left(\overline{h}^{ij}\Delta x_{ij}\Delta y_{ij}\right) &= -(\overline{hu}^{i+1/2j})\Delta y_{i+1/2j} + (\overline{hu}^{i-1/2j})\Delta y_{i-1/2j}\\ &\quad -(\overline{hv}^{ij+1/2})\Delta x_{ij+1/2} + (\overline{hv}^{ij-1/2})\Delta x_{ij-1/2}\\ \frac{d}{dt}\left(\overline{\alpha h}^{ij}\Delta x_{ij}\Delta y_{ij}\right) &= -(\overline{hu\alpha}^{i+1/2j})\Delta y_{i+1/2j} + (\overline{hu\alpha}^{i-1/2j})\Delta y_{i-1/2j}\\ &\quad -(\overline{hv\alpha}^{ij+1/2})\Delta x_{ij+1/2} + (\overline{hv\alpha}^{ij-1/2})\Delta x_{ij-1/2}\end{aligned}\right\} \tag{3.3.49}$$

Here, the overbar indicates a suitable average over the grid volume or area. It is evident that any scheme based on these finite volume equations will conserve the average mass and average mass weighted α. There are a number of choices of how this average can be carried out over this *subgrid* domain of each grid volume: one can assume that h and α are constant within the volume, or that they vary linearly, etc. A simple choice for the estimates of the average values at the center and at the walls leads naturally to the quadratically conservative differences presented above in (3.3.34) and (3.3.35):

$$\left.\begin{aligned}\overline{h}^{ij} = h_{ij} \quad \overline{hu}^{i+1/2j} &= (h_{ij} + h_{i+1j})(u_{ij} + u_{i+1j})/4\\ \overline{hu\alpha}^{i+1/2j} &= (\overline{hu}^{i+1/2j})(\alpha_{ij} + \alpha_{i+1j})/2\end{aligned}\right\} \tag{3.3.50}$$

Although in this case both methods lead to the same discretization, the finite volume approach allows additional flexibility in the choice of discretization. For example, Lin and Rood (1996) developed a combination of semi-Lagrangian and finite volume methods, in which the boundaries of the grid volume are transported to the new time step, rather than the centers of the volume as is done in the conventional semi-Lagrangian schemes (Fig. 3.3.11). Although the order of the scheme is formally low, the method seems very promising, but it requires considerable care in the detailed formulation in order both to remain conservative and to maintain the shape of the transported tracers. Lin (1997) also developed a rather simple finite volume expression to compute the horizontal pressure gradient force that can be applied to any hydrostatic vertical coordinate system. It avoids the problem of having two large terms in the pressure gradient computation that almost cancel each other, which is

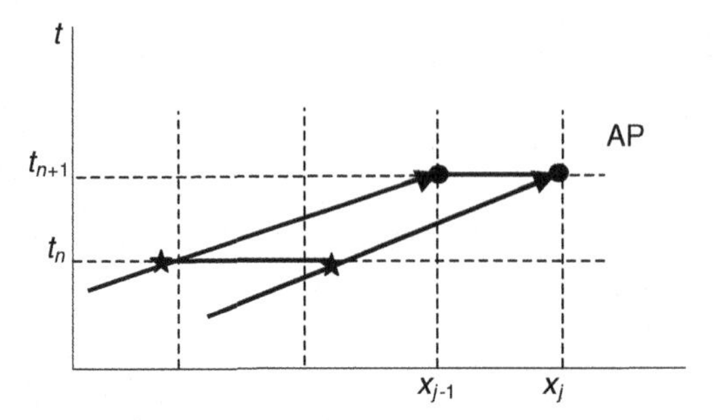

Figure 3.3.11: Schematic of the flux-form semi-Lagrangian scheme (Lin and Rood, 1996). It differs from the regular semi-Lagrangian scheme (Fig. 3.3.3) in that the walls of the volume are transported to the "arrival walls". The mass weighted average of the variables at the arrival volume is equal to the value at the departure volume (indicated by the thick segments), ensuring mass conservation. Because there is no extrapolation, the flux-form semi-Lagrangian scheme is still absolutely stable.

characteristic of the sigma and other vertical coordinate systems. There are several other variants of finite volume systems (e.g., Durran, 1999).

3.4 Boundary value problems

3.4.1 Introduction

Elliptic equations are boundary value problems, with either a fixed time, or a steady state solution at long times. Two examples of such problems arising in NWP are:

(a) Finding the new streamfunction from the vorticity after the latter has been updated to time $(n+1)\Delta t$. For example, in Section 3.3.4 we introduced the following enstrophy-conserving numerical scheme:

$$\frac{\zeta_{i,j}^{n+1} - \zeta_{i,j}^{n-1}}{2\Delta t} = -\frac{(u)_{i+1/2,j}(\zeta_{i,j} + \zeta_{i+1,j}) - (u)_{i-1/2,j}(\zeta_{i,j} + \zeta_{i-1,j})}{2\Delta x} - \frac{(v)_{i,j+1/2}(\zeta_{i,j} + \zeta_{i,j+1}) - (v)_{i,j-1/2}(\zeta_{i,j} + \zeta_{i,j-1})}{2\Delta y} \tag{3.4.1}$$

where we used the leapfrog scheme, and the right-hand side is evaluated at time $t = n\Delta t$. After solving for $\zeta_{i,j}^{n+1}$, we can obtain the streamfunction by solving the elliptic equation (Laplace) valid at $t = (n+1)\Delta t$:

$$\frac{\Psi_{i+1,j} - 2\Psi_{i,j} + \Psi_{i+1,j}}{\Delta x^2} + \frac{\Psi_{i,j+1} - 2\Psi_{i,j} + \Psi_{i,j-1}}{\Delta y^2} = \zeta_{i,j} \tag{3.4.2}$$

For this particular scheme, after solving for $\Psi_{i,j}^{n+1}$, we obtain $\overline{\Psi}_{i+1/2,j+1/2}^{n+1}$ by averaging from the four surrounding corners.

(b) Solving a semi-implicit elliptic equation for the heights also at $(n+1)\Delta t$ (Section 3.2.5):

$$\begin{aligned}\left(\delta_{2x}^2 + \delta_{2y}^2 - \frac{1}{\Phi\Delta t^2}\right)\phi^{n+1} = & -\left(\delta_{2x}^2 + \delta_{2y}^2 + \frac{1}{\Phi\Delta t^2}\right)\phi^{n-1} \\ & +2(\delta_{2x}R_u + \delta_{2y}R_v) \\ & +\frac{1}{\Delta t}(\delta_{2x}u^{n-1} + \delta_{2y}v^{n-1}) \\ & -\frac{2}{\Phi\Delta t}R_\phi = F_{i,j}^n \end{aligned} \tag{3.4.3}$$

These linear elliptic equations are easily solved with spectral methods in which the basis functions are eigenfunctions of the Laplace equation. For example, if we use spherical harmonics on the globe, and make use of

$$\begin{aligned}\nabla^2 Y_n^m &= \frac{1}{a^2}\left[\frac{1}{\cos^2\varphi}\frac{\partial^2 Y_n^m}{\partial\lambda^2} + \frac{1}{\cos\varphi}\frac{\partial}{\partial\varphi}\left(\cos\varphi\frac{\partial Y_n^m}{\partial\varphi}\right)\right] \\ &= \frac{-n(n+1)}{a^2}Y_n^m \end{aligned} \tag{3.4.4}$$

we can solve the semi-implicit equation for $\phi(\lambda,\varphi,t) = \sum_{n=0}^{N}\sum_{m=-n}^{n}\phi_n^m(t)Y_n^m(\lambda,\varphi)$ simply by writing the Helmholtz linear equation $\nabla^2\phi^{n+1} - (1/\Phi\Delta t^2)\phi^{n+1} = F$ corresponding to (3.4.3) component by component

$$\nabla^2\phi_m^n(t_{p+1})Y_n^m(\lambda,\varphi) - \frac{1}{\Phi\Delta t^2}\phi_m^n(t_{p+1})Y_n^m(\lambda,\varphi) = F_m^n(t_{p+1})Y_n^m(\lambda,\varphi) \tag{3.4.5}$$

so that the solution for each spherical harmonic coefficient is given by

$$\phi_m^n(t_{p+1}) = -\frac{1}{\left[\dfrac{n(n+1)}{a^2} + \dfrac{1}{\Phi\Delta t^2}\right]}F_m^n(t_{p+1}) \tag{3.4.6}$$

(Note that in (3.4.5) and (3.4.6) we have used p instead of n for the time step to avoid confusion with the total wavenumber n.) The simplicity with which the semi-implicit scheme can be computed is a major advantage of spectral models. For finite differences, the solution is much more involved.

The methods of solution for elliptic equations (discretized in space) are basically of two types: *direct* and *iterative*. Here we only present some simple examples of both types of methods, and refer the reader to texts such as Golub and van Loan (1996), Ferziger and Peric (2001), Dahlquist and Björk (1974) and Gustaffson *et al.* (1996) for more complete discussions of direct and iterative schemes. In the last decade, considerable work has also been done on the solution of nonsymmetric systems.

Books on computational methods for these types of problems include Barrett *et al.* (1995), Bruaset (1995), Greenbaum (1997), and Meurant (1999).

3.4.2 Direct methods for linear systems

We saw that for spectral models, the direct solution of the linear elliptic equation arising from the semi-implicit method is trivial. For finite differences, however, direct methods involve solving equations like (3.4.2) or (3.4.3), which can be written in matrix form as

$$\mathbf{A}\phi = F \tag{3.4.7}$$

using any direct solver. They are related to Gaussian elimination. If the matrix $\mathbf{A}$ is fixed (e.g., independent of the time step) the LU decomposition of $\mathbf{A} = LU$, where the diagonal of L are $l_{ii} = 1$, allows us to perform the decomposition once and then solve $LX = F$, followed by $U\Phi = X$. Here L and U are lower and upper triangular matrices.

If the matrix is tridiagonal, the direct problem is particularly easy to solve. A tridiagonal problem can be written as:

$$a_j U_{j-1} + b_j U_j + c_j U_{j+1} = d_j \tag{3.4.8}$$

with general boundary conditions

$$U_0 = A_1 U_1 + A_2 \quad U_J = B_1 U_{J-1} + B_2 \tag{3.4.9}$$

An algorithm based on Gaussian elimination is the "double sweep" method: Assume that

$$U_j = E_j U_{j+1} + F_j \tag{3.4.10}$$

Then $U_{j-1} = E_{j-1} U_j + F_{j-1}$ which can be substituted into the tridiagonal equation (3.4.8) to obtain:

$$(a_j E_{j-1} + b_j) U_j + c_j U_{j+1} = d_j - a_j F_{j-1} \tag{3.4.11}$$

From this we deduce that

$$\left.\begin{aligned} E_j &= \frac{-c_j}{a_j E_{j-1} + b_j} \\ F_j &= \frac{d_j - a_j F_{j-1}}{a_j E_{j-1} + b_j} \end{aligned}\right\} \tag{3.4.12}$$

So the method of solution is:

(a) use the lower boundary condition $U_0 = A_1 U_1 + A_2$ to determine $E_0 = A_1,\ F_0 = A_2$;

(b) sweep forward using (3.4.12) to obtain $E_j, F_j, \quad j = 1, \ldots, J-1$;

(c) determine U_J, U_{J-1} from $U_{J-1} = E_{J-1}U_J + F_{J-1}$ and the upper boundary condition $U_J = B_1 U_{J-1} + B_2$;

(d) determine U_j, $j = J-2, \ldots, 1$ using (3.4.10).

Tridiagonal matrices can thus be solved very efficiently, although problems arise when the denominator in (3.4.12) is close to zero.

3.4.3 Iterative methods for solving elliptic equations

The system $\mathbf{A}\phi = F$ can be solved iteratively by transforming it into another system,

$$\phi = (I - A)\phi + F$$

or

$$\phi = M\phi + F \tag{3.4.13}$$

choosing an initial guess ϕ^0 and then iterating (3.4.13): $\phi^{v+1} = M\phi^v + F$. The method converges if the spectral radius $\sigma(M) = \max|\lambda_i| < 1$, where λ_i are the eigenvalues of M. The asymptotic convergence rate is defined as

$$R = -\log_{10}[\sigma(M)] \tag{3.4.14}$$

We now give an example for a simple elliptic equation to provide an idea of how to attack the problem. The reader is referred to the references cited in subsection 3.4.1 for a more comprehensive discussion.

For a uniform grid with $\Delta x = \Delta y = \Delta$, an elliptic equation like (3.4.3) can be written as

$$\delta^2\phi_{i,j} - \alpha\phi_{i,j} = g_{i,j} \tag{3.4.15}$$

where the finite difference Laplace operator is

$$\delta^2\phi_{i,j} = (\phi_{i+1,j} + \phi_{i-1,j} + \phi_{i,j+1} + \phi_{i,j-1} - 4\phi_{i,j}) \tag{3.4.16}$$

Suppose we are in iteration v. Then

$$\delta^2\phi^v_{i,j} - \alpha\phi^v_{i,j} = g_{i,j} + \in^v_{i,j} \tag{3.4.17}$$

where $\in^v_{i,j}$ is the error in iteration v. If we assume at the point i, j

$$\phi^{v+1}_{i,j} = \phi^v_{i,j} + \delta\phi^v_{i,j} \tag{3.4.18}$$

and choose $\delta\phi^v_{i,j}$ to make $\in^{v+1}_{i,j} = 0$, we get

$$\phi^{v+1}_{i,j} = \phi^v_{i,j} + \frac{\delta^2\phi^v_{i,j} - \alpha\phi^v_{i,j} - g_{i,j}}{4 + \alpha} \tag{3.4.19}$$

This is the *Jacobi simultaneous relaxation method.* If we start at the southwest corner, and sweep to the right and up, by the time we reach the point i,j we have already updated the neighboring points to the west and the south, so we can use these updated values:

$$\phi_{i,j}^{v+1} = \phi_{i,j}^{v} + \frac{\phi_{i-1,j}^{v+1} + \phi_{i+1,j}^{v} + \phi_{i,j-1}^{v+1} + \phi_{i,j+1}^{v} - \alpha\phi_{i,j}^{v} - g_{i,j}}{(4+\alpha)} \tag{3.4.20}$$

This is the *Gauss–Seidel or successive relaxation method.*

If instead, we *overcorrect* by changing the sign of $\in_{i,j}^{v+1}$ rather than making it equal to zero, i.e.,

$$\phi_{i,j}^{v+1} = \phi_{i,j}^{v} + \omega\frac{\phi_{i-1,j}^{v+1} + \phi_{i+1,j}^{v} + \phi_{i,j-1}^{v+1} + \phi_{i,j+1}^{v} - \alpha\phi_{i,j}^{v} - g_{i,j}}{(4+\alpha)}$$
$$\text{with } 1 < \omega < 2 \tag{3.4.21}$$

the rate of convergence is further increased. This is the *successive overrelaxation* (SOR) method. Optimal values for ω can be obtained analytically for simple geometries such as a rectangular domain. For the equation above, the spectral radius of the Jacobi matrix **M** is

$$\lambda_1 = 1 - \varepsilon$$

where

$$\varepsilon = \sin^2\frac{\pi}{2(JM+1)} + \sin^2\frac{\pi}{2(KM+1)}$$

and *JM*, *KM* are the number of intervals in the x and y directions of the problem. Then the optimum value of the overrelaxation coefficient is

$$\omega_{opt} = \frac{2}{1+\sqrt{1-\lambda_1^2}}$$

Since the maximum error is reduced after each Jacobi iteration by the spectral radius $\lambda_1 = (1-\varepsilon)$, we can define the rate of convergence as ε.
The rates of convergence of the three methods are then:

ε = rate of convergence of the Jacobi iteration;

2ε = rate of convergence of the Gauss–Seidel iteration;

$2\sqrt{2\varepsilon}$ = rate of convergence of the SOR iteration with optimum overrelaxation.

3.4.4 Other iterative methods

We give only a simple introduction to other methods and refer the reader for further details to the references cited in Section 3.4.1.

Alternating Direction Implicit (ADI)

An efficient fractional time steps time scheme (Table 3.2.1) is used to obtain the solution of the elliptic equation as a steady state solution. For example, to solve the

Laplace equation we write the parabolic equation

$$\frac{\partial u}{\partial t} = \sigma \left(\frac{\partial^2 u}{\partial x^2} + \frac{\partial^2 u}{\partial y^2} \right) \tag{3.4.22}$$

The asymptotic long-time solution of (3.4.22) is the solution to the Laplace equation. Equation (3.4.22) is integrated numerically by separating it into two fractional steps (similar to the time scheme k in Table 3.2.1)

$$\left.\begin{aligned} \frac{u^* - u^n}{\Delta t} &= \sigma {\delta_x}^2 u^* \\ \frac{u^{n+1} - u^*}{\Delta t} &= \sigma {\delta_y}^2 u^{n+1} \end{aligned}\right\} \tag{3.4.23}$$

Since each fractional step is implicit, large time steps can be used. And since the solution of each fractional step involves only inverting tridiagonal matrices, it can be performed very efficiently (see, e.g., Hageman and Young (1981)).

Multigrid methods

The speed of convergence for iterative schemes depends on the number of grid points, and is much faster for coarser grids (see expression for λ_1 above). Moreover, the errors that take longest to converge correspond to long waves (i.e., they are smooth), whereas the shortest waves are damped fastest. Multigrid methods take advantage of this and use both coarse and fine grids (see Briggs (1987), Hackbusch (1985), Barrett *et al.* (1995)). The procedure is as follows: Several steps of a basic method on the full grid are performed in order to smooth out the error (pre-smoothing). A coarse grid is selected from a subset of the grid points, and the iterative method is used to solve the problem on this coarse grid. The coarse grid solution is then interpolated back to the original grid, and the original method applied again for a few iterations (post-smoothing). In carrying out the solution in the second step, the method can be applied recursively to coarser grids, until the number of grid points is small enough that a direct solution can be obtained.

The method of descending through a sequence of coarser grids and then ascending back to the full grid is known as a V-cycle. A W-cycle results from visiting the coarse grid twice, with some smoothing steps in between. Some multigrid methods have an (almost) optimal number of operations, i.e., almost proportional to the number of variables.

Krylov subspace methods

There are a number of iterative algorithms for solving the linear problem of (3.4.7) in the Krylov subspace, defined by

$$K_m(\mathbf{A}, r_0) = \text{span}\left\{r_0, \mathbf{A}r_0, \mathbf{A}^2 r_0, \ldots, \mathbf{A}^{m-1} r_0\right\} \tag{3.4.24}$$

where $r_0 = F - \mathbf{A}\phi_0$ is the *residual* for an arbitrary initial error ϕ_0. The approximate solution ϕ_m lies in the space $\phi_0 + K_m(\mathbf{A}, r_0)$. The residual after m steps has to satisfy certain conditions, and the choice of the condition gives rise to different types of iterative methods (e.g., Sameh and Sarin, 1999). The requirement that the residual be orthogonal to the Krylov subspace, $F - \mathbf{A}\phi_m \perp K_m(\mathbf{A}, r_0)$ leads to the *conjugate gradient* and the *Lanczos* methods. Methods like GMRES, MINRES and ORTHODIR are obtained by requiring that the residual be minimized over the Krylov subspace. The bi-conjugate gradient and QMR methods are derived requiring the residual to be orthogonal to $K_m(\mathbf{A}^T, r_0)$. The discussion of these methods applicable to nonsymmetric systems is beyond the scope of this book, but is given in the texts referred in Section 3.4.1.

3.5 Lateral boundary conditions for regional models

3.5.1 Introduction

The use of regional models for weather prediction has arisen from the desire to reduce the model errors through an increase in horizontal resolution that cannot be afforded in a global model. Operational regional models have been embedded or "nested" into coarser resolution hemispheric or global models since the 1970s. In the USA, the first regional model was the LFM model (Chapter 1). The nesting of regional models requires the use of updated lateral boundary conditions obtained from the global model.

We have seen that for pure hyperbolic equations there should be as many boundary conditions imposed at a given boundary as the number of characteristics moving *into* the domain. Parabolic equations with second order diffusion require one boundary condition at every point in the boundary for each prognostic equation. Second order elliptic equations (such as Laplace, Poisson, and Helmholtz equations) also require one boundary condition. The first forecast experiment of Charney *et al.* (1950) used the barotropic vorticity equation (conservation of absolute vorticity), and already had to deal with boundary conditions. They solved the hyperbolic equation $\partial\zeta/\partial t = -\mathbf{v} \cdot \nabla(\zeta + f)$ followed by the Poisson (elliptic) equation $\nabla^2 \partial\Psi/\partial t = \partial\zeta/\partial t$. Therefore, Charney *et al.* (1950) had to impose a boundary condition on the streamfunction at all the boundary points (needed to solve the Poisson equation) and a boundary condition for the vorticity at the inflow points. They used persistence in both cases: for the elliptic equation they used as boundary condition $\partial\Psi/\partial t = 0$ (i.e., the normal wind remains constant), and then specified that the vorticity also remained constant ($\partial\nabla^2\Psi/\partial t = 0$) at the inflow points and extrapolated the vorticity using upstream differences at the outflow points.

For the SWEs, there are three characteristics, one corresponding to a geostrophic solution, moving with the speed of the flow U, and the other two corresponding to inertia-gravity waves, moving with speed $U \pm \sqrt{f^2k^2 + \Phi}$. At the boundaries, if the speed of inertia-gravity waves is larger than U and the flow is inward, we have to specify two boundary conditions. If the flow is outward, we have to specify one boundary condition (corresponding to the inertia-gravity waves moving in). If U is greater than the speed of the inertia-gravity waves, we have to specify all three boundary conditions at the inflow points and none at the outflow points. For parabolic equations (with horizontal diffusion), each predicted variable has to be specified as well at all lateral boundaries.

Oliger and Sundstrom (1978) showed that the hydrostatic primitive equations are not purely hyperbolic (because of the loss of the time derivative of the vertical velocity), and that they do not have a well-posed set of boundary conditions. In an excellent review of the lateral boundary condition used in operational regional NWP models, McDonald (1997) pointed out that with the presence of horizontal diffusion in models there is a feeling that we can "over-specify slightly the lateral boundary conditions and not do very much damage".

In practice, boundary conditions are chosen pragmatically and tested numerically to check their appropriateness. Several methods have been tried over the years, but the most widely used is the boundary relaxation scheme introduced by Davies (1976). Davies (1983) has a very illuminating analysis of the impact of the different types of boundary conditions and their generation of spurious reflection using simple examples of wave equations and SWEs. He points out that an overspecifying boundary condition scheme is satisfactory if: (a) it transmits incoming waves from the "host" model providing boundary information without appreciable change of phase or amplitude, and (b) at the outflow boundaries, reflected waves do not reenter the domain of interest with appreciable amplitude. We follow the Davies (1983) analysis and the review by McDonald (1997) in the rest of this section. Durran (1999), Chapter 8, is also devoted to this subject.

3.5.2 Lateral boundary conditions for one-way nested models

The majority of regional models have "one-way" lateral boundary conditions, i.e., the host model, with coarser resolution, provides information about the boundary values to the nested regional model, but it is not affected by the regional model solution. This approach has some advantages: (a) it allows for independent development of the regional model, and (b) the host model can be run for long integrations without being "tainted" by problems associated with nonuniform resolution or from the regional model. Overall, the regional one-way nesting can be considered to have been successful, in the sense that the boundary information from the host model is able to penetrate the regional model, and the regional model solution is able to leave the domain without appreciable deterioration of the solutions. The success can also be measured by the fact that there have been several attempts to perform long-term integrations of nested regional models. In these long-term integrations, the initial regional information is swept out of the domain in the first day or two, and all the additional information comes from the global model integration. This approach is denoted "regional climate modeling". Takle *et al.* (1999) discuss the Project to Intercompare Regional Climate Systems (PIRCS). In these extended integrations, the regional model acts as a "magnifying glass" for the global solution, allowing the large-scale flow to interact with smaller scale forcing such as orography, variations in soil moisture and land–sea contrast, and as a result tend to give a more realistic solution. The "added value" over the global solutions empirically indicates the overall success of the one-way boundary conditions used in different models.

There are four types of "pragmatic" boundary conditions that have been formulated for one-way lateral boundary conditions:

(a) *Pseudo-radiation boundary conditions*

Orlanski (1976) proposed a finite difference approximation of the "radiation condition", i.e., specifying well-posed boundary conditions for pure hyperbolic equations. One assumes that the prognostic equations locally satisfy $\partial u/\partial t + c\partial u/\partial x = 0$ and then estimates the phase speed c using a finite difference equivalent of

$$c = -\frac{\partial u}{\partial t} \bigg/ \frac{\partial u}{\partial x} \tag{3.5.1}$$

at the points immediately inside the boundary (denoted by $b - 1$). Miller and Thorpe (1981) used first order upstream approximation

$$c' = -\frac{u_{b-1}^{n} - u_{b-1}^{n-1}}{\Delta t} \bigg/ \frac{u_{b-1}^{n-1} - u_{b-2}^{n-1}}{\Delta x} \tag{3.5.2}$$

as well as higher-order approximations. After estimating c', if it points into the domain, u_b^{n+1} is specified. If it points out, the upstream scheme is used: $u_b^{n+1} = u_b^n - c'\Delta t/\Delta x(u_b^n - u_{b-1}^n)$. If $c'\Delta t/\Delta x > 1$ because the space derivative of u is small, Orlanski (1976) suggested limiting the value of c' to $c' = \Delta x/\Delta t$. Klemp and Lilly (1978) pointed out reasons why the approximate "radiation schemes" are not completely successful in avoiding spurious reflection: there can be overspecification at the boundaries, specification of the right number of boundary conditions but not their correct values, and errors in the estimation of c'. The radiation condition has been used for research models (e.g., Durran *et al.* 1993). Klemp and Durran (1983) and Bougeault (1983) used radiation boundary conditions at the top of the model. Operational models generally do not use radiation boundary conditions and instead impose the condition that the vertical velocity be zero at the top (e.g., $\dot{\sigma} = 0$ at $\sigma = \sigma_T$ for sigma coordinates). As a result, the presence of this artificial "rigid top" leads to spurious wave reflections and even generates instabilities near the top (e.g., Kalnay and Toth, 1996, Hartman *et al.*, 1997).

(b) *Diffusive damping in a boundary zone or "sponge layer" (Burridge, 1975, Mesinger, 1977)*

In this method the global (or host) model boundary conditions are specified for all variables, and horizontal diffusion is added over a boundary zone to dissipate the noisy waves generated by the boundary conditions:

$$\frac{\partial u}{\partial t} + c\frac{\partial u}{\partial x} = \left(\frac{\partial}{\partial x}\nu\frac{\partial u}{\partial x}\right)_{BZ} \tag{3.5.3}$$

This would seem to be a natural choice for regional model boundary conditions since by increasing the order of the equation to make it parabolic within a limited boundary zone, it is possible to specify all variables at the boundary without overspecifying. However, this approach also has clear disadvantages: it damps the incoming waves from the global model (unless they are long compared to the width of the damping zone). It also produces spurious reflections of outgoing waves if ν increases abruptly, and if it increases slowly it may not be enough to damp the reflected waves. As a result, this method is not very much in use at this time.

(c) *Tendency modification scheme (Perkey and Kreitzberg, 1976)*

The wave equation is replaced by

$$\frac{\partial u}{\partial t} + c\frac{\partial u}{\partial x} = -\gamma\frac{\partial(u - \overline{u})}{\partial t} \tag{3.5.4}$$

where $\overline{u}$ is prescribed from the host model (which is assumed to be correct near the boundary), and γ is zero in the interior and increases to large values at the boundaries. Since the host model follows the wave equation

$$\frac{\partial \overline{u}}{\partial t} + c\frac{\partial \overline{u}}{\partial x} = 0 \tag{3.5.5}$$

we can write an "error" equation for the difference u' between the regional and the host model:

$$\frac{\partial u'}{\partial t} + c^*\frac{\partial u'}{\partial x} = 0 \tag{3.5.6}$$

where $c^* = c/(1 + \gamma)$. Therefore the time tendency scheme advects the error and slows it down to almost zero at the boundaries, thus avoiding overspecification. In practice this scheme is also found to produce spurious reflections.

(d) *Flow relaxation scheme (Davies 1976, 1983)*

As indicated before, this is the most widely used scheme. The forecast equations are modified by adding a Newtonian relaxation term over a boundary zone:

$$\frac{\partial u}{\partial t} + c\frac{\partial u}{\partial x} = -K(u - \overline{u}) \tag{3.5.7}$$

The "error" equation is now

$$\frac{\partial u'}{\partial t} + c\frac{\partial u'}{\partial x} = -Ku' \tag{3.5.8}$$

indicating that the error is advected to or from the boundary and damped. At the inflow boundaries only the differences between the regional and the host model are damped. Therefore this scheme mitigates the effects of overspecification at the outflow boundaries without introducing deleterious effects in the inflow boundaries.

If K increases abruptly, it can also introduce some spurious reflection. For this reason, Kallberg (1977) proposed the use of a smoothly growing function for K. Let's consider a complete prognostic equation for the regional model near the boundaries

$$\frac{\partial u}{\partial t} = F - K(u - \overline{u}) \tag{3.5.9}$$

In (3.5.9) F includes all the regular "forcing terms" in the interior time derivative (e.g., advection, sources/sinks, etc.). We can discretize it in time, using, for example, the leapfrog scheme for the regular terms and backward implicit scheme for the boundary relaxation term, as

$$\frac{u^{n+1} - u^{n-1}}{2\Delta t} = F^n - K\left(u_i^{n+1} - \overline{u}^{n+1}\right) \tag{3.5.10}$$

Here the overbar represents the host model, u^{n+1} is the updated regional model, and the subscript i indicates the regional model (internal) solution obtained *before* relaxing towards the host model values $\overline{u}^{n+1}$:

$$u_i^{n+1} = u^{n-1} + 2\Delta t F^n \tag{3.5.11}$$

From (3.5.10) and (3.5.11) we can now write

$$u^{n+1} = u_i^{n+1} - K2\Delta t u_i^{n+1} + K2\Delta t\overline{u}^{n+1} = (1-\alpha)u_i^{n+1} + \alpha\overline{u}^{n+1} \tag{3.5.12}$$

Here $\alpha = 2\Delta t K$ varies from 0 in the interior ($K = 0$), to 1 at the boundary, where the regional model solution is specified to coincide with the host model solution. McDonald (1997) mentioned three functions that have been proposed for $\alpha(j)$, where we define $j = 0, \alpha(0) = 1$ at the boundary, and assume that the boundary zone has n points so that for $j \geq n, \alpha(j) = 0$. The first function, found to be optimal in minimizing false reflection of both Rossby and gravity waves by Kallberg (1977), starts gently in the interior and has the steepest slope at the boundary: $\alpha = 1 - \tanh(j/2)$. Jones *et al.* (1995) used a linear profile $\alpha = 1 - j/n$, and McDonald and Haugen (1992) proposed a cosine profile $\alpha = [1 + \cos(j\pi/n)]/2$, which has the steepest slope at the center of the boundary zone. Benoit *et al.* (1997) in the MC2 model used $\alpha(j) = \cos^2(j\pi/2n)$ and reported good results.

3.5.3 Other examples of lateral boundary conditions

Tatsumi (1983, 1986), following an idea of Hovermale, suggested adding an "error diffusion" at the boundaries as well, which can also help to reduce the boundary errors without affecting the incoming wave:

$$\frac{\partial u}{\partial t} + c\frac{\partial u}{\partial x} = -K(u - \overline{u}) + \frac{\partial}{\partial x}\left(\nu\frac{\partial u - \overline{u}}{\partial x}\right) \tag{3.5.13}$$

This is used in the regional spectral model of the Japan Weather Service (Tatsumi, 1986).

Juang and Kanamitsu (1994) and Juang *et al.* (1997) also developed a RSM nested in the NCEP global spectral model, but they cast it as *a perturbation* model, so that the full RSM solution includes the global model solutions plus the regional perturbations. They use an "implicit" variation of the tendency modification approach with

$$\frac{\partial u}{\partial t} = F - \mu(u^{n+1} - \overline{u}^{n+1}) \tag{3.5.14}$$

where $\mu = \alpha/T$, T is an e-folding time (3 hours), and

$$u^{n+1} = u^{n-1} + 2\Delta t\frac{\partial u}{\partial t} \quad \overline{u}^{n+1} = \overline{u}^{n-1} + 2\Delta t\frac{\partial \overline{u}}{\partial t} \tag{3.5.15}$$

so that for the perturbation $u' = u - \overline{u}$, the implicit relaxation is given by

$$\frac{\partial u'}{\partial t} = \frac{F - \dfrac{\partial \bar{u}}{\partial t} - \mu u'^{n-1}}{1 + 2\mu\Delta t} \tag{3.5.16}$$

They found that the orography of the regional model also has to be blended with the global orography in the boundary zone in order to avoid spurious noise.

The Eta model at NCEP (Mesinger *et al.*, 1988, Janjic, 1994, Black, 1994) uses an "almost well-posed" approach. It uses boundary values from the NCEP global model only at the outermost row. When the flow is inwards, all the prognostic variables are prescribed from the global model. At the outflow points, the tangential velocities are extrapolated from the interior of the integration domain. The variables in the second row are a blend from the outermost and the third row. The "interior" is defined as the third row inwards, but the Eta model uses an upstream advection scheme for the five outer rows of the domain in order to minimize possible reflections at the boundary.

3.5.4 Two-way interactive boundary conditions

Finally, we note that some regional models have been developed using two-way interaction in the boundary conditions, i.e., the (presumably more accurate) regional solution, in turn, also affects the global solution. Although, in principle, this would seem a more accurate approach than the one-way boundary condition, care has to be taken that the high-resolution information does not become distorted in the coarser resolution regions, which can result in worse overall results, especially at longer time scales. There are basically two types of two-way boundary condition approaches.

The first approach corresponds to a truly nested model, with abrupt changes in the resolution, but with the inner or nested solution also used to modify the global or outer model solution. The first operational example of this type of two-way interaction was the NGM developed by Phillips (1979). Zhang *et al.* (1986) implemented two-way boundary conditions for the nesting in the MM5 model. See also Kurihara and Bender (1980), and Skamarock (1989).

The second approach is simpler, and it involves the use of continuously stretched horizontal coordinates so that only the region of interest is solved with high resolution. It is evident that with this approach, the equations in the regional high-resolution areas do not require special boundary conditions, and that they do influence the solutions in the regions more coarsely resolved, so that they can be considered as two-way interactive nesting. There have been a few methods used to obtain regional high resolution using stretched global coordinates:

(a) Uniform latitude–longitude stretching (Staniforth and Daley, 1977, Benoit *et al.*, 1989, Fox-Rabinovitz *et al.*, 1997). This method is used by the Canadian regional operational system.
(b) Stretched spherical harmonics (Courtier and Geleyn, 1988). This method is used in the French regional operational system.
(c) A regular volume (such as a cube) projected on the sphere and then stretched (Rancic *et al.*, 1996, Taylor *et al.*, 1997a). A variant of this approach is the spherical geodesic grids explored somewhat unsuccessfully during the 1960s, and now again in vogue (Williamson, 1968, Sadourny *et al.*, 1968, Masuda and Ohnishi, 1986, Heikes and Randall, 1995). The use of a regular volume to generate the grid avoids the pole problem of the convergence of the meridians in the latitude–longitude grid.

4

Introduction to the parameterization of subgrid-scale physical processes

4.1 Introduction

In Chapter 2 we derived the equations that govern the evolution of the atmosphere, and in Chapter 3 we discussed the numerical discretizations that allow the numerical integration of those equations on a computer. The discretization of the continuous governing equation is limited by the model resolution, i.e., by the size of the smallest resolvable scale. We have seen that in a finite difference scheme, the smallest scales of motion that can be (poorly) resolved are those which have a wavelength of two grid sizes. In spectral models, the motion of the smallest wave present in the solution is more accurately computed, but for these and for any type of numerical discretization there is always a minimum resolvable scale. Current *climate models* typically have a horizontal resolution of the order of several hundred kilometers, *global weather forecast models* have resolutions of 50–100 km, and regional *mesoscale models* of 10–50 km. *Storm-scale* models have even higher resolution, with grid sizes of the order of 1–10 km. In the vertical direction, model resolution and vertical extent have also been increased substantially, with current models having typically between 10 and 50 vertical levels, and extending from the surface to the stratosphere or even the mesosphere. As computer power continues to increase, so does the resolution of atmospheric models.

Despite the continued increase of horizontal and vertical resolution, it is obvious that there are many important processes and scales of motion in the atmosphere that cannot be explicitly resolved with present or future models. They include turbulent motions with scales ranging from a few centimeters to the size of the model grid, as well as processes that occur at a molecular scale, like condensation, evaporation, friction and radiation. We refer to all the processes that cannot be resolved explicitly as "subgrid-scale processes". An example of an important process that takes place at a subgrid scale is the turbulent mixing in the planetary boundary layer. During the daytime, the solar heating at the earth's surface not only warms the soil but also causes the plants to transpire and soil moisture to evaporate, thus transporting moisture into the atmosphere. Surface heating leads to turbulent motion that is on the scale of a few meters to a few hundred meters. With a horizontal grid size of 10–100 km, models cannot resolve these motions. Yet the transport of the heat and moisture into the boundary layer is very crucial to the development of afternoon thunderstorms and a host of other phenomena that are important to the resolvable atmospheric fields. Another notable example is tropical cumulus convection. The cumulus clouds in the tropics are known to be extremely important to the global energy balance, yet each cloud typically occupies only a few kilometers of space horizontally and vertically (Pan, 1999).

Although these processes occur at small scales, they depend on, and in turn, affect the larger-scale fields and processes that are explicitly resolved by a numerical model. For example, condensation of water vapor on a subgrid scale occurs if the resolved-scale humidity field is sufficiently high, and, in turn, condensation releases latent heat that warms the grid-scale temperature field. For this reason, it is not possible to ignore the effect of the subgrid processes on the resolvable-scale fields without degrading the quality of the forecast. To reproduce the interaction of the grid and subgrid-scale processes, the subgrid-scale phenomena are "parameterized", i.e., their effect is formulated in terms of the resolved fields. Fig. 4.1.1, adapted from Arakawa (1997), indicates schematically the resolved processes (usually referred to as the "dynamics of the model"), and the processes that must be parameterized ("the model physics"),

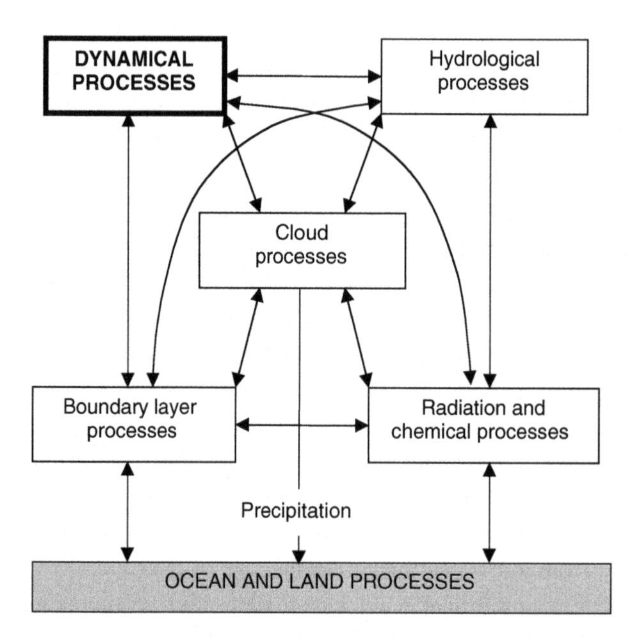

Figure 4.1.1: Physical processes in the atmosphere and their interactions. The dynamical processes for resolvable scales, in bold, are explicitly computed by the model "dynamics" (discussed in Chapters 2 and 3). The other subgrid-scale processes are parameterized in terms of the resolved-scale fields. (Adapted from Arakawa, 1997.)

and their interactions. Arakawa (1997) points out that some subgrid-scale processes can be interpreted as *adjustment* processes. For example, the atmosphere adjusts to the surface conditions through boundary layer adjustment processes, which are very efficient if the planetary boundary layer is unstable. Radiative fluxes occur because temperature tends to adjust towards radiative equilibrium. Convective processes occur in the presence of an unstable stratification and adjust the field towards a more neutrally stable state. Because radiative equilibrium is convectively unstable for the lower troposphere, radiative–convective adjustment is a dominant process controlling the vertical thermal structure of the troposphere.

The details of the parameterizations have a profound effect on the model forecast, especially at longer time scales, and are the subject of very intense research. In this chapter we provide only a very elementary introduction to model parameterizations. A short but inspiring introduction is presented in Arakawa (1997). An overview of different subgrid processes, and their parameterizations in atmospheric models appears in Haltiner and Williams (1980), and a more recent review, including ocean and land models, is available in *Climate system modeling* edited by Trenberth (1992). Stull (1988) and Garrart (1994) are texts on the atmospheric boundary layer processes. Emanuel and Raymond (1993) edited a volume including detailed discussions of a number of cumulus parameterizations. Pan (1999) discusses the philosophy that guides modelers in the development of parameterizations. Randall (2000) has edited a book honoring Akio Arakawa on the occasion of his retirement, which includes many review papers on areas related to physical parameterizations (as well as numerical modeling).

4.2 Subgrid-scale processes and Reynolds averaging

Consider the prognostic equation for water vapor written in flux form in z-coordinates (Section 2.5):

$$\frac{\partial \rho q}{\partial t} = -\frac{\partial \rho u q}{\partial x} - \frac{\partial \rho v q}{\partial y} - \frac{\partial \rho w q}{\partial z} + \rho E - \rho C \tag{4.2.1}$$

In the real atmosphere, both u and q contain scales that are resolved by the grid of the model, and smaller, subgrid scales. We write then

$$\left.\begin{aligned} u &= \overline{u} + u' \\ q &= \overline{q} + q' \end{aligned}\right\} \tag{4.2.2}$$

where the overbar represents the spatial average over a grid, and the primes, the subgrid-scale perturbation. We can neglect the subgrid-scale variations of ρ. By definition, the grid-box average of all quantities linear in the perturbations is zero, e.g., $\overline{q'} = 0$, $\overline{u'\overline{q}} = 0$. Also, averaging a grid-average quantity does not change it, e.g., $\overline{\overline{u}\overline{q}} = \overline{u}\overline{q}$. These are the rules for *Reynolds averaging*, a method originally developed by Reynolds in 1895 for use in time averages, but that we apply to grid-box averages. We can substitute (4.2.2) in the moisture equation (4.2.1), take a grid average, and obtain:

$$\frac{\partial \rho \bar{q}}{\partial t} = -\frac{\partial \rho \bar{u}\bar{q}}{\partial x} - \frac{\partial \rho \bar{v}\bar{q}}{\partial y} - \frac{\partial \rho \bar{w}\bar{q}}{\partial z} - \frac{\partial \rho \overline{u'q'}}{\partial x} - \frac{\partial \rho \overline{v'q'}}{\partial y} - \frac{\partial \rho \overline{w'q'}}{\partial z} + \rho E - \rho C \tag{4.2.3}$$

The first three terms of the right-hand side are the grid-scale (resolved) advection terms, whose numerical discretization we have studied in the Chapter 3. They are included in the "dynamical processes" box of Fig. 4.1.1. The next three terms are *the divergences of the eddy fluxes of moisture* or turbulent moisture transports. The last two terms (evaporation and condensation) are subgrid-scale processes that occur at a molecular scale and that we still need to parameterize. Both the molecular-scale processes and eddy fluxes that occur at scales much larger than molecular, but smaller than the grid resolution, are denoted collectively as "subgrid-scale processes". As indicated in the introduction, the impact of at least some of these physical processes on the larger scales explicitly represented in the model must be included. Without the parameterization of at least the most important subgrid-scale processes, the model integrations cease to be realistic in a very short period, from a day or two for large-scale flow, to less than an hour for storm-scale simulations.

There are several choices for the parameterization of the effect of turbulent transport terms in terms of the resolved scales. Consider, for example, the vertical turbulent flux of moisture (which, because of the strong vertical gradients, especially in the planetary boundary layer, is by far the dominant component of the eddy fluxes). We can choose to:

(a) Neglect the vertical turbulent flux, assuming that, in the boundary layer, the grid-scale field is well mixed:

$$-\rho \overline{w'q'} = 0 \tag{4.2.4}$$

This is known as a "zeroth order" closure, in which only the average properties are sought. An example is the bulk parameterization of the mixed boundary layer (Deardorff, 1972), in which the potential temperature, water vapor, and wind are assumed to be well mixed, and only the depth of the layer is forecast.

(b) Parameterize the vertical flux as a "turbulent diffusion process" in terms of $\bar{q}$ and the other grid-scale variables (this is a first order closure, and is the most commonly used):

$$-\rho \overline{w'q'} = K\frac{\partial \bar{q}}{\partial z} \tag{4.2.5}$$

This represents the effect of turbulent mixing due to parcels moving up or down, bringing with them the moisture from their original level, and mixing with the environment at the new level. The main problem in "K-theory", as this approach is also known, is to find a suitable formulation of the eddy diffusivity K, which also depends on the grid-average fields and the stability of the flow.

(c) Obtain a prognostic equation for $\overline{w'q'}$ by multiplying the vertical equation of motion by ρq and adding it to (4.2.1) multiplied by w. We obtain an equation with many terms like

$$\frac{\partial \rho w q}{\partial t} = -\frac{\partial \rho u w q}{\partial x} - \cdots \tag{4.2.6}$$

We can then take its Reynolds average and subtract it from (4.2.6), and derive a prognostic equation for the turbulent fluxes $\partial \rho \overline{w'q'}/\partial t = \cdots - \partial \rho \overline{w'w'q'}/\partial z \cdots$. This equation can be included as an additional model equation. Since it contains triple products of turbulent terms, these terms, in turn, have to be parameterized in terms of the double products:

$$-\rho \overline{w'w'q'} = K' \frac{\partial \rho \overline{w'q'}}{\partial z} \tag{4.2.7}$$

This is a second order closure. Second order closure models have many additional prognostic equations (for all the products of turbulent variables) but are an alternative to high-resolution models to obtain an estimate of turbulent transports (e.g., Moeng and Wyngaard, 1989). Mellor and Yamada (1974, 1982) show how to construct a hierarchy of closures for vertical fluxes and provide simplifying assumptions.

If an important physical process that occurs in the real atmosphere on a scale unresolved by the model is not parameterized, it may still appear in the model integration "aliased" into the resolved scales. For example, primitive equation model integrations will be ruined by dry convective instability if it is not parameterized. In the real atmosphere, if the potential temperature decreases with height, the unstable convective circulation that takes place occurs at very small horizontal scales, of the same order as the depth of the unstable layer, typically 1 km or less. Since this cannot be resolved with horizontal grids of the order of 10–100 km, models with unstable layers develop an unrealistic appearance of "vertical noodles", with narrow columns moving up and down side by side. In order to handle this problem, Manabe *et al.* (1965) developed the *dry convective adjustment*, a simple parameterization of dry convection still used in most present-day models. In this parameterization, when the grid-scale atmosphere lapse rate exceeds the dry adiabatic lapse rate $\Gamma_d = g/C_p \approx 10$ K/km, the unstable atmospheric column is instantaneously adjusted to an adiabatic or very slightly stable profile, while keeping constant the layer total enthalpy. Moist (cumulus) convection that occurs when there is grid-scale saturation and the temperature gradient exceeds the moist adiabatic lapse rate also results in a "wet noodles" circulation. This led to the *moist convective adjustment*, the first parameterization of cumulus convection (Manabe *et al.*, 1965), adjusting to a moist adiabatic profile. The moist convective adjustment was not found to be a sufficiently realistic cumulus parameterization, and has since been replaced by other convective parameterizations by Kuo, 1974, Arakawa and Schubert, 1974, Betts and Miller, 1986, Kain and Fritsch, 1990. See the volume edited by Emanuel and Raymond (1993) for a detailed review of cumulus parameterizations, and some updates in Randall (2000). Cumulus convection is one of the most important parameterizations in determining the characteristics of the model climatology (e.g., Miyakoda and Sirutis, 1977).

When a process occurs at scales not much smaller than the grid size, it presents an additional difficulty: the resolved scales and the unresolved scales to be parameterized are not well separated. An example of a process only marginally resolved in present-day models, which therefore appears aliased into the shortest waves present in the solution, is the sea-breeze circulation. A model with a grid size of 50–100 km (or more) cannot resolve the real sea-breeze circulation that takes place, for example, over a distance of the order of 1–20 km in the Florida peninsula on summer days. Therefore, in the model, the sea-breeze coastal circulation becomes distorted into a $2\Delta x$ circulation, and because the scales are not well separated, its effects on the large scales are difficult to parameterize. Similar effects are observed near heated mountain slopes when they are not properly resolved. The same problem of lack of scale separation complicates cumulus convection parameterization in models with a resolution of the order of 10 km, which is close to the horizontal scale of the convection, but not high enough to resolve convection explicitly.

4.3 Overview of model parameterizations

In a typical hydrostatic model on pressure coordinates, the governing equations (Chapter 2), including subgrid-scale processes, denoted with a tilde, are written as:

$$\frac{d\overline{\mathbf{v}}}{dt} = -\nabla_p \overline{\phi} - f\,k \times \overline{\mathbf{v}} - g\frac{\partial \tilde{\tau}}{\partial p} \tag{4.3.1}$$

for the two horizontal equations of motion, including the effect of eddy fluxes of momentum,

$$\frac{\partial \overline{\phi}}{\partial p} = -\overline{\alpha} \tag{4.3.2}$$

the hydrostatic equation,

$$\nabla_p \cdot \overline{\mathbf{v}} + \frac{\partial \overline{\omega}}{\partial p} = 0 \tag{4.3.3}$$

the continuity equation,

$$\frac{\partial \overline{p}_s}{\partial t} + \overline{\mathbf{v}} \cdot \nabla \overline{p}_s = -\int_0^\infty \boldsymbol{\nabla}_p \cdot \overline{\mathbf{v}} dp \tag{4.3.4}$$

the rate of change of the surface pressure,

$$\overline{p}\overline{\alpha} = R\overline{T} \tag{4.3.5}$$

the equation of state,

$$C_p \frac{\overline{T}}{\overline{\theta}} \frac{d\overline{\theta}}{dt} = \tilde{Q} = \tilde{Q}_{rad} - g\frac{\partial \tilde{F}_\theta}{\partial p} + L(\tilde{C} - \tilde{E}) \tag{4.3.6}$$

and the first law of thermodynamics, which includes radiative heating and cooling, sensible heat fluxes and condensation and evaporation, and

$$\frac{d\overline{q}}{dt} = \tilde{E} - \tilde{C} - g\frac{\partial \tilde{F}_q}{\partial p} \tag{4.3.7}$$

the conservation equation for water vapor. Condensation takes place when the grid average value oversaturates (stable or grid-scale condensation), or when there is moist convective instability and cumulus convection. The condensed water falls as precipitation, and may evaporate if the layers below are not saturated. Additional conservation equations can be written for cloud and rain water in models with prognostic (rather than diagnostic) clouds, and for other substances such as ozone.

In these equations the quantities with an overbar are the grid-averaged quantities computed by the model dynamics, and the terms with the tilde represent the terms that are parameterized. In a typical model, the vertical eddy flux of momentum $\tilde{\tau} = \rho\overline{w'u'}\,\mathbf{i} + \rho\overline{w'v'}\,\mathbf{j}$ (also known as eddy stress) of sensible heat $\tilde{F}_T = \rho C_p \overline{w'T'}$ and of moisture $\tilde{F}_q = \rho\overline{w'q'}$ may be represented using K-theory in the boundary layer and neglected in the free atmosphere above the boundary layer (using $K = 0$ or a very small value). The vertical derivatives of the turbulent fluxes that appear in the right-hand sides of (4.3.1), (4.3.6), and (4.3.7) introduce a requirement for lower boundary conditions for the surface fluxes of heat, moisture, and momentum. These surface fluxes are computed using a bulk parameterization based on the Monin–Obukhov (1954) similarity theory. This theory concludes that the profiles of wind and temperature in the turbulent *surface layer* can be described by a set of equations that depends only on a few parameters, including the surface roughness length z_0. The hypothesis of similarity, based on many observational studies, suggests that the fluxes of momentum and heat are nearly constant with height in the *surface layer* (of depth 10–100 m, which is much thinner than the planetary boundary layer). The fluxes in the surface or *constant flux layer* are usually represented with bulk aerodynamic formulas:

$$\left.\begin{aligned} \boldsymbol{\tau} &= -\rho C_D |\mathbf{v}|\mathbf{v} \\ F_\theta &= -\rho C_H |\mathbf{v}| C_p(\theta - \theta_S) \\ F_q &= -\rho C_E |\mathbf{v}| \beta(q - q_S) \end{aligned}\right\} \tag{4.3.8}$$

Here, $\mathbf{v}, \theta, q$ are the velocity, potential temperature, and mixing ratio in the surface layer, respectively, and the variables with an S subscript are the corresponding values at the underlying ocean or land surface ($\mathbf{v}_S = 0$). C_D, C_H, and C_E are transfer coefficients (C_D is known as "drag coefficient") and they depend on the stability of the surface layer (measured by the bulk Richardson number $Ri_B = gz\left[(\theta - \theta_S)/\overline{\theta}\right]/\mathbf{v}^2$, the height z and the surface roughness length). They are nondimensional and have typical values of the order of 10^{-3} for stable conditions and 10^{-2} for unstable conditions (Louis, 1979). β is a coefficient representing the degree of saturation of the underlying surface (1 for oceans, 0–1 for land depending on the degree of saturation in the soil moisture content). The surface layer values are either obtained through the use of a thin (order 10 m) prognostic layer or diagnosed.

The radiative heating in (4.3.6) is determined from the vertical divergence of the upward and downward fluxes of *short- and long-wave radiation*, obtained using the radiative transfer equation. See Kiehl (1992) for a review of the parameterization of radiation. The interaction between *clouds and radiation* is very complex, and is a major area of research. Early models specified clouds climatologically (Manabe *et al.*, 1965). In the 1980s the cloud cover was specified diagnostically, based on relative humidity (Slingo, 1987, Campana, 1994). More recently, cloud and rain water were predicted using budget equations and cloud cover was deduced from the amount of cloud water (e.g., Zhao *et al.*, 1997). The cloud properties are also important: rather than plane slabs, as generally assumed, clouds have a fractal structure, which effectively reduces their albedo and increases atmospheric absorption of solar radiation (Cahalan *et al.*, 1994).

An important area of research is the effect of subgrid-scale mountains. Wallace *et al.* (1983) proposed representing the blocking effect of subgrid-scale hills and valleys by increasing the effective height of mountains above its grid average by a factor of order one times the standard deviation of the subgrid-scale orography. This approach has been denoted "envelope orography". Similarly, Mesinger *et al.* (1988) chose a method essentially that defines the grid mountain height by the tallest peaks ("silhouette orography"). Lott and Miller (1997) formulated a new parameterization using developments in the nonlinear theory of stratified flows around obstacles, paying special attention to the parameterization of the blocked flow when the effective height of the subgrid-scale orography is high enough. They showed that this method can duplicate the results using envelope orography. In addition to its blocking effect, under stable conditions, small-scale orography generates internal gravity waves that propagate upwards, increase their amplitude, and eventually break at upper levels, depositing their low-level momentum (Lilly and Kennedy, 1973). The net result is a deceleration due to surface orography at upper levels. Modelers have introduced a *gravity-wave parameterization* following Palmer *et al.* (1986), McFarlane (1987), and Lindzen (1988). Kim and Arakawa (1995) developed a parameterization of the drag due to gravity waves.

Other areas of research in the parameterization of subgrid processes are related to the fact that the underlying surfaces (ocean and land) have their own evolution and therefore provide a "longer memory" to the forecast model which cannot be represented diagnostically. Equation (4.3.8) indicates that over ocean it is necessary to know the surface stress $\boldsymbol{\tau}$ and the SST. Short-range forecasts are performed with observed SSTs, under the assumption that they do not change significantly with time, but this is clearly not a reasonable assumption for medium-range or longer forecasts (e.g., Peña *et al.*, 2002).

For *seasonal and interannual predictions*, the SST is predicted using an ocean model coupled to the atmospheric model (Ji *et al.*, 1994, Trenberth, 1992). In addition, the surface fluxes over the ocean depend on the surface waves, which are driven by the wind. Currently most models use the Charnock (1955) parameterization relating an effective roughness length to the surface stress. In an iterative procedure the stress and the roughness length are obtained and bulk-aerodynamical formulas used to deduce the sensible and latent heat fluxes. However, in reality, ocean waves have a memory of their previous interactions with the atmosphere: swell ("old sea") is smoother than "new sea" where waves are driven by sudden changes in the wind, and, in turn, this affects the surface stress and the fluxes of heat and moisture. To take this effect into account, it is necessary to couple atmospheric models with ocean wave models.

Over land, similarly, the surface fluxes of heat and moisture are strongly dependent on the vegetation and soil moisture. Older models followed Manabe *et al.* (1965) by representing the effect of available soil moisture with a simple 15-cm "bucket" model, whose content was reduced by evaporation and increased by precipitation, with overflow representing river runoff. The surface temperature was obtained diagnostically assuming zero heat capacity for the land. Current models include coupling the atmospheric model with multilevel soil models with prognostic equations for the soil temperature and moisture, and include the very important controlling effect of plants on evapotranspiration (see reviews by Sellers (1992), Dickinson (1992), Pan (1990)).

5

Data assimilation

5.1 Introduction

In previous chapters we saw that NWP is an initial/boundary value problem: given an estimate of the present state of the atmosphere (initial conditions), and appropriate surface and lateral boundary conditions, the model simulates (forecasts) the atmospheric evolution. Obviously, the more accurate the estimate of the initial conditions, the better the quality of the forecasts. Currently, operational NWP centers produce initial conditions through *a statistical combination of observations and short-range forecasts.* This approach has become known as "data assimilation", whose purpose is defined by Talagrand (1997) as "using all the available information, to determine as accurately as possible the state of the atmospheric (or oceanic) flow."

There are several excellent reviews of this subject, which has become an important science in itself. The book *Atmospheric data analysis* by Daley (1991) is a comprehensive description of methods for atmospheric data analysis and assimilation. Ghil and Malanotte-Rizzoli (1991) have written a rigorous discussion of present data assimilation methods with special emphasis on sequential methods. Talagrand (1997) gives an elegant introductory overview of current methods of data assimilation, and Zupanski and Kalnay (1999) also provide a short introduction to the subject. The book *Data assimilation in meteorology and oceanography: Theory and practice* (Ghil *et al.*, editors, 1997) contains a wealth of important papers on current methods for data assimilation. An earlier but still useful book is *Dynamic meteorology: Data assimilation methods* (Bengtsson *et al.*, editors, 1981). Thiebaux and Pedder (1987) provided a description of spatial interpolation methods applied to meteorology. Several workshops on data assimilation have taken place at ECMWF, and their proceedings are extremely useful.

In the early NWP experiments, Richardson (1922) and Charney *et al.* (1950) performed hand interpolations of the available observations to a regular grid, and these fields of initial conditions were then manually digitized, which was a very time consuming procedure. The need for an automatic "objective analysis" became quickly apparent (Charney, 1951), and interpolation methods fitting observations to a regular grid were soon developed. Panofsky (1949) developed the first objective analysis algorithm based on two-dimensional polynomial interpolation, a procedure that can be considered "global" since the same function is used to fit all the observations.

Gilchrist and Cressman (1954) developed a "local polynomial" interpolation scheme for the geopotential height (Fig. 5.1.1). A quadratic polynomial in x and y was defined at each grid point:

$$z(x, y) = a_{00} + a_{10}x + a_{01}y + a_{20}x^2 + a_{11}xy + a_{02}y^2 \tag{5.1.1}$$

The six coefficients were determined by minimizing the mean square difference between the polynomial and observations close to the grid point (within *a radius of influence of the grid point*):

$$\min_{a_{ij}} E = \min_{a_{ij}} \sum_{k=1}^{K_z} p_k\left(z_k^o - z(x_k, y_k)\right)^2 + \sum_{k=1}^{K_v} q_k\left\{\left[u_k^o - u_g(x_k, y_k)\right]^2 + \left[v_k^o - v_g(x_k, y_k)\right]^2\right\} \tag{5.1.2}$$

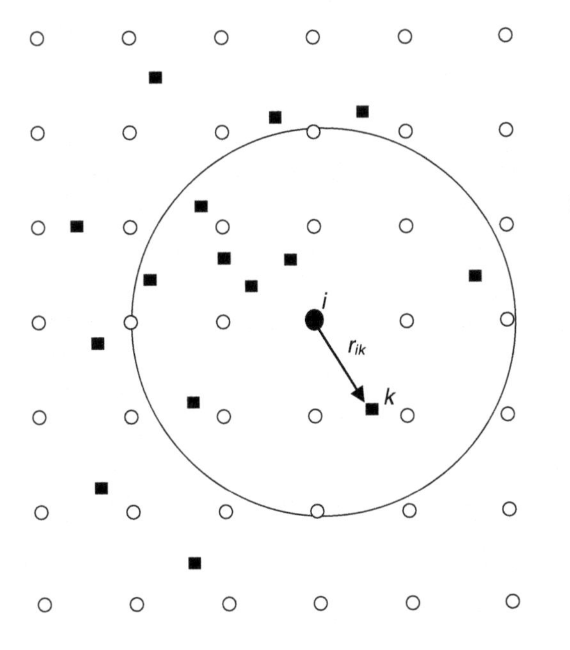

Figure 5.1.1: Schematic of grid points (circles), irregularly distributed observations (squares), and a radius of influence around a grid point i marked with a black circle. In 4DDA, the grid-point analysis is a combination of the forecast at the grid point (first guess) and the observational increments (observation minus first guess) computed at the observational points k. In certain analysis schemes, like SCM, only observations within the radius of influence, indicated by a circle, affect the analysis at the black grid point.

Here p_k, q_k are empirical weighting coefficients, and u_g, v_g are the geostrophic wind components computed from the gradient of the geopotential height $z(x, y)$ at the observation point k, and K is the total number of observations within the radius of influence. Note that although the field being analyzed is just the geopotential height, the wind observations are useful as well because they provide additional information about its gradient.

However, for operational primitive equation models, it is not enough to perform spatial interpolation of observations into regular grids, because not enough data are available to initialize current models. As pointed out in the introduction, the number of degrees of freedom in a modern NWP model is of the order of 10^7, whereas the total number of conventional observations of the variables used in the models (e.g., from rawinsondes) is of the order of 10^4. There are many new types of data currently available, including remotely sensed data such as satellite and radar observations, but they do not measure directly the variables used in the models (wind, temperature, moisture, and surface pressure). Moreover, their distribution in space and time is very nonuniform (Fig. 1.4.1), with regions like North America and Eurasia that are relatively data-rich, and others that are much more poorly observed.

For this reason, it became clear rather early in the history of NWP that, in addition to the observations, it was necessary to have a complete *first guess* estimate of the state of the atmosphere at all the grid points in order to generate the initial conditions for the forecasts (Bergthorsson and Döös, 1955). The first guess (also known as *background field* or *prior information*) should be our best estimate of the state of the atmosphere prior to the use of the observations. Initially climatology, or a combination of climatology and a short forecast were used as a first guess (e.g., Gandin, 1963, Bergthorsson and Döös, 1955). As forecasts became better, the use of short-range forecasts as a first guess was universally adopted in operational systems in what is called an "analysis cycle" (Fig. 5.1.2).

The analysis cycle is an intermittent data assimilation system that continues to be used in most global operational systems, which typically use a 6-h cycle performed four times a day. The model forecast plays a very important role. Over data-rich regions, the analysis is dominated by the information contained in the observations. In data-poor regions, the forecast benefits from the information upstream. For example, 6-h forecasts over the North Atlantic Ocean are very good, even in the absence of satellite data, because of the information coming from North America. The forecast is thus able to transport information from data-rich to data-poor areas, and for this reason, data assimilation using a short-range forecast as a first guess has become known as four-dimensional data assimilation (4DDA).

In Section 5.2 we describe empirical analysis schemes (SCM and nudging), and Sections 5.3 *et seq.* are devoted to statistical interpolation schemes.

(a)

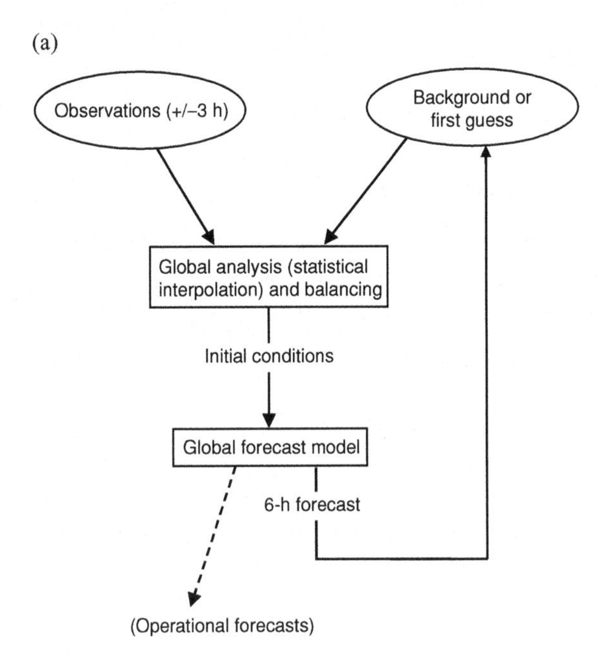

(b)

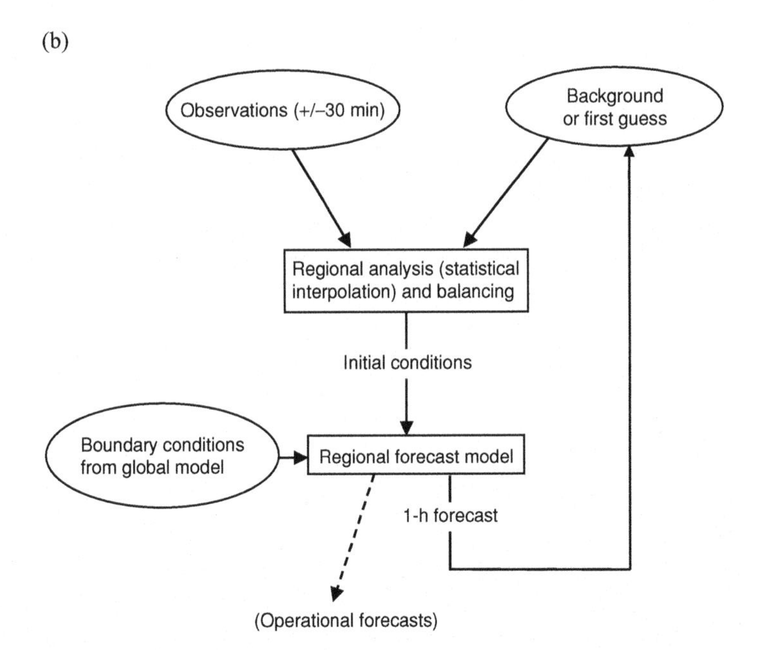

Figure 5.1.2: (a) Typical global 6-h analysis cycle performed at 00, 06, 12, and 18 UTC. The observations should be valid for the same time as the first guess. In the global analysis this has usually meant the rawinsondes are launched mostly at the main observing times (00 and 12 UTC), and satellite data are lumped into windows centered at the main observing times. The observations can be direct observations of variables used by the model, or indirect observations of geophysical parameters, such as radiances, that depend on the variables used in the model. (b) Typical regional analysis cycle. The main difference with the global cycle is that boundary conditions coming from global forecasts are an additional requirement for the regional forecasts.

5.2 Empirical analysis schemes

5.2.1 Successive corrections method (SCM)

The first analysis method used in 4DDA was based on an *empirical* approach known as the SCM, developed by Bergthorsson and Doos (1955) in Sweden and by Cressman (1959) of the US Weather Service. In SCM the first estimate of the gridded field is given by the background (or first guess) field:

$$f_i^0 = f_i^b \tag{5.2.1}$$

where f_i^b is the background field evaluated at the ith grid point, and f_i^0 the corresponding zeroth iteration estimate of the gridded field (Fig. 5.1.1).

After this first estimate, the following iterations are obtained by "successive corrections":

$$f_i^{n+1} = f_i^n + \frac{\sum_{k=1}^{K_i^n} w_{ik}^n \left(f_k^O - f_k^n\right)}{\sum_{k=1}^{K_i^n} w_{ik}^n + \varepsilon^2} \tag{5.2.2}$$

where f_i^n is the nth iteration estimation at the grid point i, f_k^O is the kth observation surrounding the grid point i, f_k^n is the value of the nth field estimate evaluated at the observation point k (obtained by interpolation from the surrounding grid points), and ε^2 is an estimate of the ratio of the observation error variance to the background error variance. The weights w_{ik}^n can be defined in different ways. Cressman (1959) defined the weights in the SCM as

$$\left.\begin{aligned} w_{ik}^n &= \frac{R_n^2 - r_{ik}^2}{R_n^2 + r_{ik}^2} \qquad \text{for} \qquad r_{ik}^2 \le R_n^2 \\ w_{ik}^n &= 0 \qquad\qquad\qquad \text{for} \qquad r_{ik}^2 > R_n^2 \end{aligned}\right\} \tag{5.2.3}$$

where r_{ik}^2 is the square of the distance between an observation point $\mathbf{r}_k$ and a grid point at $\mathbf{r}_i$.

The radius of influence R_n is allowed to vary with the iteration, and K_i^n is the number of observations within a distance R_n of the grid point i. For example, in the 1980s the Swedish operational system used $R_1 = 1500$ km, $R_2 = 900$ km for upper air analyses, and $R_1 = 1500$ km, $R_2 = 1200$ km, $R_3 = 750$ km, $R_4 = 300$ km for the surface pressure analysis. The reduction of the radius of influence results in a field that reflects the large scales after the first iteration and converges towards the smaller scales after the additional iterations.

In the Cressman SCM, the coefficient ε^2 is assumed to be zero. This results in a "credulous" analysis that more faithfully reflects the observations, and for a very small radius of influence the analysis *converges to the observation values* if the observations are located at the grid points. If the data are noisy (e.g., if an observation has gross errors, or if it contains an unrepresentative sample of subgrid-scale variability), this can lead to "bull's eyes" (many isolines around an unrealistic grid-point value) in the analysis. Including $\varepsilon^2 > 0$ assumes that the observations have errors, and gives some weight to the background field.

Barnes (1964, 1978) developed another empirical version of the SCM that has been widely used for analyses where there is no available background or first guess field, such as the analysis of radar data or other small-scale observations. Since we have no information on the background field, its error variance can be considered to be very large, so that $\varepsilon^2 = 0$. The weights are given by $w_{ik}^n = e^{-r_{ik}^2/2R_n^2}$. The radii of influence are changed by a constant factor at each iteration: $R_{n+1}^2 = \gamma R_n^2$. If $\gamma = 1$, only the large scales are captured. For $\gamma < 1$ more details in the observations are reproduced in the analysis as more iterations are performed.

Although the SCM method is empirical, it is simple and economical, and it provides reasonable analyses. Bratseth (1986) showed that if the weights are chosen appropriately instead of using the empirical formulas presented above, the SCM can be made to converge to a proper statistical interpolation (OI) (Section 5.3).

5.2.2 Nudging

Another empirical and fairly widely used method for data assimilation is *Newtonian relaxation* or *nudging* (Hoke and Anthes, 1976, Kistler, 1974). This consists of adding to the prognostic equations a term that nudges the solution towards the observations (interpolated to the model grid). For example, for a primitive equation model, the zonal velocity forecast equation is written as

$$\frac{\partial u}{\partial t} = -\mathbf{v} \cdot \nabla u + f v - \frac{\partial \phi}{\partial x} + \frac{u_{obs} - u}{\tau_u} \tag{5.2.4}$$

and similarly for the other equations.

The relaxation time scale, τ, is chosen based on empirical considerations and may depend on the variable. If τ is very small, the solution converges towards the observations too fast, and the dynamics do not have enough time to adjust. If τ is too large, the errors in the model can grow too much before the nudging becomes effective. Hoke and Anthes indicated that τ should be chosen so that the last term is similar in magnitude to the less dominant terms. They used a very short time scale, about 20 minutes, in their experiments. Stauffer and Seaman (1990) used about one hour in experiments assimilating synoptic observations, and reported a fair amount of success. Zou *et al.* (1992) made optimal parameter estimations of the nudging time scale. Kaas *et al.* (1999) performed an interesting experiment, nudging a model towards a 15-y reanalysis from the ECMWF, and by averaging the mean forcing introduced by nudging, empirically determined corrections to reduce model deficiencies.

Although this method is not generally used for large-scale assimilation, some groups use it for assimilating small-scale observations (e.g., radar observations) when there are no available statistics to perform a statistical interpolation.

5.3 Introduction to least squares methods

We have described in Section 5.2 several empirical methods for data assimilation. In this section we present methods that are based on statistical estimation theory. According to Talagrand (1997):

> Assimilation of meteorological or oceanographical observations can be described as the process through which all the available information is used in order to estimate as accurately as possible the state of the atmospheric or oceanic flow. The available information essentially consists of the *observations* proper, and of the *physical laws* that govern the evolution of the flow. The latter are available in practice under the form of a *numerical model.* The existing assimilation algorithms can be described as either *sequential* or *variational.*

5.3.1 Least squares method

In this section we give "baby examples" of both sequential and variational approaches. The methodology and results derived from this simple case carry over to multivariate OI, Kalman filtering, and 3D-Var and 4D-Var assimilation.

The best estimate of the state of the atmosphere (analysis) is obtained, as indicated by Talagrand (1997), from combining prior information about the atmosphere (background or first guess) with observations, but in order to combine them optimally we also need *statistical information* about the errors in these "pieces of information." A classic example of determining the best estimate of the true value of a scalar (e.g., the true temperature T_t) given two independent observations (or pieces of information), T_1 and T_2, serves as an introduction to statistical estimation:

$$\left.\begin{aligned} T_1 &= T_t + \varepsilon_1 \\ T_2 &= T_t + \varepsilon_2 \end{aligned}\right\} \tag{5.3.1}$$

The observations have errors ε_i that we don't know. Let $E()$ represent the *expected value,* i.e., the average that one would obtain if making many similar measurements. We assume that the instruments that measure T_1 and T_2 are unbiased: $E(T_1 - T_t) = E(T_2 - T_t) = 0$, or equivalently,

$$E(\varepsilon_1) = E(\varepsilon_2) = 0 \tag{5.3.2}$$

and that we know the variances of the observational errors:

$$E\left(\varepsilon_1^2\right) = \sigma_1^2 \quad E\left(\varepsilon_2^2\right) = \sigma_2^2 \tag{5.3.3}$$

We also assume that the errors of the two measurements are uncorrelated:

$$E(\varepsilon_1\varepsilon_2) = 0 \tag{5.3.4}$$

Equations (5.3.2), (5.3.3) and (5.3.4) represent the statistical information that we need about the actual observations. We try to estimate T_t from a linear combination of the two observations since they represent all the information that we have about the true value of T:

$$T_a = a_1 T_1 + a_2 T_2 \tag{5.3.5}$$

The "analysis" T_a should be unbiased:

$$E(T_a) = E(T_t) \tag{5.3.6}$$

which implies

$$a_1 + a_2 = 1 \tag{5.3.7}$$

T_a will be the *best estimate* of T_t if the coefficients are chosen to minimize the mean squared error of T_a:

$$\sigma_a^2 = E[(T_a - T_t)^2] = E[(a_1(T_1 - T_t) + a_2(T_2 - T_t))^2] \tag{5.3.8}$$

subject to the constraint (5.3.7). Substituting $a_2 = 1 - a_1$, the minimization of σ_a^2 with respect to a_1 gives

$$a_1 = \frac{1/\sigma_1^2}{1/\sigma_1^2 + 1/\sigma_2^2} \qquad a_2 = \frac{1/\sigma_2^2}{1/\sigma_1^2 + 1/\sigma_2^2} \tag{5.3.9}$$

or

$$a_1 = \frac{\sigma_2^2}{\sigma_1^2 + \sigma_2^2} \qquad a_2 = \frac{\sigma_1^2}{\sigma_1^2 + \sigma_2^2} \tag{5.3.10}$$

i.e., the weights of the observations are proportional to the "precision" or accuracy of the measurements (defined as the inverse of the variances of the observational errors). Moreover, substituting the coefficients (5.3.10) in (5.3.8), we obtain a relationship between the analysis variance and the observational variances:

$$\frac{1}{\sigma_a^2} = \frac{1}{\sigma_1{}^2} + \frac{1}{\sigma_2{}^2} \tag{5.3.11}$$

i.e., *if the coefficients are optimal, and the statistics of the errors are exact, then the "precision" of the analysis (defined as the inverse of the variance) is the sum of the precisions of the measurements.*

Exercise 5.3.1: Derive equations (5.3.9), (5.3.10), and (5.3.11).

5.3.2 Variational (cost function) approach

We can also obtain the same best estimate of T_t by minimizing a function of the temperature defined as the *sum of the square of the distance (or misfit) of the estimate T to the two observations*, weighted by their observational error precisions:

$$J(T) = \frac{1}{2}\left[\frac{(T-T_1)^2}{\sigma_1^2} + \frac{(T-T_2)^2}{\sigma_2^2}\right] \tag{5.3.12}$$

Exercise 5.3.2: Show that the minimum of the *cost function J* is obtained for $T = T_a$ defined in (5.3.5) with the same weights as in (5.3.10). Hint: $\partial J/\partial T = 0$ for $T = T_a$.

One may ask the motivation for defining a cost function as in (5.3.12). We now show that (5.3.12) can be formulated using the *maximum likelihood* approach, where we ask the question: Given the two independent observations T_1 and T_2, which are assumed to have normally distributed errors with standard deviations σ_1 and σ_2, what is the most likely value of the true temperature T? We *define the analysis as the most likely value of T given the observations and their statistical errors.*

The probability distribution of an observation T_1 given a true value T and an observational standard deviation σ_1, is given by the gaussian distribution

$$p_{\sigma_1}(T_1|T) = \frac{1}{\sqrt{2\pi}\sigma_1} e^{-\frac{(T_1-T)^2}{2\sigma_1^2}}$$

Conversely, the likelihood (Edwards, 1984) of a true value T given an observation T_1 with a standard deviation σ_1 is given by

$$L_{\sigma_1}(T||T_1) = p_{\sigma_1}(T_1|T) = \frac{1}{\sqrt{2\pi}\sigma_1} e^{-\frac{(T_1-T)^2}{2\sigma_1^2}}$$

Similarly, the likelihood of a true value T given an observation T_2 with a standard deviation σ_2 is

$$L_{\sigma_2}(T||T_2) = p_{\sigma_2}(T_2|T) = \frac{1}{\sqrt{2\pi}\sigma_2} e^{-\frac{(T_2-T)^2}{2\sigma_2^2}}$$

Therefore the most likely value of T given the two independent measurements T_1 and T_2 is the one that maximizes the joint probability, i.e., their product:

$$\max_T L_{\sigma_1,\sigma_2}(T||T_1, T_2) = p_{\sigma_1}(T_1|T)p_{\sigma_2}(T_2|T) = \frac{1}{2\pi\sigma_1\sigma_2} e^{-\frac{(T_1-T)^2}{2\sigma_1^2} - \frac{(T_2-T)^2}{2\sigma_2^2}}$$

Since the logarithm is a monotonic function, we can take the logarithm of the likelihood and obtain the same maximum likelihood temperature:

$$\max_T \ln L_{\sigma_1,\sigma_2}(T||T_1, T_2) = \max_T \left[const. - \frac{(T_1-T)^2}{2\sigma_1^2} - \frac{(T_2-T)^2}{2\sigma_2^2}\right] \tag{5.3.13}$$

The standard deviations are constant so that the maximum likelihood is attained when the cost function (5.3.12) is minimized.

Alternatively (Purser, 1984), the Bayesian derivation of (5.3.12) assumes we made the observation T_1 (the background forecast in the data assimilation problem), which implies a *prior* probability distribution of the truth $p_{T_1,\sigma_1}(T) = (1/\sqrt{2\pi}\sigma_1)e^{-(T_1-T)^2/2\sigma_1^2}$, prior to the second observation. Then Bayes formula for the *a posteriori* probability of the truth given observation T_2 is

$$p_{\sigma_2}(T|T_2) = \frac{p_{\sigma_2}(T_2|T)p_{T_1,\sigma_1}(T)}{p_{\sigma_2}(T_2)} = \frac{\frac{1}{\sqrt{2\pi}\sigma_2}e^{-\frac{(T_2-T)^2}{2\sigma_2^2}}\frac{1}{\sqrt{2\pi}\sigma_1}e^{-\frac{(T_1-T)^2}{2\sigma_1^2}}}{p_{\sigma_2}(T_2)} \tag{5.3.14}$$

The denominator

$$p_{\sigma_2}(T_2) = \int_{T'} \frac{1}{\sqrt{2\pi}\sigma_2}e^{-\frac{(T_2-T')^2}{2\sigma_2^2}}\,dT'$$

is independent of T. The estimate of the truth that maximizes the *a posteriori* probability (5.3.14) is obtained by maximizing the logarithm of the numerator, and is given, once again, by the minimum of the cost function (5.3.12).

Note that the *control variable* for the minimization of (5.3.12) (i.e., the variable with respect to which we are minimizing the cost function J) is now the *temperature itself, not the weights*. The *equivalence* between the minimization of the analysis error variance (finding the optimal weights through a least squares approach), and the variational cost function approach (finding the optimal analysis that minimizes the distance to the observations weighted by the inverse of the error variance) is an important property. This equivalence also holds true for multidimensional problems (in which case we use the covariance matrix rather than the scalar variance), and it indicates that OI (Gandin, 1963) and 3D-Var (e.g., Sasaki, 1970, Parrish and Derber, 1992) are solving the same problem (Lorenc, 1986).

Figure 5.3.1 illustrates the probability distribution for a simple case. Note that the analysis (the most likely value of the truth that maximizes the joint probability of T_1 and T_2) has a probability distribution with a maximum closer to T_2, and a smaller standard deviation (higher precision) than either observation.

5.3.3 Simplest sequential assimilation and Kalman filtering for a scalar

This is a prototype of the full multivariate OI. Assume that one of the two pieces of information $T_1 = T_b$ is the forecast (or any other "background" value) and the other

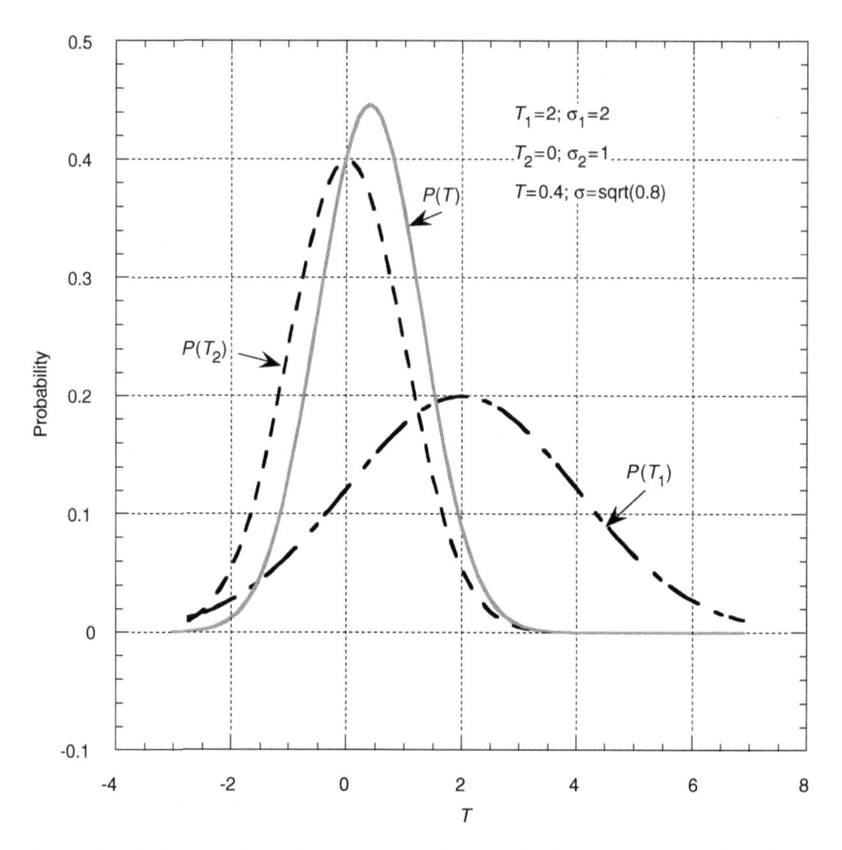

Figure 5.3.1: Illustration of the properties of the probability distribution of the analysis T, given observations T_1 and T_2, using either the least squares approach or the Bayesian approach (after Purser, 1984).

is an observation $T_2 = T_o$. From (5.3.5) and (5.3.10), we can write the analysis as

$$T_a = T_b + W(T_o - T_b) \tag{5.3.15}$$

where $(T_o - T_b)$ is defined as the *observational "innovation"* , i.e., the new information brought by the observation. It is also known as the *observational increment* (with respect to the background); W is the optimal weight, given by

$$W = \sigma_b^2 (\sigma_b^2 + \sigma_o^2)^{-1} \tag{5.3.16}$$

and the analysis error variance is, as before,

$$\sigma_a^2 = (\sigma_b^{-2} + \sigma_o^{-2})^{-1} \tag{5.3.17}$$

The analysis variance can in turn be written as $\sigma_a^2 = \sigma_b^2 \sigma_o^2 / (\sigma_b^2 + \sigma_o^2)$, or

$$\sigma_a^2 = (1 - W)\sigma_b^2 \tag{5.3.18}$$

Exercise 5.3.3: Derive equations (5.3.15)–(5.3.18).

Equations (5.3.15)–(5.3.18) have been derived for the simplest scalar case, but they are important for the problem of data assimilation because they have exactly the same form as the least squares sequential estimation methods used for real multidimensional problems (OI, interpolation, 3D-Var and even Kalman filtering). Therefore we interpret these equations in detail:

Equation (5.3.15) says: "The analysis is obtained by adding to the first guess (background) the innovation (difference between the observation and first guess) weighted by the optimal weight."

Equation (5.3.16) says: "The optimal weight is the background error variance multiplied by the inverse of the total error variance (the sum of the background and the observation error variances)." Note that the larger the background error variance, the larger the correction to the first guess.

Equation (5.3.17) says: "The precision of the analysis (inverse of the analysis error variance) is the sum of the precisions of the background and the observation."

Equation (5.3.18) says: "The error variance of the analysis is the error variance of the background, reduced by a factor equal to one minus the optimal weight."

All these statements are important because they also hold true for sequential data assimilation systems (OI and Kalman filtering) for multidimensional problems. In these problems, in which T_b and T_a are three-dimensional fields of size order 10^7 and T_o is a set of observations (typically of size 10^5 or 10^6), we have to replace the expression "error variance" by "error covariance matrix", and the "optimal weight" by an "optimal gain matrix".

Note also from (5.3.16) that there is one essential "tuning" parameter in OI: the ratio of the *a priori* estimate of the observational to the background error variances $(\sigma_o/\sigma_b)^2$.

Moreover, if the background is a forecast, we can use equations (5.3.15), (5.3.16), and (5.3.18) to create a simple sequential "analysis cycle", in which the observation is used once at the time it appears and then discarded. Assume that we have completed the analysis at time t_i (e.g., at 12 UTC), and we want to proceed to the next cycle (time t_{i+1}, or 18 UTC in the example). The analysis cycle has two phases, a *forecast* phase to update the background T_b and its error variance σ_b^2, and an *analysis* phase, to update the analysis T_a and its error variance σ_a^2.

In the *forecast phase of the analysis cycle*, the background is first obtained through a forecast:

$$T_b(t_{i+1}) = M\left[T_a(t_i)\right] \tag{5.3.19}$$

where M represents a forecast model (which could be a dynamical model, persistence, climatology, extrapolation, etc.). We also need to estimate the error variance of the background. In OI, this is done by making some suitable simple assumption, such as that the model integration increases the initial error variance by a fixed amount, e.g., a factor a not much greater than 1 (such as 1.5 or 2).

$$\sigma_b^2(t_{i+1}) = a\sigma_a^2(t_i) \tag{5.3.20}$$

This allows the new weight $W(t_{i+1})$ to be estimated using (5.3.16).

In Kalman filtering, (5.3.19) is the same as in OI, but instead of *assuming* a value for $\sigma_b^2(t_{i+1})$ as in (5.3.20) we *compute the forecast error covariance using the forecast model itself.* If we applied the model (5.3.19) to update the true temperature, there would be an error, since the model is not perfect: $T_t(t_{i+1}) = M\,[T_t(t_i)] - \varepsilon_M$.

The model error is assumed to be unbiased (unfortunately this is not in general a good assumption) with an error variance $Q^2 = E(\varepsilon_M^2)$. Then,

$$\varepsilon_{b,i+1} = (T_b - T_t)_{i+1} = M(T_a)_i - M(T_t)_i + \varepsilon_M = \mathbf{M}\varepsilon_{a,i} + \varepsilon_M \tag{5.3.21}$$

where $\mathbf{M} = \partial M/\partial T$ is the linearized or tangent linear model operator, and the forecast for the background error covariance at the new time level is:

$$\sigma_{b,i+1}^2 = E\left(\varepsilon_{b,i+1}^2\right) = \mathbf{M}^2\sigma_{a,i}^2 + Q^2 \tag{5.3.22}$$

Exercise 5.3.4: Derive (5.3.21) and (5.3.22).

In the *analysis phase of the analysis cycle* (for both OI and Kalman filtering) we get the new observation $T_o(t_{i+1})$, and we derive the new analysis $T_a(t_{i+1})$ using (5.3.15), the estimates of σ_b^2 from either (5.3.20) for OI, or (5.3.22) for Kalman filtering, and the new analysis error variance $\sigma_a^2(t_{i+1})$ using (5.3.18). After the analysis, the cycle for time t_{i+1} is completed, and we can proceed to the next cycle.

Remarks 5.3.1

In general, we cannot directly observe the model variables that we want to analyze (i.e., temperature, moisture, wind, and surface pressure at the grid points of the model). Instead we have rawinsondes (which were designed to provide these desirable variables) but at locations which are different from the analysis grid points, so that we have to perform horizontal and vertical interpolations. A more complex problem is that we may have remote sensing instruments (like satellites and radars) that measure quantities *influenced* by the desired variables, like radiances, reflectivities, refractivities, and Doppler shifts, rather than the variables themselves. Typically, then, we have to use an *observation operator* $H(T_b)$ (also known as an observational *forward* operator) to obtain from the first guess gridded field *a first guess of the observations*. The observation operator H includes spatial interpolations (or spectral to physical space transformation) from the first guess to the location of the observations. It also includes transformations based on physical laws, such as the radiative transfer equations that go from a model vertical profile of temperature and moisture to "observed" first guess satellite radiances.

Instead of using the observation operator, the operational assimilation of remotely sensed data used to be done following the "retrieval" approach. For example, TOVS (TIROS-N Operational Vertical Sounder) is an instrument that measures radiances in the infrared and microwave range of the spectrum. The forward operator model *H(T,q,clouds)* was inverted by first filtering the clouds and then retrieving "observed" profiles of temperature and moisture, $T(p)$ and $q(p)$. The retrieved profiles (that "looked" like rawinsonde observations) were then assimilated into the models. As indicated in Chapter 1, the direct assimilation of radiances, using the forward observational model H to convert the first guess into "first guess TOVS radiances" and then the assimilation of the "radiances innovations" (observed minus first guess radiances) has resulted in major improvements in the forecasts in both hemispheres (e.g., Fig. 1.4.3). This will be discussed further in Section 5.5, but we remark here that the improvements obtained by direct assimilation of the radiances are due basically to two reasons:

(1) There are fewer independent radiance observations than vertical levels of T and q in the model, which means that the problem of deriving a "retrieval" using only radiances is underdetermined. Therefore, in order to "retrieve" (invert the forward observational operator) it is necessary to introduce additional and less accurate statistical information into the problem. The introduction of this ancillary information (usually based on climatology, and generally less accurate than a short-range forecast used as a first guess) is unnecessary with the direct assimilation of radiance innovations.

(2) The observation error covariance of the retrieved T and q profiles is very difficult to determine, since it involves strong error correlations among retrievals in different latitude and longitude locations introduced by the use of ancillary information. On the other hand, observed radiances have "cleaner" error covariances, since they depend only on instrument errors, and not on how the data were processed. As a result, the observational error covariance for the radiances is usually diagonal.

5.4 Multivariate statistical data assimilation methods

We now generalize the least squares method to obtain the OI equations for vectors of observations and background fields. These equations were derived originally by Eliassen (1954, reproduced in Bengtsson *et al.*, 1981). However, Gandin (1963) derived the multivariate OI equations independently and applied them to objective analysis in the Soviet Union. Gandin's work had a profound influence upon the research and operational community, and OI became the operational analysis scheme of choice during the 1980s and early 1990s. In this discussion we generally follow the notation proposed by Ide *et al.* (1997) for data assimilation methods. This short paper, although ostensibly devoted to notation, is also an excellent overview of data assimilation. Later in this section we show that 3D-Var is equivalent to the OI method, although the method for solving it is quite different and advantageous for operational systems.

5.4.1 Optimal interpolation (OI)

In Section 5.3 we studied the formulation of the "optimal" analysis of a scalar at a single point. We now consider the complete NWP operational problem of finding an optimum analysis of a field of model variables $\mathbf{x}_a$, given a background field $\mathbf{x}_b$ available at grid points in two or three dimensions, and a set of p observations $\mathbf{y}_o$ available at irregularly spaced points $\mathbf{r}_i$ (Fig. 5.1.1).

The unknown analysis and the known background can be two-dimensional fields of a single variable like the temperature analysis $\mathbf{T}_a(x, y)$, or the three-dimensional field of the initial conditions for all the model prognostic variables: $\mathbf{x}=(p_s, T, q, u, v)$. These model variables are ordered by grid point and by variable, forming a single vector of length n, where n is the product of the number of points times the number of variables. The (unknown) "truth" $\mathbf{x}_t$, discretized at the model points, is also a vector of length n.

Note that we have used a different variable $\mathbf{y}_o$ for the observations than for the field we want to analyze. This is to emphasize that the observed variables are, in general, different from the model variables by: (a) being located in different points, and (b) possibly being *indirect* measures of the model variables. Examples of these measurements are radar reflectivities and Doppler shifts, satellite radiances, and global positioning system (GPS) atmospheric refractivities.

As we did in (5.3.15) for a scalar, the analysis is cast as the background plus the innovation weighted by optimal weights which we will determine from statistical interpolation,

$$\begin{aligned} \mathbf{x}_t - \mathbf{x}_b &= \mathbf{W}[\mathbf{y}_o - H(\mathbf{x}_b)] - \varepsilon_a = \mathbf{W}\mathbf{d} - \varepsilon_a \\ \varepsilon_a &= \mathbf{x}_a - \mathbf{x}_t \end{aligned} \tag{5.4.1}$$

but now the truth, the analysis, and the background are vectors of length n (the total number of grid points times the number of model variables) and the weights are given by a matrix of dimension $(n \times p)$. The forward observational operator H converts the background field into "first guesses of the observations." H can be nonlinear (e.g., the radiative transfer equations that go from temperature and moisture vertical profiles to the satellite observed radiances). The observation field $\mathbf{y}_o$ is a vector of length p, the number of observations. The vector $\mathbf{d}$, also of length p, is the "innovation" or "observational increments" vector:

$$\mathbf{d} = \mathbf{y}_o - H(\mathbf{x}_b) \tag{5.4.2}$$

Remarks 5.4.1

(a) The weight matrix $\mathbf{W}$ is also called the *gain matrix* $\mathbf{K}$, the same matrix that appears in Kalman filtering.

(b) An error covariance matrix is obtained by multiplying a vector error

$$\varepsilon = \begin{bmatrix} e_1 \\ e_2 \\ \vdots \\ e_n \end{bmatrix}$$

by its transpose $\varepsilon^T = \begin{bmatrix} e_1 \; e_2 \dots e_n \end{bmatrix}$, and averaging over many cases, to obtain the expected value:

$$\mathbf{P} = \overline{\varepsilon\varepsilon^T} = \begin{bmatrix} \overline{e_1e_1} & \overline{e_1e_2} & \cdots & \overline{e_1e_n} \\ \overline{e_2e_1} & \overline{e_2e_2} & \cdots & \overline{e_2e_n} \\ \vdots & \vdots & & \vdots \\ \overline{e_ne_1} & \overline{e_ne_2} & \cdots & \overline{e_ne_n} \end{bmatrix} \tag{5.4.3}$$

where the overbar represents the expected value (i.e. is the same as $E(\,)$). A covariance matrix is symmetric and positive definite. The diagonal elements are the variances of the vector error components $\overline{e_ie_i} = \sigma_i^2$. If we normalize the covariance matrix, dividing each component by the product of the standard deviations $\overline{e_ie_j}/\sigma_i\sigma_j = \text{corr}(e_i, e_j) = \rho_{ij}$, we obtain a correlation matrix

$$\mathbf{C} = \begin{bmatrix} 1 & \rho_{12} & \cdots & \rho_{1n} \\ \rho_{12} & 1 & \cdots & \rho_{2n} \\ \vdots & \vdots & & \vdots \\ \rho_{1n} & \rho_{12} & \cdots & 1 \end{bmatrix} \tag{5.4.4}$$

and if

$$\mathbf{D} = \begin{bmatrix} \sigma_1^2 & 0 & \cdots & 0 \\ 0 & \sigma_2^2 & \cdots & 0 \\ \vdots & \vdots & & \vdots \\ 0 & 0 & \cdots & \sigma_n^2 \end{bmatrix}$$

is the diagonal matrix of the variances, then we can write

$$\mathbf{P} = \mathbf{D}^{1/2}\mathbf{C}\mathbf{D}^{1/2} \tag{5.4.5}$$

(c) The transpose of matrix products is given by the product of the transposes, but in reverse order: $[\mathbf{AB}]^T = \mathbf{B}^T\mathbf{A}^T$; a similar rule applies to the inverse of a product: $[\mathbf{AB}]^{-1} = \mathbf{B}^{-1}\mathbf{A}^{-1}$

(d) The general form of a quadratic function is $F(\mathbf{x}) = \frac{1}{2}\mathbf{x}^T\mathbf{A}\mathbf{x} + \mathbf{d}^T\mathbf{x} + c$, where $\mathbf{A}$ is a symmetric matrix, $\mathbf{d}$ is a vector and c a scalar. To find the gradient of this scalar function $\nabla_{\mathbf{x}}F = \partial F/\partial\mathbf{x}$ (a column vector) we use the following properties of the gradient with respect to $\mathbf{x}$: $\nabla(\mathbf{d}^T\mathbf{x}) = \nabla(\mathbf{x}^T\mathbf{d}) = \mathbf{d}$ (since $\nabla_{\mathbf{x}}\mathbf{x}^T = \mathbf{I}$, the identity matrix), and $\nabla(\mathbf{x}^T\mathbf{A}\mathbf{x}) = 2\mathbf{A}\mathbf{x}$. Therefore,

$$\nabla F(\mathbf{x}) = \mathbf{A}\mathbf{x} + \mathbf{d} \qquad \nabla^2 F(\mathbf{x}) = \mathbf{A} \qquad \text{and} \qquad \delta F = (\nabla F)^T \delta x \tag{5.4.6}$$

(e) Multiple regression or best linear unbiased estimation (BLUE).
Assume we have two time series of vectors

$$\mathbf{x}(t) = \begin{bmatrix} x_1(t) \\ x_2(t) \\ \vdots \\ x_n(t) \end{bmatrix} \qquad \mathbf{y}(t) = \begin{bmatrix} y_1(t) \\ y_2(t) \\ \vdots \\ y_p(t) \end{bmatrix}$$

centered about their mean value, $E(\mathbf{x}) = 0$, $E(\mathbf{y}) = 0$, i.e., vectors of anomalies. We derive now the best linear unbiased estimation of $\mathbf{x}$ in terms of $\mathbf{y}$, i.e., the optimal value of the weight matrix $\mathbf{W}$ in the multiple linear regression

$$\mathbf{x}_a(t) = \mathbf{W}\mathbf{y}(t) \tag{5.4.7}$$

which approximates the true relationship

$$\mathbf{x}(t) = \mathbf{W}\mathbf{y}(t) - \boldsymbol{\varepsilon}(t) \tag{5.4.8}$$

Here $\boldsymbol{\varepsilon}(t) = \mathbf{x}_a(t) - \mathbf{x}(t)$ is the linear regression ("analysis") error, and $\mathbf{W}$ is an $n \times p$ matrix that minimizes the mean squared error $E(\boldsymbol{\varepsilon}^T\boldsymbol{\varepsilon})$. To derive $\mathbf{W}$ we write the regression equation matrix components explicitly:

$$x_i(t) = \sum_{k=1}^{p} w_{ik} y_k(t) - \varepsilon_i(t)$$

Then

$$\sum_{i=1}^{n} \varepsilon_i^2(t) = \sum_{i=1}^{n} \left[\sum_{k=1}^{p} w_{ik} y_k(t) - x_i(t) \right]^2$$

and the derivative with respect to the weight matrix components is

$$\frac{\partial \sum_{i=1}^{n} \varepsilon_i^2(t)}{\partial w_{ij}} = 2\left[\sum_{k=1}^{p} w_{ik} y_k(t) - x_i(t)\right]\left[y_j(t)\right]$$
$$= 2\left[\sum_{k=1}^{p} w_{ik} y_k(t) y_j(t) - x_i(t) y_j(t)\right] \quad (5.4.9)$$

In matrix form, this is

$$\frac{\partial \varepsilon^T \varepsilon}{\partial w_{ij}} = 2\left\{\left[\mathbf{W}\mathbf{y}(t)\mathbf{y}^T(t)\right]_{ij} - \left[\mathbf{x}(t)\mathbf{y}^T(t)\right]_{ij}\right\}$$

so that if we take a long time mean, and choose $\mathbf{W}$ to minimize the mean squared error, we get the normal equation

$$\mathbf{W}E\left(\mathbf{y}\mathbf{y}^T\right) - E\left(\mathbf{x}\mathbf{y}^T\right) = 0$$

or

$$\mathbf{W} = E\left(\mathbf{x}\mathbf{y}^T\right)\left[E\left(\mathbf{y}\mathbf{y}^T\right)\right]^{-1} \quad (5.4.10)$$

which gives the best linear unbiased estimation $\mathbf{x}_a(t) = \mathbf{W}\mathbf{y}(t)$.

Statistical assumptions

We define the background error and the analysis error as vectors of length n:

$$\boldsymbol{\varepsilon}_b(x, y) = \mathbf{x}_b(x, y) - \mathbf{x}_t(x, y) \quad (5.4.11a)$$
$$\boldsymbol{\varepsilon}_a(x, y) = \mathbf{x}_a(x, y) - \mathbf{x}_t(x, y) \quad (5.4.11b)$$

The p observations available at irregularly spaced points $\mathbf{y}_o(\mathbf{r}_i)$ have observational errors

$$\boldsymbol{\varepsilon}_{oi} = \mathbf{y}_o(\mathbf{r}_i) - \mathbf{y}_t(\mathbf{r}_i) = \mathbf{y}_o(\mathbf{r}_i) - H[\mathbf{x}_t(\mathbf{r}_i)] \quad (5.4.12)$$

We don't know the truth $\mathbf{x}_t$, thus we don't know the errors of the available background and observations, but we can make a number of assumptions about their statistical properties. The background and observations are assumed to be unbiased:

$$\left.\begin{aligned} E\{\boldsymbol{\varepsilon}_b(x, y)\} &= E\{\mathbf{x}_b(x, y)\} - E\{\mathbf{x}_t(x, y)\} = 0 \\ E\{\boldsymbol{\varepsilon}_o(\mathbf{r}_i)\} &= E\{\mathbf{y}_o(\mathbf{r}_i)\} - E\{\mathbf{y}_t(\mathbf{r}_i)\} = 0 \end{aligned}\right\} \quad (5.4.13)$$

If the forecasts (background) and the observations are biased, in principle we can and should correct the bias before proceeding. Dee and Da Silva (1998) show how the model bias can actually be estimated as part of the analysis cycle.

We define the error covariance matrices for the analysis, background and observations respectively:

$$\left.\begin{aligned}\mathbf{P}_a &= \mathbf{A} = E\{\boldsymbol{\varepsilon}_a\boldsymbol{\varepsilon}_a^T\}\\ \mathbf{P}_b &= \mathbf{B} = E\{\boldsymbol{\varepsilon}_b\boldsymbol{\varepsilon}_b^T\}\\ \mathbf{P}_o &= \mathbf{R} = E\{\boldsymbol{\varepsilon}_o\boldsymbol{\varepsilon}_o^T\}\end{aligned}\right\} \tag{5.4.14}$$

The nonlinear observation operator H that transforms model variables into observed variables can be linearized as

$$H(\mathbf{x} + \delta\mathbf{x}) = H(\mathbf{x}) + \mathbf{H}\delta\mathbf{x} \tag{5.4.15}$$

where $\mathbf{H}$ is a $p \times n$ matrix, denoted the linear observation operator with elements $h_{i,j} = \partial H_i/\partial x_j$. We also assume that the background (usually a model forecast) is a good approximation of the truth, so that the analysis and the observations are equal to the background values plus small increments. Therefore, the innovation vector (5.4.2) can be written as

$$\begin{aligned}\mathbf{d} &= \mathbf{y}_o - H(\mathbf{x}_b) = \mathbf{y}_o - H(\mathbf{x}_t + (\mathbf{x}_b - \mathbf{x}_t))\\ &= \mathbf{y}_o - H(\mathbf{x}_t) - \mathbf{H}(\mathbf{x}_b - \mathbf{x}_t) = \boldsymbol{\varepsilon}_o - \mathbf{H}\boldsymbol{\varepsilon}_b\end{aligned} \tag{5.4.16}$$

The $\mathbf{H}$ matrix transforms vectors in model space into their corresponding values in observation space. Its transpose or adjoint $\mathbf{H}^T$ transforms vectors in observation space to vectors in model space.

The background error covariance $\mathbf{B}$ (a matrix of size $n \times n$) and the observation error covariance $\mathbf{R}$ (a matrix of size $p \times p$) are assumed to be known. In addition, we assume that the observation and background errors are uncorrelated:

$$E\{\boldsymbol{\varepsilon}_o\boldsymbol{\varepsilon}_b^T\} = 0 \tag{5.4.17}$$

We will now use the best linear unbiased estimation formula (5.4.10) to derive the optimal weight matrix $\mathbf{W}$ in (5.4.1). $\mathbf{x}_a - \mathbf{x}_b = \mathbf{W}\mathbf{d}$, which approximates the true relationship $\mathbf{x}_t - \mathbf{x}_b = \mathbf{W}\mathbf{d} - \boldsymbol{\varepsilon}_a$.

From (5.4.16), $\mathbf{d} = \mathbf{y}_o - H(\mathbf{x}_b) = \boldsymbol{\varepsilon}_o - \mathbf{H}\boldsymbol{\varepsilon}_b$, and from (5.4.10) the optimal weight matrix $\mathbf{W}$ (also known as the gain matrix $\mathbf{K}$) that minimizes $\boldsymbol{\varepsilon}_a^T\boldsymbol{\varepsilon}_a$ is given by

$$\begin{aligned}\mathbf{W} &= E\{(\mathbf{x}_t - \mathbf{x}_b)[\mathbf{y}_o - H(\mathbf{x}_b)]^T\}(E\{[\mathbf{y}_o - H(\mathbf{x}_b)][\mathbf{y}_o - H(\mathbf{x}_b)]^T\})^{-1}\\ &= E[(-\boldsymbol{\varepsilon}_b)(\boldsymbol{\varepsilon}_o - \mathbf{H}\boldsymbol{\varepsilon}_b)^T]\{E[(\boldsymbol{\varepsilon}_o - \mathbf{H}\boldsymbol{\varepsilon}_b)(\boldsymbol{\varepsilon}_o - \mathbf{H}\boldsymbol{\varepsilon}_b)^T]\}^{-1}\end{aligned} \tag{5.4.18}$$

Recall that in (5.4.17) we assumed that the background errors are not correlated with the observational errors, i.e., that their covariance is equal to zero. Substituting the definitions of background error covariance $\mathbf{B}$ and observational error covariance $\mathbf{R}$

(5.4.14) into (5.4.18) we obtain the optimal weight matrix

$$\mathbf{W} = \mathbf{B}\mathbf{H}^T(\mathbf{R} + \mathbf{H}\mathbf{B}\mathbf{H}^T)^{-1} \tag{5.4.19}$$

Finally, we derive the analysis error covariance.

$$\begin{aligned}\mathbf{P}_a = E\{\boldsymbol{\varepsilon}_a\boldsymbol{\varepsilon}_a^T\} &= E\{\boldsymbol{\varepsilon}_b\boldsymbol{\varepsilon}_b^T + \boldsymbol{\varepsilon}_b(\boldsymbol{\varepsilon}_o - \mathbf{H}\boldsymbol{\varepsilon}_b)^T\mathbf{W}^T \\ &\quad + \mathbf{W}(\boldsymbol{\varepsilon}_o - \mathbf{H}\boldsymbol{\varepsilon}_b)\boldsymbol{\varepsilon}_b^T + \mathbf{W}(\boldsymbol{\varepsilon}_o - \mathbf{H}\boldsymbol{\varepsilon}_b)(\boldsymbol{\varepsilon}_o - \mathbf{H}\boldsymbol{\varepsilon}_b)^T\mathbf{W}^T\} \\ &= \mathbf{B} - \mathbf{B}\mathbf{H}^T\mathbf{W}^T - \mathbf{W}\mathbf{H}\mathbf{B} + \mathbf{W}\mathbf{R}\mathbf{W}^T + \mathbf{W}\mathbf{H}\mathbf{B}\mathbf{H}^T\mathbf{W}^T\end{aligned}$$

and substituting (5.4.19) we obtain

$$\mathbf{P}_a = (\mathbf{I} - \mathbf{W}\mathbf{H})\mathbf{B} \tag{5.4.20}$$

Exercise 5.4.1: Derive (5.4.20).

For convenience, we repeat the basic equations of OI, and express in words their interpretation, which is similar to that for a scalar least square problem from the last section:

$$\mathbf{x}_a = \mathbf{x}_b + \mathbf{W}[\mathbf{y}_o - H(\mathbf{x}_b)] = \mathbf{x}_b + \mathbf{W}\mathbf{d} \tag{5.4.1}$$

$$\mathbf{W} = \mathbf{B}\mathbf{H}^T(\mathbf{R} + \mathbf{H}\mathbf{B}\mathbf{H}^T)^{-1} \tag{5.4.19a}$$

We will see in Section 5.5 (where we derive the variational approach or 3D-Var) that the weight matrix (5.4.19) can be written in an alternative equivalent form as

$$\mathbf{W} = (\mathbf{B}^{-1} + \mathbf{H}^T\mathbf{R}^{-1}\mathbf{H})^{-1}\mathbf{H}^T\mathbf{R}^{-1} \tag{5.4.19b}$$

(see (5.5.11) in Section 5.5)

$$\mathbf{P}_a = (\mathbf{I}_n - \mathbf{W}\mathbf{H})\mathbf{B} \tag{5.4.20}$$

where the subscript n is a reminder that the identity matrix is in the analysis or model space.

The interpretation of these equations is very similar to the scalar case discussed in Section 5.3:

Equation (5.4.1) says: "The analysis is obtained by adding to the first guess (background) the product of the optimal weight (or gain) matrix and the innovation (the difference between the observation and the first guess). The first guess of the observations is obtained by applying the observation operator H to the background vector." Also, note that from (5.4.15), $H(\mathbf{x}_b) = H(\mathbf{x}_t) + \mathbf{H}(\mathbf{x}_b - \mathbf{x}_t) = H(\mathbf{x}_t) + \mathbf{H}\boldsymbol{\varepsilon}_b$, where the matrix $\mathbf{H}$ is the linear tangent perturbation of H.

Equation (5.4.19a) says: "The optimal weight (or gain) matrix is given by the background error covariance in the observation space ($\mathbf{B}\mathbf{H}^T$) multiplied by the inverse of the total error covariance (the sum of the background and the observation error covariances)." Note that the larger the background error covariance compared with the observation error covariance, the larger the correction to the first guess.

Equation (5.4.20) says: "The error covariance of the analysis is given by the error covariance of the background, reduced by a matrix equal to the identity matrix ($n \times n$) minus the optimal weight matrix."

Finally we derive an alternative formulation for the analysis error covariances showing (as in the scalar case) the additive properties of the "precisions" (if all the statistical assumptions hold true). From (5.4.1), (5.4.16), and (5.4.19b) we can show that

$$\begin{aligned}\boldsymbol{\varepsilon}_a &= \boldsymbol{\varepsilon}_b + [\mathbf{B}^{-1} + \mathbf{H}^T\mathbf{R}^{-1}\mathbf{H}]^{-1}\mathbf{H}^T\mathbf{R}^{-1}(\boldsymbol{\varepsilon}_0 - \mathbf{H}\boldsymbol{\varepsilon}_b) \\ &= [\mathbf{B}^{-1} + \mathbf{H}^T\mathbf{R}^{-1}\mathbf{H}]^{-1}[\mathbf{B}^{-1}\boldsymbol{\varepsilon}_b + \mathbf{H}^T\mathbf{R}^{-1}\boldsymbol{\varepsilon}_o]\end{aligned} \tag{5.4.21}$$

If we again compute $\mathbf{P}_a = E\{\boldsymbol{\varepsilon}_a\boldsymbol{\varepsilon}_a^T\}$ from (5.4.21), and make use of $E\{\boldsymbol{\varepsilon}_b\boldsymbol{\varepsilon}_o^T\} = 0$, $\mathbf{P}_b = \mathbf{B} = E\{\boldsymbol{\varepsilon}_b\boldsymbol{\varepsilon}_b^T\}$, $\mathbf{P}_o = \mathbf{R} = E\{\boldsymbol{\varepsilon}_o\boldsymbol{\varepsilon}_o^T\}$, we obtain

$$\mathbf{P}_a^{-1} = \mathbf{B}^{-1} + \mathbf{H}^T\mathbf{R}^{-1}\mathbf{H} \tag{5.4.22}$$

Equation (5.4.22) says: "The analysis precision, defined as the inverse of the analysis error covariance, is the sum of the background precision and the observation precision projected onto the model space."

Note that all these statements are dependent on the assumption that the statistical estimates of the errors are accurate. If the observations and/or background error covariances are poorly known, if there are biases, or if the observations and background errors are correlated, the analysis precision can be considerably worse than implied by (5.4.20) or (5.4.22).

Remarks 5.4.2

(a) It is important to note that the observation error variances come from two different sources: one is the instrumental error variances proper, the second is the presence in the observations of subgrid-scale variability not represented in the grid-average values of the model and analysis. The second type of error is denoted "*error of representativeness*". By performing a grid average similar to the Reynolds average discussed in Chapter 4, we obtain that the observational error variance $\mathbf{R}$ is the sum of the instrument error variance $\mathbf{R}_{instr}$ and the representativeness error variance $\mathbf{R}_{repr}$, assuming that these errors are not correlated. If in addition we allow for errors in the observation operator H with observation error covariance $\mathbf{R}_H$, these can also be included in the observation error covariance (Lorenc, 1986):

$$\mathbf{R} = \mathbf{R}_{instr} + \mathbf{R}_{repr} + \mathbf{R}_H \tag{5.4.23}$$

(b) The equations for Kalman filtering, discussed in detail in Section 5.5, are very similar as those for OI. The main difference is that the background error covariance **B**, instead of being assumed to be constant in time as in OI or 3D-Var, is updated (forecasted) from the previous analysis time t_n to the new analysis time t_{n+1}. The model forecast starts from the analysis at time t_n, $\mathbf{x}_b^{n+1} = M(\mathbf{x}_a^n)$, where M is the nonlinear model. Therefore, subtracting $\mathbf{x}_t^{n+1} = M(\mathbf{x}_t^n) - \varepsilon_M$ from both sides,

$$\varepsilon_b^{n+1} = \mathbf{M}\varepsilon_a^n + \varepsilon_M \tag{5.4.24}$$

where ε_M is the model error. From (5.4.24) we obtain the Kalman filter new forecast error covariance

$$\mathbf{B} = \mathbf{P}_f(t_{n+1}) = \boldsymbol{\varepsilon}_b^{n+1}\left(\boldsymbol{\varepsilon}_b^{n+1}\right)^T = \mathbf{M}(t_n)\mathbf{P}_a(t_n)\mathbf{M}^T(t_n) + \mathbf{Q}(t_n) \tag{5.4.25}$$

Here $\mathbf{Q} = E(\boldsymbol{\varepsilon}_m\boldsymbol{\varepsilon}_m^T)$ is the forecast model error covariance, **M** is the linear tangent model and $\mathbf{M}^T$ its adjoint. With this change, the weight matrix becomes the Kalman gain matrix **K.** Although this is apparently a small change from OI, the matrix multiplications by **M** in (5.4.20) are approximately equivalent to integrating the forecast model $n/2$ times, where n is the number of degrees of freedom of the model.

5.4.2 Approximations made in the practical implementation of OI

We have seen that, in matrix form, the analysis is obtained from

$$\mathbf{x}_a = \mathbf{x}_b + \mathbf{W}[\mathbf{y}_o - H(\mathbf{x}_b)] \tag{5.4.26}$$

or if we define increments from the background as $\delta\mathbf{x} = \mathbf{x} - \mathbf{x}_b$, then the analysis increment is

$$\delta\boldsymbol{x}_a = \boldsymbol{W}\delta\boldsymbol{y}_o \tag{5.4.27}$$

The optimal weight matrix **W** that minimizes the analysis error covariance is given by

$$\mathbf{W} = \mathbf{B}\mathbf{H}^T(\mathbf{H}\mathbf{B}\mathbf{H}^T + \mathbf{R})^{-1} \tag{5.4.28}$$

If all the statistical assumptions are accurate, i.e., the background error covariance **B** and the observations error covariance **R** are known exactly, then the formulas (5.4.26) or (5.4.27), and (5.4.28) provide the OI analysis. In that case, the analysis error covariance is given by

$$\mathbf{P}_a = \mathbf{A} = (\mathbf{I} - \mathbf{W}\mathbf{H})\mathbf{B} \tag{5.4.29}$$

If, as occurs in reality, the statistics are only approximations of the true statistics, then (5.4.26) and (5.4.28) provide a "*statistical interpolation*", not necessarily "optimal interpolation".

OI is typically performed in physical space, either grid point by grid point (e.g., McPherson *et al.*, 1979, DiMego *et al.*, 1985) or over limited volumes (Lorenc, 1981). In the implementation of OI, (5.4.27) and (5.4.28) are solved point by point (or volume by volume) in grid-point space. To make the formulation in physical space clearer, we expand the matrix equations:

$$\left.\begin{aligned} \mathbf{B} &= \begin{bmatrix} b_{11} & \dots & b_{1n} \\ \vdots & & \vdots \\ b_{n1} & \dots & b_{nn} \end{bmatrix} & \mathbf{H} &= \begin{bmatrix} h_{11} & \dots & h_{1n} \\ \vdots & & \vdots \\ h_{p1} & \dots & h_{pn} \end{bmatrix} \\ \mathbf{R} &= \begin{bmatrix} r_{11} & \dots & r_{1p} \\ \vdots & & \vdots \\ r_{p1} & \dots & r_{pp} \end{bmatrix} & \mathbf{W} &= \begin{bmatrix} w_{11} & \dots & w_{1p} \\ \vdots & & \vdots \\ w_{n1} & \dots & w_{np} \end{bmatrix} \end{aligned}\right\} \tag{5.4.30}$$

$\mathbf{H}$ is the linear perturbation (Jacobian) of the forward observational model H, and $\mathbf{H}^T$ is its *transpose* or *adjoint*. Multiplying by $\mathbf{H}$ on the left transforms grid-point increments into observation increments (e.g., by linear interpolation), and $\mathbf{H}^T$ transforms from observation points back to grid points. There are n grid points, or if we are considering several variables, n is the product of the number of grid points and the variables. Consider a specific grid point with the subscript g. The subscripts j and k represent particular *observations* affecting the grid point g, and there are p such observations. Recall that $\mathbf{B}$ is the background error covariance, so that the background error is $\varepsilon_b(x, y) = \mathbf{x}_b(x, y) - \mathbf{x}_t(x, y)$, and $b_{jk} = E[\varepsilon_b(x_j, y_j)\varepsilon_b^T(x_k, y_k)]$, and the expected value is the average over many cases.

We can rewrite the equation for the weights (5.4.28) as

$$\mathbf{W}(\mathbf{HBH}^T + \mathbf{R}) = \mathbf{BH}^T \tag{5.4.31}$$

Consider again the OI equations in matrix form:

$$\mathbf{x}_a = \mathbf{x}_b + \mathbf{W}[\mathbf{y}_o - H(\mathbf{x}_b)]$$

where the optimal weight matrix is obtained from the system of equations (5.4.31).

As an illustration, let us write (5.4.31) for the simple case of three grid points e, f, g, and two observations, 1 and 2 (Fig. 5.4.1).

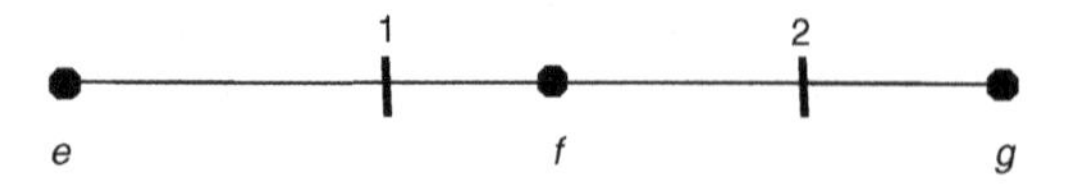

Figure 5.4.1: Simple system with three grid points (black dots) e, f, g, and two observation points, 1 and 2.

In this case, $\mathbf{x}^a = (x_e^a, x_f^a, x_g^a)^T$, and similarly for the background $\mathbf{x}^b = (x_e^b, x_f^b, x_g^b)^T$. The observation vector is $\mathbf{y}^o = (y_1^o, y_2^o)^T$, and the background values at the observation points are

$$\mathbf{H}\mathbf{x}^b = \begin{pmatrix} h_{1e} & h_{1f} & h_{1g} \\ h_{2e} & h_{2f} & h_{2g} \end{pmatrix} \begin{pmatrix} x_e^b \\ x_f^b \\ x_g^b \end{pmatrix} = \mathbf{y}^b \tag{5.4.32}$$

The coefficients of the observation operator $\mathbf{H}$ are obtained from linear or higher order interpolation of the grid location to the observation location. We are assuming that the observed and analyzed variables are the same, so that the coefficients of the matrix $\mathbf{H}$ are simply interpolation coefficients. For example, if we used linear interpolation, $\mathbf{H}$ would be

$$\mathbf{H} = \begin{pmatrix} \dfrac{x_f - x_1}{x_f - x_e} & \dfrac{x_1 - x_e}{x_f - x_e} & 0 \\ 0 & \dfrac{x_g - x_2}{x_g - x_f} & \dfrac{x_2 - x_f}{x_g - x_f} \end{pmatrix}$$

The background error covariance matrix elements are the covariances between grid points:

$$\mathbf{B} = \begin{pmatrix} b_{ee} & b_{ef} & b_{eg} \\ b_{fe} & b_{ff} & b_{fg} \\ b_{ge} & b_{gf} & b_{gg} \end{pmatrix}$$

so that

$$\mathbf{B}\mathbf{H}^T = \begin{pmatrix} b_{e1} & b_{e2} \\ b_{f1} & b_{f2} \\ b_{g1} & b_{g2} \end{pmatrix}$$

is an approximation by interpolation of the background error covariances between grid to observation points, e.g., $b_{g2} = b_{ge}h_{2e} + b_{gf}h_{2f} + b_{gg}h_{2g}$. Then

$$\mathbf{H}\mathbf{B}\mathbf{H}^T = \begin{pmatrix} b_{11} & b_{12} \\ b_{21} & b_{22} \end{pmatrix}$$

is an approximation by back interpolation of the background error covariance between observation points. The observation error covariance for this case is

$$\mathbf{R} = \begin{pmatrix} r_{11} & r_{12} \\ r_{21} & r_{22} \end{pmatrix}$$

It is usually a reasonable assumption that measurement errors made at different locations are uncorrelated, in which case $\mathbf{R}$ is a diagonal matrix. (Measurement errors could be correlated, but only within small groups of observations made by the same instrument, in which case $\mathbf{R}$ is a block diagonal matrix, where the blocks are still easily invertible.) From this simple example, it is apparent that, in general, we can write the OI equation for a *particular grid point g influenced by p observations* as:

$$x_g^a = x_g^b + \sum_{j=1}^{p} w_{gj} \delta y_j \tag{5.4.33}$$

where the weights are the solution of the linear system

$$\sum_{j=1}^{p} w_{gj}(b_{jk}+r_{jk}) = b_{gk} \qquad k = 1, \ldots, p \tag{5.4.34}$$

In (5.4.34), w_{gj} is the weight that multiplies the observation increment δy_j to contribute to the analysis increment δx_g^a, r_{jk} is the observation error covariance between two observation points j and k, b_{jk} is the *background error covariance between the observation points j and k*, and b_{gk} is the background error covariance between the grid point g and the observation point k. We are assuming that the observation vector has already been transformed into the same type of variables as the model, i.e., that a retrieval method is used, and the interpolation matrix **H** has already been multiplied by **B,** generating grid-to-observation and observation-to-observation background correlations as in the simple example above. There are equations like (5.4.34) and (5.4.33) for each grid point, and in the case of multivariate analysis (e.g., the geopotential height and the two horizontal wind components z, u, v), for each variable at each grid point.

Equations (5.4.33) and (5.4.34) constitute the OI scheme written for each grid point g. Note the fundamental role that the background error covariance plays in the determination of the optimal weights (5.4.34). The background error covariance determines the scale and the structure of the corrections to the background. In the practical implementation of (5.4.33) and (5.4.34), there are a number of commonly made additional simplifications, especially in the background error covariance elements b.

Remarks 5.4.3

(a) Perhaps the most important advantage of statistical interpolation schemes such as OI and 3D-Var over empirical schemes such as SCM is that the correlation between observational increments is taken into account. Recall that in SCM, the weights of the observational increments depend only on their distance to the grid point. Therefore, in SCM all observations will be given similar weight even if a number of them are "bunched up" in one quadrant (Fig. 5.1.1), with just a single observation in a different quadrant. In OI (or 3D-Var), by contrast, the isolated observational increment will be given more weight in the analysis than observations that are close together and therefore less independent. The fact that isolated observations have more independent information than observations close together is a result of the fact that the forecast error correlations $b_{jk}/\sqrt{b_{jj}b_{kk}}$ at the observation points j, k are large if the observation points are close together.

(b) When several observations are too close together, then the solution of (5.4.34) becomes an ill-posed problem. In those cases, it is common to compute a "superobservation" combining the close individual observations. This has the advantage of removing the ill-posedness, while at the same time reducing by averaging the random errors of the individual observations. The superobservation should be a weighted average that takes into account the relative observation errors of the original close observations.

In order to develop an OI-based data assimilation, we need to make actual estimates of the prior error covariances, **B** and **R** and the observation operator **H**. We saw before that, for a simple system in which the observations are the same as the model variables, the observation operator is simply an interpolator from the model to the observation location. If the variables are different, it has to include not only the interpolation to the observation location, but also the "forward model" that represents the observations that would be obtained if the model was true. For example, if the observations are satellite radiances, the observation operator interpolates from the model grid to the radiance observation location, and then uses radiative transfer theory to convert a model column of temperature and moisture as a function of pressure into synthetic "radiances". The observational error covariance is obtained from instrument error estimates. If the measurements are independent, the matrix **R** is diagonal, which is a major advantage. The forecast error covariance **B** is the most difficult error covariance to estimate, and has a crucial impact on the results.

Most of the rest of this section is devoted to a brief review of the methods that were in use in the 1980s to estimate **B** in OI applications. They were based on estimations of the horizontal and vertical correlations between forecast errors, estimated as the differences between the short-range forecasts and the rawinsonde observations (Thiebaux and Pedder, 1987, Hollingsworth and Lönnberg, 1986). In contrast, for 3D-Var, the method that has been almost universally adopted does not depend on measurements at all, but on the difference between forecasts verifying at the same time. This is known as the "NMC method" and is discussed in Section 5.4.8.

In OI it is common to standardize the background error covariance with **D,** the diagonal matrix of the variances:

$$\mathbf{B} = \mathbf{D}^{1/2}\mathbf{C}\mathbf{D}^{1/2} \quad \mathbf{D} = \begin{bmatrix} \sigma_1^2 & 0 & \dots & 0 \\ 0 & \sigma_2^2 & \dots & 0 \\ \vdots & \vdots & & \vdots \\ 0 & 0 & \dots & \sigma_n^2 \end{bmatrix} \quad \mathbf{C} = \begin{bmatrix} \mu_{11} & \mu_{12} & \dots & \mu_{1p} \\ \mu_{12} & \mu_{22} & \dots & \mu_{2p} \\ \vdots & \vdots & & \vdots \\ \mu_{1p} & \mu_{2p} & \dots & \mu_{pp} \end{bmatrix} \tag{5.4.35}$$

Here

$$\mu_{ij} = b_{ij}/(\sqrt{b_{ii}}\sqrt{b_{jj}}) = b_{ij}/\left(\sqrt{\sigma_i^2}\sqrt{\sigma_j^2}\right) \tag{5.4.36}$$

are the correlations of the background errors at two observational points i, j, and σ_i^2 are the error variances.

It has been commonly assumed in OI implementations that the background error correlations can be separated into the product of the horizontal correlation and the vertical correlation. As we will see, these simplified correlations are typically defined as functions of distance only.

We present an example of simplifications frequently made in practice by considering the two-dimensional analysis of z, u, v, at a single pressure level p, using rawinsondes as the only observations. Since the observation errors of two separate rawinsondes at points i and j are not correlated (although the geopotential errors are correlated in the vertical), we can assume that the observation error covariance matrix is a diagonal matrix:

$$\left.\begin{aligned} r_{ij} &= 0 \text{ if } i \neq j \\ r_{ii} &= \sigma_{oi}^2 \end{aligned}\right\} \tag{5.4.37}$$

We also assume that the background error variance is constant for each variable, $b_{ii} = \sigma_{bg}^2$ (equal to the background error variance at the grid point and assumed to be the same for all the grid points).

With these assumptions, (5.4.34) for a grid point g becomes

$$\sum_{j=1}^{p} w_{gj}\mu_{jk} + \eta_k w_{gk} = \mu_{gk} \qquad k = 1, \ldots, p \tag{5.4.38}$$

Here $\eta_k = \sigma_{ok}^2/\sigma_{bg}^2$ is the mean square of the relative error of observations compared with the background error, a parameter frequently "tuned" to give more or less weight to the observations. We can also show from (5.4.29) that the relative analysis error at the grid point is given by:

$$\frac{\sigma_{ag}^2}{\sigma_{bg}^2} = 1 - \sum_{k=1}^{p} w_{gk}\mu_{gk} \tag{5.4.39}$$

where the analysis error has been scaled by the background error variance at the grid point. Note that (5.4.39) is the equivalent of (5.4.20) in grid-point space.

We now further assume that the background error correlation between two points in the same horizontal surface is homogeneous and isotropic (i.e., it doesn't change with a rigid translation or rotation of the two points). In that case the background error correlation of the geopotential height depends only on the distance between the two points. Gandin (1963), Schlatter (1975), Thiebaux and Pedder (1987) and others have used a Gaussian exponential function for the geopotential error correlation:

$$\mu_{ij} = e^{-r_{ij}^2/2L_\phi^2} \tag{5.4.40}$$

where $r_{ij}^2 = (x_i - x_j)^2 + (y_i - y_j)^2$ is the square of the distance between two points i and j, and L_ϕ, typically of the order of 500 km, defines the background error correlation scale. Gaussian functions have also been used for the vertical correlation functions. These assumptions are clearly crude, and only qualitatively begin to reflect the true structure of the background error correlation. For example, in the real atmosphere the background error correlation length should depend on the Rossby radius of deformation, and therefore should be a function of latitude, with longer horizontal error correlations in the tropics than in the extratropics (Balgovind *et al.*, 1983, Baker *et al.*, 1987). It should also depend on the data density: at the boundaries between data-rich and data-poor regions, the correlations of the forecast errors should not be isotropic (Cohn and Parrish, 1986). We refer the reader to the discussions in Daley (1991) and Thiebaux and Pedder (1987) for further details and references.

Another important assumption usually made in the OI analysis of large-scale flow is that the background wind error correlations are geostrophically related to the geopotential height error correlations. This has two advantages: it avoids having to estimate independently the wind error correlation, and it imposes an approximate geostrophic balance of the wind and height analysis increments, and therefore improves the balance of the analysis (see Remark 5.4.3(c)). Once a functional assumption for the background error correlation of the height like (5.4.40) is made, then the multivariate correlation between heights and winds can be obtained from the height correlations. For example, consider the background error correlation between two horizontal wind components:

$$E(\delta u_i \delta v_j) = -\frac{g}{f_i}\frac{g}{f_j}E\left(\frac{\partial \delta z_i}{\partial y_i}\frac{\partial \delta z_j}{\partial x_j}\right) \tag{5.4.41}$$

Now, since the geopotential error at the point x_j is independent of y_i and vice versa, we can combine the derivatives, use (5.4.40) and write

$$\begin{aligned} E(\delta u_i \delta v_j) &= -\frac{g}{f_i}\frac{g}{f_j}\frac{\partial^2 E(\delta z_i \delta z_j)}{\partial y_i \partial x_j} = -\frac{g}{f_i}\frac{g}{f_j}\frac{\partial^2 b_{ij}}{\partial y_i \partial x_j} \\ &= -\frac{g^2}{f_i}\frac{\sigma_z^2}{f_j}\frac{\partial^2 \mu_{ij}}{\partial y_i \partial x_j} \end{aligned} \tag{5.4.42}$$

The standard deviation of the wind increments can also be derived from the geostrophic relationship, $E(\delta u_i^2)^{1/2} = (g\sigma_z/f_i)$, $E(\delta v_j^2)^{1/2} = (g\sigma_z/f_j)$, so that we obtain the correlation of the increments of the two wind components by dividing (5.4.42) by these standard deviations: $\rho_{u,v} = -\partial^2 \mu_{ij}/\partial y_i \partial x_j$. Similarly, we can obtain the correlations between the increments of any two of the three variables at two points i, j: $\rho_{h,h} = \mu_{ij}$, $\rho_{h,u} = -\frac{\partial \mu_{ij}}{\partial y_i}$, $\rho_{u,h} = -\frac{\partial \mu_{ij}}{\partial y_j}$, etc.

Fig. 5.4.2 shows schematically the shape of typical wind/height correlation functions used in OI (e.g., Schlatter, 1975). Note that the *u*–*h* correlations have the opposite sign than the *h*–*u* correlations because the first and second variables correspond to the first and second points *i* and *j* respectively. For example, a height observation leading to a positive analysis increment of *h* will result in positive increments of *u* to its north. Conversely, a positive increment of *u* will lead to negative increments of *h* to its north.

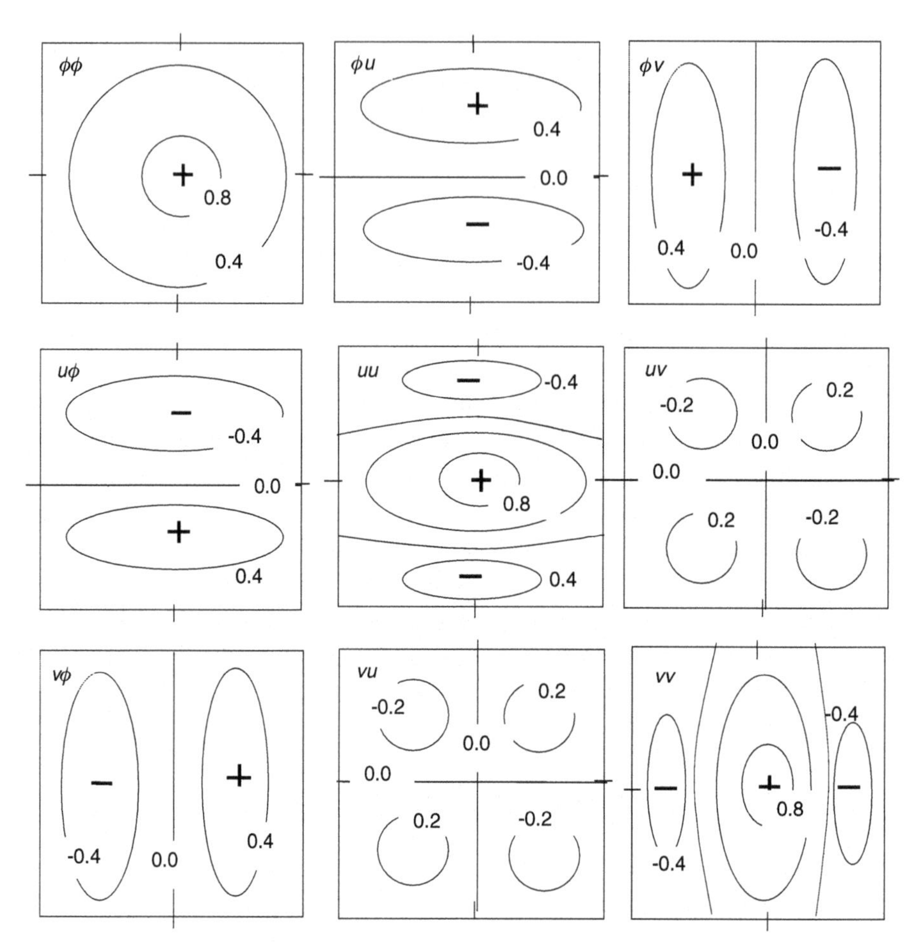

Figure 5.4.2: Schematic illustration of the correlation and cross-correlation functions for multivariate OI analysis derived using the geostrophic increment assumption (after Gustaffson, 1981). Both x- and y-axes go from $-\sqrt{2}L_\phi$ to $+\sqrt{2}L_\phi$.

Equations (5.4.42) are not valid at the equator, and additional approximations have to be made in the tropics to allow for a smooth decoupling of wind and height increments (Lorenc, 1981).

In addition, it is common to select the observations to be included in solving the linear system for the weight coefficients (5.4.38), depending on the computer resources available for the analysis, allowing for a maximum number of observations affecting each grid point. Rules for the selection of the subset of observations to be used typically depend on the distance to the grid point (within a maximum radius of influence), the types of observations (giving priority to the most accurate) and their distribution (trying to "cover" all quadrants, and choosing the closest stations).

Lorenc (1981) gave a comprehensive description of an OI as implemented at the ECMWF. Although later improvements were implemented in the error covariances (Hollingsworth and Lönnberg, 1986, Lönnberg and Hollingsworth, 1986, Hollingsworth, 1989) and other components of the analysis cycle, it remained the backbone of the ECMWF analysis until its replacement by 3D-Var in 1996. A more recent operational implementation of OI including advanced three-dimensional estimations of the three-dimensional background error covariance is that of Mitchell *et al.* (1990) in the Canadian operational system.

5.4.3 Bratseth's iterative method for OI

Bratseth (1986) proposed a variation of the SCM that converges to OI analysis (see also Daley (1991), Appendix F). It is based on the convergence of the geometric series

$$\sum_{k=0}^{\infty} (\mathbf{I} - \mathbf{A})^k = \mathbf{A}^{-1} \text{ if } (\mathbf{I} - \mathbf{A})^k \to 0 \text{ as } k \to \infty \tag{5.4.43}$$

The OI algorithm is written as

$$\mathbf{x}_g^a = \mathbf{x}_g^b + (\mathbf{BH}^T)_g \mathbf{M}^{-1}\mathbf{d} \tag{5.4.44}$$

where $\mathbf{d}$ is a correction vector which is determined by successive iterations. $\mathbf{M}$ is a diagonal (and hence easily invertible) matrix which is chosen to speed convergence.

Bratseth's algorithm consists of computing a series of corrections (without inverting a matrix to determine the optimal weights, as required by OI):

$$\left.\begin{aligned} \mathbf{d}_0 &= \delta\mathbf{y} \\ \mathbf{d}_\nu &= [\mathbf{I} - (\mathbf{HBH}^\mathrm{T} + \mathbf{R})\mathbf{M}^{-1}]\mathbf{d}_{\nu-1} \end{aligned}\right\} \tag{5.4.45}$$

Using the iteration formula (5.4.45), the corrections are

$$\mathbf{d}_\nu = \sum_{k=0}^{\nu} [\mathbf{I} - (\mathbf{HBH}^\mathrm{T} + \mathbf{R})\mathbf{M}^{-1}]^k \delta\mathbf{y} \tag{5.4.46}$$

The summation in (5.4.46) is a geometric series. Because $\mathbf{M}$ is chosen to ensure convergence, in the limit of large j, the series converges to

$$\mathbf{d}_\infty = [(\mathbf{HBH}^\mathrm{T} + \mathbf{R})\mathbf{M}^{-1}]^{-1}\delta\mathbf{y} = \mathbf{M}(\mathbf{HBH}^\mathrm{T} + \mathbf{R})^{-1}\delta\mathbf{y} \tag{5.4.47}$$

After a sufficient number of iterations, the correction (5.4.47) is substituted into (5.4.44), which results in the desired OI solution (independent of $\mathbf{M}$):

$$\mathbf{x}_g^a = \mathbf{x}_g^b + (\mathbf{BH}^\mathrm{T})_\mathbf{g} (\mathbf{HBH}^\mathrm{T} + \mathbf{R})^{-1}\delta\mathbf{y} \tag{5.4.48}$$

Since the diagonal matrix $\mathbf{M}$ is arbitrary, Bratseth suggested a choice that speeds convergence:

$$m_{jj} = \sum_{k=1}^{p} |b_{jk} + r_{jk}| \tag{5.4.49}$$

where b_{jk} are the elements of $\mathbf{HBH}^{\mathrm{T}}$, and r_{jk} the elements of $\mathbf{R}$. This method is further illustrated in the following examples.

5.4.4 One-dimensional example of OI and comparison with Bratseth's scheme

Consider a simple example in one-dimensional grid space, with just two geopotential observations along the x-axis, at the points $x_1 = 0$, and $x_2 = \alpha L_\phi$, where L_ϕ is the correlation length of geopotential observations. The grid points are also on the x-axis: $x_g = g\Delta x$. For each grid point we can rewrite equations (5.4.33) and (5.4.34) as

$$\phi_g^a = \phi_g^b + w_{g1}(\phi_1^o - \phi_1^b) + w_{g2}(\phi_2^o - \phi_2^b) \tag{5.4.50}$$

where the weights are obtained from the solution of the linear system

$$\left.\begin{aligned}(b_{11} + r_{11})w_{g1} + b_{12}w_{g2} &= b_{1g}\\ b_{21}w_{g1} + (b_{22} + r_{22})w_{g2} &= b_{2g}\end{aligned}\right\} \tag{5.4.51}$$

As noted before, we have assumed that the observational errors are uncorrelated at different points, and that they are uncorrelated with the background errors.

We assume as before that the background error correlation is Gaussian,

$$b_{ij} = \sigma_\phi^2 \mu_{ij} = \sigma_\phi^2 e^{-(x_i - x_j)^2/2L_\phi^2} \tag{5.4.52}$$

and that the ratio of the observation and background error variances is

$$\eta = r_{ii}/b_{ii} = \sigma_o^2/\sigma_\phi^2 \tag{5.4.53}$$

If one of the observations is of geopotential and the second is a wind observation v, the term b_{ij} can be computed as

$$\begin{aligned}b_{ij} &= \frac{1}{f}\frac{\partial E(\delta\phi_i \delta v_j)}{\partial x_j} = \frac{\sigma_\phi^2}{f}\frac{\partial \mu_{ij}}{\partial x_j} = \frac{\sigma_\phi^2}{f}\frac{\partial e^{-(x_i - x_j)^2/2L_\phi^2}}{\partial x_j}\\ &= \frac{\sigma_\phi^2 (x_i - x_j)}{f L_\phi^2} e^{-(x_i - x_j)^2/2L_\phi^2}\end{aligned} \tag{5.4.54}$$

In Exercise 5.4.2, this OI problem is solved directly. We now write Bratseth's iterative scheme for the same grid point and allow for more than two observations.

First we need to compute correction vector $\mathbf{d}$ by performing enough successive iterations. Assume p observations are influencing the grid point g. The first iteration is simply the vector of observational increments:

$$\mathbf{d}_0 = \begin{pmatrix} \phi_1^o - \phi_1^b \\ \vdots \\ \phi_p^o - \phi_p^b \end{pmatrix}$$

The following iterations are given by

$$\mathbf{d}_\nu = \mathbf{A}\mathbf{d}_{\nu-1} + \mathbf{d}_0 \tag{5.4.55}$$

The elements a_{ij} of the matrix $\mathbf{A}$ are obtained as in (5.4.45), from

$$a_{ij} = \delta_{ij} - (b_{ij} + \delta_{ij} r_{ij})/m_{jj} \tag{5.4.56}$$

where $\delta_{ij} = 1$ if $i = j$ and zero otherwise, and the scaling diagonal matrix $\mathbf{M}^{-1}$ has elements

$$m_{jj} = \sum_{k=1}^{p} |b_{jk} + \delta_{jk} r_{jk}| \tag{5.4.57}$$

chosen by Bratseth to speed up convergence.

Once enough iterations have been performed for $\mathbf{d}$, the grid-point analysis is obtained from

$$\phi_g^a = \phi_g^b + \begin{pmatrix} b_{g1} & \cdots & b_{gp} \end{pmatrix} \begin{pmatrix} 1/m_{11} & 0 & \cdots & 0 \\ 0 & 1/m_{22} & \cdots & 0 \\ \vdots & \vdots & & \vdots \\ 0 & 0 & \cdots & 1/m_{pp} \end{pmatrix} \begin{pmatrix} d_1 \\ \vdots \\ d_p \end{pmatrix} \tag{5.4.58}$$

As shown before, $\mathbf{d}$ converges with large ν towards

$$\mathbf{d}_\nu \to \mathbf{M}(B + R)^{-1}\mathbf{d}_0$$

so that

$$\phi_g^a \to \phi_g^b + \begin{pmatrix} b_{g1} & \cdots & b_{gp} \end{pmatrix} \begin{pmatrix} b_{11} + r_{11} & b_{12} & \cdots & b_{1p} \\ b_{21} & b_{22} + r_{22} & \cdots & b_{2p} \\ \vdots & \vdots & & \vdots \\ b_{p1} & b_{p2} & \cdots & b_{pp} + r_{pp} \end{pmatrix}^{-1} \begin{pmatrix} d_1 \\ \vdots \\ d_p \end{pmatrix}_0 \tag{5.4.59}$$

which is the OI analysis.

Exercise 5.4.2: (adapted from Bratseth, 1986)

(1) Write a FORTRAN or MATLAB code to solve an OI analysis (5.4.59) with two observations. Assume that the background is zero, i.e., compute just the analysis increments for the geopotential, using the following default values: $\sigma_\phi = 200\ \text{m}^2/\text{s}^2; \sigma_o = 100\ \text{m}^2/\text{s}^2$; $f = 10^{-4}\ \text{s}^{-1}$; $L_\phi = 500$ km; $\Delta x = 50$ km; distance between the observations $x_2 - x_1 = \alpha L_\phi, \alpha = 1$; the observational increments are $(\phi_1^o - \phi_1^b) = 500\ \text{m}^2/\text{s}^2$, $(\phi_2^o - \phi_2^b) = 0$.

Compute the geopotential analysis increments at all the grid points between $-2L_\phi$ and $2L_\phi$.

Plot the analysis at the grid points and the two observational points. Compute and plot the estimated analysis error at each point.

(2) Same as (1) but varying the distance between the observations ($\alpha = 1, \sqrt{2}, 2$).

(3) Do the default exercise again but change the ratio of the observation and background error variances, $\eta = 0.0, 0.5, 1.0, 2.0$.

(4) Assume that one has a "credulous" analysis that believes that $\eta = 0.0$, whereas the real ratio between observation and background error variance is $\eta = 1.0$. Compare the true analysis error with the estimated analysis error.

Exercise 5.4.3: Adapt the program written in Exercise 5.4.2 to two multivariate observations, ϕ_1, v_1; ϕ_2, v_2. Take the default values in Exercise 5.4.2, and assume

$$\left(\phi_1^o - \phi_1^b\right) = 500\ \mathrm{m}^2/\mathrm{s}^2,\ v_1^o - v_1^b = 0$$

$$\left(\phi_2^o - \phi_2^b\right) = 0,\ v_2^o - v_2^b = 0$$

Perform the analysis over the grid points as in Exercise 5.4.2.(1)

Exercise 5.4.4: Perform a Bratseth iterative OI and determine how many iterations are required to obtain a satisfactory approximation to OI (compare with the exact OI solution obtained in Exercise 5.4.2).

5.5 3D-Var, the physical space analysis scheme (PSAS), and their relationship to OI

We saw in Section 5.3 that there is an important equivalence between the formulation of the optimal analysis of a scalar by minimizing the analysis error variance (finding the *optimal weights* through a least squares approach) and solving the same problem through a variational approach (finding the *analysis* that minimizes a cost function measuring its distance to the background and to the observations). The same is true when the analysis involves a full three-dimensional field, as in Section 5.4.1. In the derivation of OI, *we found the optimal weight matrix* $\mathbf{W}$ *that minimized the analysis error covariance (a matrix).* Lorenc (1986) showed that this solution is equivalent to a specific variational assimilation problem: *Find the optimal analysis* $\mathbf{x}_a$ *field that minimizes a (scalar) cost function,* where the cost function is defined as the distance between $\mathbf{x}$ and the background $\mathbf{x}_b$, weighted by the inverse of the background error covariance, plus the distance to the observations $\mathbf{y}_o$, weighted by the inverse of the observations error covariance:

$$2J(\mathbf{x}) = (\mathbf{x} - \mathbf{x}_b)^T \mathbf{B}^{-1}(\mathbf{x} - \mathbf{x}_b) + [\mathbf{y}_o - H(\mathbf{x})]^T \mathbf{R}^{-1}[\mathbf{y}_o - H(\mathbf{x})] \tag{5.5.1}$$

We saw in Section 5.3.2 for the case of a simple scalar with two measurements that the variational cost function can be derived through a *maximum likelihood* approach, i.e., the analysis is the most likely state of the atmosphere, given the two independent measurements. Similarly we can define here the likelihood (Edwards, 1984) of the true state given the background field (6-h forecast) and the new observations:

$$L_{\mathbf{B}}(\mathbf{x}||\mathbf{x}_b) = p_{\mathbf{B}}(\mathbf{x}_b|\mathbf{x}) = \frac{1}{(2\pi)^{n/2}\,|\mathbf{B}|^{1/2}} e^{-\frac{1}{2}[(\mathbf{x}_b-\mathbf{x})^T\mathbf{B}^{-1}(\mathbf{x}_b-\mathbf{x})]} \tag{5.5.2a}$$

$$L_{\mathbf{R}}(\mathbf{x}||\mathbf{y}_o) = p_{\mathbf{R}}(\mathbf{y}_o|\mathbf{x}) = \frac{1}{(2\pi)^{p/2}\,|\mathbf{R}|^{1/2}} e^{-\frac{1}{2}[(\mathbf{y}_o-H(\mathbf{x}))^T\mathbf{R}^{-1}(\mathbf{y}_o-H(\mathbf{x}))]} \tag{5.5.2b}$$

Since the background and new observations are independent, their joint probability is the product of the two Gaussian probabilities. The most likely state $\mathbf{x}$ of the atmosphere (analysis) maximizes the joint probability. This maximum is also attained when the logarithm of the joint probability is maximized, which is the same as minimizing the cost function (5.5.1).

Alternatively, the cost function (5.5.1) can also be derived based on a Bayesian approach.[1] In this case we assume that the true field is a realization of a random process defined by the prior probability distribution function (given the background field)

$$p_{\mathbf{B}}(\mathbf{x}) = \frac{1}{(2\pi)^{n/2}\,|\mathbf{B}|^{1/2}} e^{-\frac{1}{2}[(\mathbf{x}_b-\mathbf{x})^T\mathbf{B}^{-1}(\mathbf{x}_b-\mathbf{x})]}$$

Bayes theorem indicates that given the new observations $\mathbf{y}_o$, the *a posteriori* probability distribution of the true field is

$$P(\mathbf{x}|\mathbf{y}_o) = \frac{p(\mathbf{y}_o|\mathbf{x})\,p_{\mathbf{B}}(\mathbf{x})}{p(\mathbf{y}_o)} \tag{5.5.3}$$

The Bayesian estimate of the true state is the one that maximizes the *a posteriori* probability (5.5.3). The denominator in (5.5.3) is the "climatological" distribution of observations, and does not depend on the current true state $\mathbf{x}$. Therefore the maximum of the *a posteriori* probability is attained when the numerator is maximum, and is given by the minimum of the cost function (5.5.1).

The minimum of $J(\mathbf{x})$ in (5.5.1) is attained for $\mathbf{x} = \mathbf{x}_a$, i.e., the analysis is given by the solution of

$$\nabla_{\mathbf{x}} J(\mathbf{x}_a) = 0 \tag{5.5.4}$$

An exact solution can be obtained in the following way. As we did in Section 5.4.1, we can expand the second term of (5.5.1), the observational differences, assuming that the analysis is a close approximation to the truth and therefore to the observations, and linearizing H around the background value:

$$\mathbf{y}_o - H(\mathbf{x}) = \mathbf{y}_o - H[\mathbf{x}_b + (\mathbf{x} - \mathbf{x}_b)] = \{\mathbf{y}_o - H(\mathbf{x}_b)\} - \mathbf{H}(\mathbf{x} - \mathbf{x}_b) \tag{5.5.5}$$

1 I am grateful to Peter Lyster for a discussion about this topic.

Substituting (5.5.5) into (5.5.1) we get

$$2J(\mathbf{x}) = (\mathbf{x} - \mathbf{x}_b)^T \mathbf{B}^{-1}(\mathbf{x} - \mathbf{x}_b) + [\{\mathbf{y}_o - H(\mathbf{x}_b)\} - \mathbf{H}(\mathbf{x} - \mathbf{x}_b)]^T \mathbf{R}^{-1}[\{\mathbf{y}_o - H(\mathbf{x}_b)\} - \mathbf{H}(\mathbf{x} - \mathbf{x}_b)] \tag{5.5.6}$$

Expanding the products, and using the rules to transpose matrix products, we then get

$$\begin{aligned} 2J(\mathbf{x}) = {} & (\mathbf{x} - \mathbf{x}_b)^T \mathbf{B}^{-1}(\mathbf{x} - \mathbf{x}_b) + (\mathbf{x} - \mathbf{x}_b)^T \mathbf{H}^T \mathbf{R}^{-1} \mathbf{H}(\mathbf{x} - \mathbf{x}_b) \\ & - \{\mathbf{y}_o - H(\mathbf{x}_b)\}^T \mathbf{R}^{-1} \mathbf{H}(\mathbf{x} - \mathbf{x}_b) - (\mathbf{x} - \mathbf{x}_b)^T \mathbf{H}^T \mathbf{R}^{-1} \{\mathbf{y}_o - H(\mathbf{x}_b)\} \\ & + \{\mathbf{y}_o - H(\mathbf{x}_b)\}^T \mathbf{R}^{-1} \{\mathbf{y}_o - H(\mathbf{x}_b)\} \end{aligned} \tag{5.5.7}$$

The cost function is a quadratic function of the analysis increments $(\mathbf{x} - \mathbf{x}_b)$ and therefore we can use Remark 5.4.1(d): Given a quadratic function $F(\mathbf{x}) = \frac{1}{2}\mathbf{x}^T \mathbf{A}\mathbf{x} + \mathbf{d}^T \mathbf{x} + c$, where $\mathbf{A}$ is a symmetric matrix, $\mathbf{d}$ is a vector and c a scalar, the gradient is given by $\nabla F(\mathbf{x}) = \mathbf{A}\mathbf{x} + \mathbf{d}$. The gradient of the cost function J with respect to $\mathbf{x}$ (or with respect to $(\mathbf{x} - \mathbf{x}_b)$) is

$$\nabla J(\mathbf{x}) = \mathbf{B}^{-1}(\mathbf{x} - \mathbf{x}_b) + \mathbf{H}^T \mathbf{R}^{-1} \mathbf{H}(\mathbf{x} - \mathbf{x}_b) - \mathbf{H}^T \mathbf{R}^{-1} \{\mathbf{y}_o - H(\mathbf{x}_b)\} \tag{5.5.8}$$

We now set $\nabla J(\mathbf{x}_a) = 0$ to ensure that J is a minimum, and obtain an equation for $(\mathbf{x}_a - \mathbf{x}_b)$

$$(\mathbf{B}^{-1} + \mathbf{H}^T \mathbf{R}^{-1} \mathbf{H})(\mathbf{x}_a - \mathbf{x}_b) = \mathbf{H}^T \mathbf{R}^{-1} \{\mathbf{y}_o - H(\mathbf{x}_b)\} \tag{5.5.9}$$

or

$$\mathbf{x}_a = \mathbf{x}_b + (\mathbf{B}^{-1} + \mathbf{H}^T \mathbf{R}^{-1} \mathbf{H})^{-1} \mathbf{H}^T \mathbf{R}^{-1} \{\mathbf{y}_o - H(\mathbf{x}_b)\} \tag{5.5.10}$$

which, in incremental form, is

$$\delta \mathbf{x}_a = (\mathbf{B}^{-1} + \mathbf{H}^T \mathbf{R}^{-1} \mathbf{H})^{-1} \mathbf{H}^T \mathbf{R}^{-1} \delta \mathbf{y}_o$$

Formally, this is the solution of the variational (3D-Var) analysis problem, but in practice the solution is obtained through minimization algorithms for $J(\mathbf{x})$ using iterative methods for minimization such as the conjugate gradient or quasi-Newton methods.

Note that the *control variable* for the minimization (i.e., the variable with respect to which we are minimizing the cost function J) is now the *analysis*, not the *weights* as in OI. The *equivalence* between the minimization of the analysis error variance (finding the optimal weights through a least squares approach), and the three-dimensional variational cost function approach (finding the optimal analysis that minimizes the distance to the observations weighted by the inverse of the error variance) is an important property.

5.5.1 Equivalence between the OI and 3D-Var statistical problems

We now demonstrate the equivalence of the 3D-Var solution and the OI analysis solution obtained in Section 5.4.1 (Lorenc, 1986). We have to show that the weight matrix that multiplies the innovation $\{\mathbf{y}_o - H(\mathbf{x}_b)\} = \delta\mathbf{y}_o$ in (5.5.10) is the same as the weight matrix obtained with OI, i.e., that

$$\mathbf{W} = (\mathbf{B}^{-1} + \mathbf{H}^T\mathbf{R}^{-1}\mathbf{H})^{-1}\mathbf{H}^T\mathbf{R}^{-1} = (\mathbf{B}\mathbf{H}^T)(\mathbf{R} + \mathbf{H}\mathbf{B}\mathbf{H}^T)^{-1} \tag{5.5.11}$$

This identity is a variant of the Sherman–Morrison–Woodbury formula (Golub and Van Loan, 1996). If the variables that we are observing are the same as the model variables, i.e., if $\mathbf{H} = \mathbf{H}^T = \mathbf{I}$, then it is rather straightforward to prove (5.5.11), using the rules for the inverse and transpose of a matrix product. However, in general $\mathbf{H}$ is rectangular, and noninvertible. The equality (5.5.11) can be proven[2] by considering the following block matrix equation:

$$\begin{bmatrix} \mathbf{R} & \mathbf{H} \\ \mathbf{H}^T & -\mathbf{B}^{-1} \end{bmatrix} \begin{bmatrix} \mathbf{w} \\ \delta\mathbf{x}_a \end{bmatrix} = \begin{bmatrix} \delta\mathbf{y}_o \\ \mathbf{0} \end{bmatrix} \tag{5.5.12}$$

where $\mathbf{w}$ is a vector which will be discussed further in Section 5.5.2. We want to derive from (5.5.12) an equation of the form $\delta\mathbf{x}_a = \mathbf{W}\delta\mathbf{y}_o$. Eliminating $\mathbf{w}$ from both block rows in (5.5.12) we find that $\delta\mathbf{x}_a = (\mathbf{B}^{-1} + \mathbf{H}^T\mathbf{R}^{-1}\mathbf{H})^{-1}\mathbf{H}^T\mathbf{R}^{-1}\delta\mathbf{y}_o$, the 3D-Var version of the weight matrix. On the other hand, eliminating $\delta\mathbf{x}_a$ from both block rows, we obtain an equation for the vector $\mathbf{w}$, $\mathbf{w} = (\mathbf{R} + \mathbf{H}\mathbf{B}\mathbf{H}^T)^{-1}\delta\mathbf{y}_o$. From this, substituting $\mathbf{w}$ in the second block row of (5.5.12), we obtain the OI version of the weight matrix: $\delta\mathbf{x}_a = \mathbf{B}\mathbf{H}^T(\mathbf{H}\mathbf{B}\mathbf{H}^T + \mathbf{R})^{-1}\delta\mathbf{y}_o$. This demonstrates the formal equivalence of the problems solved by 3D-Var and OI. Because the methods of solution are different, however, their results are different, and most centers have adopted the 3D-Var approach.

Exercise 5.5.1: Prove (5.5.11), using the rules for the inverse and transpose of a matrix product assuming $\mathrm{H} = 1$.

5.5.2 Physical space analysis system (PSAS)

Da Silva *et al.* (1995) introduced another scheme related to 3D-Var and OI, in which the minimization is performed in the (physical) space of the observations, rather than in the model space as in the 3D-Var scheme (spectral variables in the NCEP scheme (Parrish and Derber, 1992)). They solved the same OI/3D-Var equation (5.5.10),

2 I am very grateful to Jim Purser for suggesting this elegant proof of (5.5.11) and for pointing out the relationship (5.5.17).

written as in the OI approach,

$$\delta\mathbf{x}_a = (\mathbf{BH}^T)(\mathbf{R} + \mathbf{HBH}^T)^{-1}\delta\mathbf{y}_o \tag{5.5.13}$$

but separated it into two steps:

$$\mathbf{w} = (\mathbf{R} + \mathbf{HBH}^T)^{-1}\delta\mathbf{y}_o \tag{5.5.14}$$

followed by

$$\delta\mathbf{x}_a = (\mathbf{BH}^T)\mathbf{w} \tag{5.5.15}$$

The first step is the most computer intensive, and is solved by minimization of a cost function:

$$J(\mathbf{w}) = \frac{1}{2}\mathbf{w}^T(\mathbf{R} + \mathbf{HBH}^T)\mathbf{w} - \mathbf{w}^T[\mathbf{y}_o - H(\mathbf{x}_b)] \tag{5.5.16}$$

If the number of observations is much smaller than the number of degrees of freedom in the model, this is a more efficient method for achieving results similar to those of 3D-Var.

Exercise 5.5.1 indicates that the intermediate solution vector $\mathbf{w}$ is also given by

$$\mathbf{w} = \mathbf{R}^{-1}(\delta\mathbf{y}_o - \mathbf{H}\delta\mathbf{x}_a) = \mathbf{R}^{-1}[\mathbf{y}_o - H(\mathbf{x}_a)] \tag{5.5.17}$$

i.e., it is the misfit of the observations to the analysis weighted by the inverse of the observation covariance matrix.

5.5.3 Final comments on the relative advantages of 3D-Var, PSAS and OI

Although the three statistical interpolation methods, 3D-Var, OI, and PSAS, have been shown to formally solve the same problem, there are important differences in the methods of solution. As indicated before, in practice OI requires the introduction of a number of approximations, and local solution of the analysis, grid point by grid point, or small volume by small volume (Lorenc, 1981). This in turn requires the use of a "radius of influence" and selection of only the stations closest to the grid point or volume being analyzed. The background error covariance matrix also has to be locally approximated.

Despite their formal equivalence, 3D-Var and the closely related PSAS have several important advantages with respect to OI, because the cost function (5.5.1) is minimized using global minimization algorithms, and as a result it makes unnecessary many of the simplifying approximations required by OI (Parrish and Derber, 1992, Derber *et al.*, 1991, Courtier *et al.*, 1998, Rabier *et al.*, 1998, Andersson *et al.*, 1998):

(a) In 3D-Var (and PSAS) there is no data selection, all available data are used simultaneously. This avoids jumpiness in the boundaries between regions that have selected different observations.

(b) In OI the background error covariance has been crudely obtained assuming, for example, separability of the correlation into products of horizontal and vertical gaussian correlations, and that the background errors are in geostrophic balance. The background error covariance matrix for 3D-Var, although it may still require simplifying assumptions, can be defined with a more general, global approach, rather than the local approximations used in OI. In particular, most centers have adopted the "NMC method" (Parrish and Derber, 1992) for estimating the forecast error covariance:

$$\mathbf{B} \approx \alpha E\{[\mathbf{x}_f(48\text{ h}) - \mathbf{x}_f(24\text{ h})][\mathbf{x}_f(48\text{ h}) - \mathbf{x}_f(24\text{ h})]^T\} \tag{5.5.18}$$

As indicated in (5.5.18), in the "NMC" (now NCEP) method, the structure of the forecast or background error covariance is estimated as the average over many (e.g., 50) differences between two short-range model forecasts verifying at the same time. The magnitude of the covariance is then appropriately scaled. In this approximation, rather than estimating the structure of the forecast error covariance from differences with rawinsondes (Thiebaux and Pedder, 1987, Hollingsworth and Lönnberg, 1986), the model–forecast differences themselves provide a multivariate global forecast difference covariance. The forecast covariance (5.5.18) strictly speaking is the covariance of the forecast differences and is only a proxy for the structure of forecast errors. Nevertheless, it has been shown to produce better results than previous estimates computed from forecast minus observation estimates. An important reason for this improvement is that the rawinsonde network does not have enough density to allow a proper estimate of the global structures (Parrish and Derber, 1992, Rabier *et al.*, 1998), whereas (5.5.18) provides a global representation of the forecast error structures. In the NCEP system, the analysis variables are based on the spectral model forecast variables. This allows a major simplification: the assumption of horizontal homogeneity and isotropy of the error covariance imply that the spectral model errors are uncorrelated, i.e., the background error covariance in spectral space is diagonal. In the vertical, Parrish and Derber (1992) use an empirical orthogonal function expansion of (5.5.18).

(c) The background error covariance $\mathbf{B}$ has a fundamental impact in determining the characteristics of the OI analysis increment. Essentially, *the analysis increment can only occur within the subspace spanned by* $\mathbf{B}$.

This can be easily demonstrated if we assume that the background error covariance is spanned by a single vector $\mathbf{b}$, i.e., $\mathbf{B} = \mathbf{bb}^T$. This assumes that the forecast error can take place only in the direction of $\mathbf{b}$. Assume also for simplicity that $\mathbf{H} = \mathbf{I}$, i.e., that the model variables are observed at all the model grid points, and that $\mathbf{R} = \alpha^2\mathbf{I}$, i.e., that the observational errors are uncorrelated and have equal variance. Then the solution of the OI problem $\delta\mathbf{x}_a = \mathbf{x}_a - \mathbf{x}_b = \mathbf{B}\mathbf{H}^T\left[\mathbf{H}\mathbf{B}\mathbf{H}^T + \mathbf{R}\right]^{-1}[\mathbf{y}_o - H(\mathbf{x}_b)]$ can be written exactly as $\delta\mathbf{x}_a = \mathbf{b}\mathbf{b}^T\delta\mathbf{y}_o/(\mathbf{b}^T\mathbf{b} + \alpha^2)$.

Note that *the analysis increment has the direction of* $\mathbf{b}$, and that its magnitude is proportional to the projection of the observational increment upon the subspace of the vector $\mathbf{b}$ (a similar formula with $\alpha^2 = 0$ was used by Kalnay and Toth, (1994)).

If $\mathbf{H} \neq \mathbf{I}$, we can write $\tilde{\mathbf{b}} = \mathbf{H}\mathbf{b}$. Then $\mathbf{H}\delta\mathbf{x} = \mathbf{H}\mathbf{B}\mathbf{H}^T[\mathbf{H}\mathbf{B}\mathbf{H} + \mathbf{R}]^{-1} \times [\mathbf{y}_o - H(\mathbf{x}_b)]$, and from the previous formula we obtain

$$\delta\mathbf{x}_a = \frac{\mathbf{b}\mathbf{b}^T\mathbf{H}^T[\mathbf{y}_o - H(\mathbf{x}_b)]}{\mathbf{b}^T\mathbf{H}^T\mathbf{H}\mathbf{b} + \alpha^2} \tag{5.5.19}$$

again showing that *the analysis increment takes place in the direction of* $\mathbf{b}$, with an amplitude proportional to the projection of innovation onto the subspace of $\mathbf{b}$.

Exercise 5.5.2: Prove (5.5.19) for a vector $\mathbf{b}$ of dimension 2.

If $\mathbf{B} = \sum_{i=1}^{k}\mathbf{b}_i$, spanning a subspace of dimension $k < n$, the dimension of the model, then the 3D-Var cost function $2J(\mathbf{x}) = (\mathbf{x} - \mathbf{x}_b)^T\mathbf{B}^{-1}(\mathbf{x} - \mathbf{x}_b) + [\mathbf{y}_o - H(\mathbf{x})]^T\mathbf{R}^{-1}[\mathbf{y}_o - H(\mathbf{x})]$ can be used to show again that the analysis increment has to be within the k-dimensional subspace spanned by the vectors $\mathbf{b}_i$. This is because outside this subspace, the inverse of the covariance matrix is infinitely large, and therefore increments not within the k-dimensional subspace are forbidden because they would result in large increases of the value of the cost function.

(d) It is possible to add constraints to the cost function without increasing the cost of the minimization. For example, Parrish and Derber (1992) included a "penalty" term in the cost function (5.5.1) forcing *simultaneously* the analysis increments to approximately satisfy the linear global balance equation. In OI, the imposition of the geostrophic constraint on the increments ensured only an approximate balance in the analysis. In practice it was found necessary to follow the OI analysis with a *nonlinear normal mode initialization* (Section 5.7). With the global balance equation added as a weak constraint to the cost function, the NCEP global model spin up (indicated for example by the change of precipitation over the first 12 hours of integration) was reduced by more than an order of magnitude compared with the results obtained with OI. In other words, with the implementation of 3D-Var it became unnecessary to perform a separate initialization step in the analysis cycle.

(e) It is also possible to incorporate important nonlinear relationships between observed variables and model variables in the H operator in the minimization of the cost function (5.5.1) by performing "inner" iterations with the linearized $\mathbf{H}$ observation operator kept constant and "outer" iterations in which it is updated. This is harder to do in the OI approach.

(f) The introduction of 3D-Var has allowed three-dimensional variational assimilation of radiances (Derber and Wu, 1998). In this approach, there is no attempt to perform retrievals and, instead, each satellite sensor is taken as an independent observation with uncorrelated errors. As a result, for each satellite observation spot, even if some channel measurements are rejected because of cloud contamination, others may still be used. In addition, because all the data are assimilated simultaneously, information from one channel at a certain location can influence the use of satellite data at a different geographical location. The quality control of the observations becomes easier and more reliable when it is made in the space of the observations than in the space of the retrievals.[3]

(g) It is also possible to include quality control of the observations within the 3D-Var analysis (Section 5.6).

(h) Cohn *et al.* (1998) have shown that the observation-space form of 3D-Var (PSAS) offers opportunities, through the grouping of data, to improve the preconditioning of the problem in an iterative solution.

5.6 Advanced data assimilation methods with evolving forecast error covariance

In Section 5.4 we have discussed OI, which minimizes the expected analysis error covariance, and its practical implementations, and in Section 5.5 the closely related 3D-Var and PSAS methods, solving essentially the same problem but minimizing a cost function. In these methods, the forecast error covariance matrix is estimated once and for all, as if the forecast errors were statistically stationary.

From Fig. 5.6.1 we can evaluate whether this is indeed a good approximation. It shows the 6-h forecast errors in the western and eastern thirds of the USA from

3 Joiner and Da Silva (1998) pointed out that the use of retrievals from remotely sensed observations is a viable option within the variational analysis approach, as long as the innovation vector is computed consistently with the use of retrievals from radiances. If $\mathbf{D}y^o$ is the (linearized) retrieval algorithm applied to satellite radiances to obtain, e.g., temperature and moisture profiles, then the innovation vector should be computed consistently as $\mathbf{D}y^o - \mathbf{DF}x^b$, where $\mathbf{F}$ is the forward (linearized radiative transfer) algorithm that converts model variables into model radiances. In other words, the forward observational operator in (5.5.13) is $\mathbf{H} = \mathbf{DF}$, and the observational error covariance for the retrievals becomes $E[(\mathbf{D}\delta y^o)(\mathbf{D}\delta y^o)^T] = \mathbf{D}E[\delta y^o \delta y^{oT}]\mathbf{D}^T = \mathbf{DRD}^T$ instead of $\mathbf{R}$, which is the observation error covariance when radiances are directly assimilated. If this method is applied to OI, the weight matrix becomes $\mathbf{K} = \mathbf{BF}^T\mathbf{B}^T(\mathbf{DFBF}^T\mathbf{D}^T + \mathbf{DRD}^T)^{-1}$.

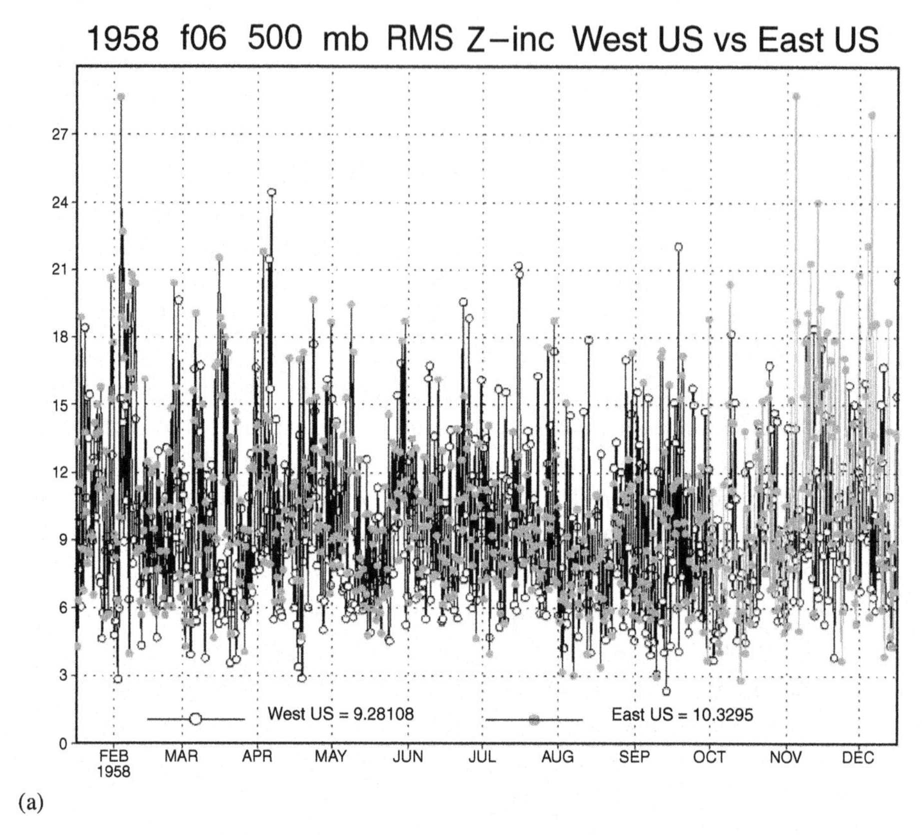

(a)

Figure 5.6.1: Daily variation of the rms increment between the 6-h forecast and the analysis over two data-rich regions in the USA, from the NCEP-NCAR reanalysis (Kistler *et al.*, 2001): (a) 1958; (b) 1996. The average error is indicated in the box.

the NCEP/NCAR reanalysis (Kistler *et al.*, 2001) estimated from the difference between the forecast and the analysis or analysis increments. Since in these data-rich regions the analysis is close to the truth, the analysis increment is a good estimate of the forecast error. Figure 5.6.1(a) corresponds to 1958 and Fig. 5.6.1(b) is calculated for 1996. The NCEP/NCAR reanalysis used a 3D-Var data assimilation system unchanged in time, so that the difference between the figures is due only to the changes in the observing system. Over these four decades the improvements in the observing system in the Northern Hemisphere show a positive impact on the 6-h forecast errors of about 20%, with the average analysis increment reduced from about 10 m to 8 m. Note that the NCEP/NCAR reanalysis used satellite temperature retrievals, not direct assimilation of radiances, which would probably have resulted in larger improvements. However, the most striking result apparent in the figure is that the day-to-day variability in the forecast error (with a time scale of a few days) *is about as large as the average error*, not just in 1958 but even in 1996. This figure emphasizes the importance of the "errors of the day" (Kalnay *et al.*, 1997), which

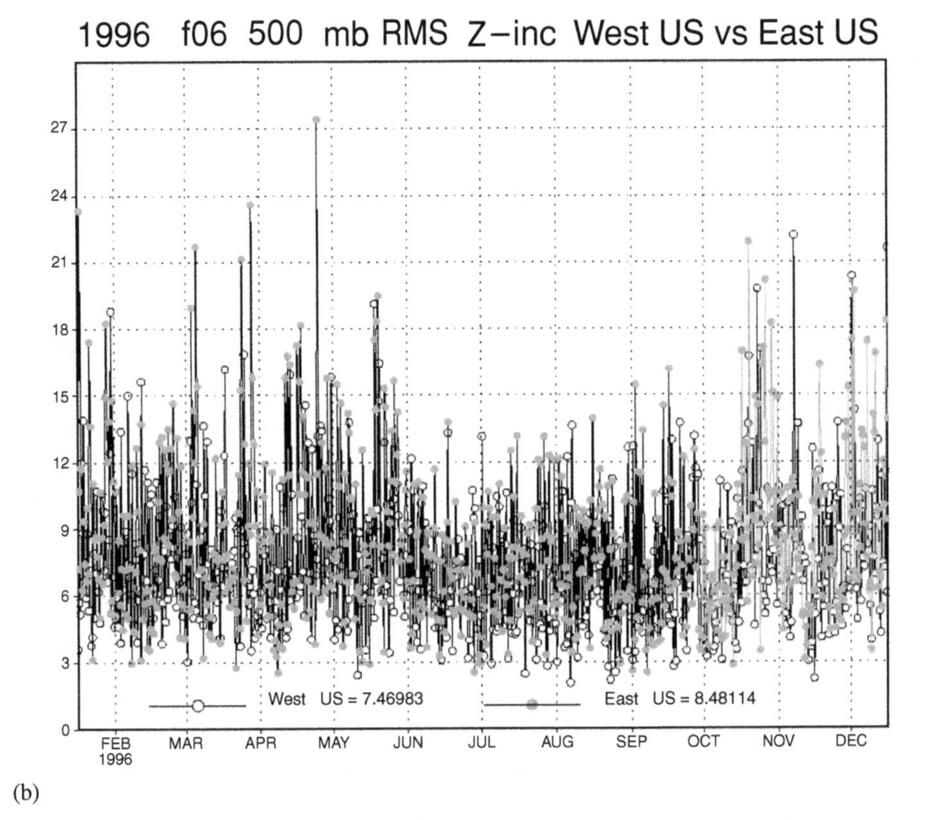

(b)

Figure 5.6.1: *(cont.)*

in these areas are dominated presumably by baroclinic instabilities of synoptic time scales, and which are ignored when the forecast error covariance is assumed to be constant.

In this section we give a brief introduction to more advanced (and much costlier) schemes that include, at least implicitly, the evolution of the forecast error covariance. A number of papers in Ghil *et al.* (1997) provide more details about the theory and practice of some of these methods. Ide *et al.* (1997) is a brief but extremely clear overview.

5.6.1 Extended Kalman filtering

As we discussed in Section 5.3 for the simple case of a scalar analysis, the Kalman filter (KF) is formally very similar to OI, but with one major difference: the forecast or background error covariance $\mathbf{P}^f(t_i)$ is advanced using the model itself, rather than estimating it as a constant covariance matrix $\mathbf{B}$.

Following the notation of Ide *et al.* (1997), let $\mathbf{x}^f(t_i) = M_{i-1}[\mathbf{x}^a(t_{i-1})]$ represent the (nonlinear) model forecast that advances from the previous analysis time t_{i-1}

to the current t_i. The model is imperfect (in particular, it has been discretized, so that subgrid processes are not included). Therefore, we assume that for the true atmosphere

$$\mathbf{x}^t(t_i) = M_{i-1}[\mathbf{x}^t(t_{i-1})] + \eta(t_{i-1}) \tag{5.6.1}$$

where η is a noise process with zero mean and covariance matrix $\mathbf{Q}_{i-1} = E(\eta_{i-1}\eta_{i-1}^T)$ (in other words, when starting from perfect initial conditions, the forecast error is given by $-\eta_{i-1}$, where the negative sign is chosen for convenience). Although we are assuming that the mean error is zero, in reality model errors have significant biases that should be taken into account. Dee and DaSilva (1998) showed how to estimate and remove these model biases.

In the extended Kalman filter, the forecast error covariance is obtained linearizing the model about the nonlinear trajectory of the model between t_{i-1} and t_i, so that if we introduce a perturbation in the initial conditions, the final perturbation is given by

$$\begin{aligned}\mathbf{x}(t_i) + \delta\mathbf{x}(t_i) &= M_{i-1}\left[\mathbf{x}(t_{i-1}) + \delta\mathbf{x}(t_{i-1})\right] \\ &= M_{i-1}\left[\mathbf{x}(t_{i-1})\right] + \mathbf{L}_{i-1}\delta\mathbf{x}(t_{i-1}) + O(|\delta\mathbf{x}|^2)\end{aligned} \tag{5.6.2}$$

The linear tangent model $\mathbf{L}_{i-1}$ is a matrix that transforms an initial perturbation at time t_{i-1} to the final perturbation at time t_i (Lorenz, 1965). The linear tangent model and its transpose or adjoint model $\mathbf{L}_{i-1}^T$ will be discussed in more detail in Chapter 6, which is devoted to predictability, and in Appendix B. We point out here that if there are several steps in a time interval $t_0 - t_i$, the linear tangent model that advances a perturbation from t_0 to t_i is given by the product of the linear tangent model matrices that advance it over each step:

$$\mathbf{L}(t_0, t_i) = \prod_{j=i-1}^{0} \mathbf{L}(t_j, t_{j+1}) = \prod_{j=i-1}^{0} \mathbf{L}_j = \mathbf{L}_{i-1}\mathbf{L}_{i-2}\cdots\mathbf{L}_0 \tag{5.6.3}$$

Therefore, the adjoint model (transpose of the linear tangent model) is given by

$$\mathbf{L}(t_i, t_0)^T = \prod_{j=0}^{i-1} \mathbf{L}(t_{j+1}, t_j)^T = \prod_{j=0}^{i-1} \mathbf{L}_j^T \tag{5.6.4}$$

Equation (5.6.4) shows that the adjoint model "advances" a perturbation backwards in time, from the final to the initial time. Adjoint models are discussed in more detail in Chapter 6 and in Appendix B.

As we did in OI and 3D-Var, observations are assumed to have random errors with zero mean and an observational error covariance matrix $\mathbf{R}_i = E(\varepsilon_i^o \varepsilon_i^{oT})$, where

$$\mathbf{y}_i^o = H(\mathbf{x}^t(t_i)) + \varepsilon_i^o \tag{5.6.5}$$

and H is the forward or observation operator.

Note that the forecast error over a 6-h forecast depends on the initial (analysis) error and on the errors introduced by the forecast model during that period:

$$\varepsilon_i^f = M_{i-1}\left(\mathbf{x}_{i-1}^t\right) + \eta_i - M_{i-1}\left(\mathbf{x}_{i-1}^a\right) = M_{i-1}\left(\mathbf{x}_{i-1}^a + \mathbf{x}_{i-1}^t - \mathbf{x}_{i-1}^a\right)$$
$$+\eta_i - M_{i-1}\left(\mathbf{x}_{i-1}^a\right) \approx \mathbf{L}_{i-1}\varepsilon_{i-1}^a + \eta_i \qquad (5.6.6)$$

where we have neglected higher order terms.

The analysis and forecast error covariances are defined, as usual, from their corresponding errors at the appropriate time:

$$P_i = E\left(\varepsilon_i \varepsilon_i^T\right) \qquad (5.6.7)$$

From these equations we can define extended Kalman filtering which consists of a "forecast step" that advances the forecast and the forecast error covariance, followed by an "analysis" or update step, a sequence analogous to OI. After the forecast step, an optimal weight matrix or Kalman gain matrix is calculated as in OI, and this matrix is used in the analysis step.

The forecast step is

$$\left.\begin{aligned} \mathbf{x}^f(t_i) &= M_{i-1}[\mathbf{x}^a(t_{i-1})] \\ \mathbf{P}^f(t_i) &= \mathbf{L}_{i-1}\mathbf{P}^a(t_{i-1})\mathbf{L}_{i-1}^T + \mathbf{Q}(t_{i-1}) \end{aligned}\right\} \qquad (5.6.8)$$

The analysis step is written as in OI, with

$$\left.\begin{aligned} \mathbf{x}^a(t_i) &= \mathbf{x}^f(t_i) + \mathbf{K}_i\mathbf{d}_i \\ \mathbf{P}^a(t_i) &= (\mathbf{I} - \mathbf{K}_i\mathbf{H}_i)\mathbf{P}^f(t_i) \end{aligned}\right\} \qquad (5.6.9)$$

where

$$\mathbf{d}_i = \mathbf{y}_i^o - H[\mathbf{x}^f(t_i)] \qquad (5.6.10)$$

is the observational increment or innovation.

The formula for the Kalman gain or weight matrix in (5.6.10), computed after completing the forecast step, is obtained by minimizing the analysis error covariance $\mathbf{P}_i^a$. It is given by the same formula derived for OI, but with the constant background error covariance $\mathbf{B}$ replaced by the evolved forecast error covariance $\mathbf{P}^f(t_i)$:

$$\mathbf{K}_i = \mathbf{P}^f(t_i)\mathbf{H}_i^T[\mathbf{R}_i + \mathbf{H}_i\mathbf{P}^f(t_i)\mathbf{H}^T]^{-1} \qquad (5.6.11)$$

The extended Kalman filter is the "gold standard" of data assimilation. Even if a system starts with a poor initial guess of the state of the atmosphere, the extended Kalman filter may go through an initial transient period of a week or so, after which it should provide the best linear unbiased estimate of the state of the atmosphere *and* its error covariance. However, if the system is very unstable, and the observations are not frequent enough, it is possible for the linearization to become inaccurate, and the extended Kalman filter may drift away from the true solution (Miller *et al.*, 1994).

The updating of the forecast error covariance matrix ensures that the analysis takes into account the "errors of the day". Unfortunately the extended Kalman filter is exceedingly expensive, since the linear model matrix $\mathbf{L}_{i-1}$ has size n, the number of degrees of freedom of a modern model (more than 10^6) and updating the error covariance is equivalent to performing $O(n)$ model integrations. For this reason, this step has been replaced by the use of simplifying assumptions (e.g., a lower order model and/or infrequent updating).

5.6.2 Ensemble Kalman filtering

One promising simplification of Kalman filtering is ensemble Kalman filtering. In this approach, an ensemble of K data assimilation cycles is carried out simultaneously (Houtekamer *et al.*, 1996, Houtekamer and Mitchell, 1998, 2001, Hamill and Snyder, 2000, Hamill *et al.*, 2001, Anderson, 2001). All the cycles assimilate the same real observations, but in order to maintain them realistically independent, different sets of *random perturbations are added to the observations* assimilated in each member of the ensemble data assimilations. This ensemble of data assimilation systems can be used to estimate the forecast error covariance (Evensen, 1994, Evensen and van Leewen, 1996, Houtekamer and Mitchell, 1998, Hamill and Snyder, 2000). After completing the ensemble of analyses at time t_{i-1}, and the K forecasts $\mathbf{x}_k^f(t_i) = M_{i-1}^k[\mathbf{x}_k^a(t_{i-1})]$, one can obtain an estimate of the forecast error covariance from the K forecasts $\mathbf{x}_k^f(t_i)$. For example one could assume

$$\mathbf{P}^f \approx \frac{1}{K-1}\sum_{k=1}^{K}\left(\mathbf{x}_k^f - \overline{\mathbf{x}}^f\right)\left(\mathbf{x}_k^f - \overline{\mathbf{x}}^f\right)^T$$

where the overbar represents the ensemble average, but this would tend to underestimate the variance of the forecast errors because every forecast is used to compute the estimate of its own error covariance. Houtekamer and Mitchell (1998) and Hamill and Snyder (2000) suggested instead to compute the forecast error covariance for ensemble member i from an ensemble that excludes the forecast l:

$$\mathbf{P}_l^f \approx \frac{1}{K-2}\sum_{k \neq l}\left(\mathbf{x}_k^f - \overline{\mathbf{x}}_l^f\right)\left(\mathbf{x}_k^f - \overline{\mathbf{x}}_l^f\right)^T \tag{5.6.12}$$

Hamill and Snyder (2000) also suggested a hybrid between 3D-Var and ensemble Kalman filtering, where the forecast error covariance is obtained from a linear combination of the (constant) 3D-Var covariance $\mathbf{B}_{3D\text{-}Var}$:

$$P_l^{f(hybrid)} = (1-\alpha)P_l^f + \alpha\mathbf{B}_{3D\text{-}Var} \tag{5.6.13}$$

where α is a tunable parameter that varies from 0, pure ensemble Kalman filtering from (5.6.12) to 1, pure 3D-Var. In (5.6.12) the ensemble Kalman filtering covariance is estimated from only a limited sample of ensemble members $K-1$, compared with a much larger number of degrees of freedom of the model, it is rank deficient. The combination with the 3D-Var, computed from many estimated forecast errors (using for example the method of Parrish and Derber (1992)) may ameliorate this sampling problem and "fill out" the error covariance. In the experiments of Hamill and Snyder (2000) the best results were obtained for low values of α, between 0.1 and 0.4, indicating good impact of the use of the ensemble-evolved forecast error covariance. They found that 25–50 ensemble members were enough to provide the benefit of ensemble Kalman filtering (but this may be different when using a more complex model than the quasi-geostrophic model used here).

The ensemble Kalman filtering approach has several advantages: (a) K is of the order of 10–100, so that the computational cost (compared with OI or 3D-Var) is increased by a factor of 10–100. Although this increased cost may seem large, it is small compared to extended Kalman filtering, which requires a cost increase of the order of the number of degrees of freedom of the model. (b) Ensemble Kalman filtering does not require the development of a linear and adjoint model. (c) It does not require the linearization of the evolution of the forecast error covariance. (d) It may provide excellent initial perturbations for ensemble forecasting. Despite these advantages, at the time of writing no operational center has yet implemented this system, although Canada has plans to do so. Ott *et al.* (2002) proposed to do ensemble Kalman filtering based on the bred vectors available from operations, and taking advantage of the local low dimensionality observed by Patil *et al.* (2001). Ensemble Kalman filtering appears at the present time to be one of the most promising approaches for the future.

5.6.3 4D-Var

This is an important extension of the 3D-Var which allows for observations distributed within a time interval (t_0, t_n) (e.g., Lewis and Derber, 1985, Courtier and Talagrand, 1990, Derber, 1989, Daley, 1991, Zupanski, 1993, Bouttier and Rabier, 1997). The cost function includes a term measuring the distance to the background *at the beginning of the interval*, and a summation over time of the cost function for each observational increment computed with respect to the model integrated to the time of the observation:

$$J[\mathbf{x}(t_0)] = \frac{1}{2}[\mathbf{x}(\mathrm{t}_0) - \mathbf{x}^b(t_0)]^T \mathbf{B}_0^{-1}[\mathbf{x}(\mathrm{t}_0) - \mathbf{x}^b(t_0)] + \frac{1}{2}\sum_{i=0}^{N}\left[H(\mathbf{x}_i) - \mathbf{y}_i^0\right]^T \mathbf{R}_i^{-1}\left[H(\mathbf{x}_i) - \mathbf{y}_i^o\right] \tag{5.6.14}$$

The control variable (the variable with respect to which the cost function is minimized) is the *initial* state of the model with the time interval $\mathbf{x}(t_0)$, whereas the analysis at the end of the interval is given by the *model integration* from the solution $\mathbf{x}(t_n) = M_0[\mathbf{x}(t_0)]$. Thus, the model is used as *a strong constraint* (Sasaki, 1970), i.e., the analysis solution has to satisfy the model equations. In other words, 4D-Var seeks an initial condition such that the forecast best fits the observations within the assimilation interval. The fact that the 4D-Var method assumes a perfect model is a disadvantage since, for example, it will give the same credence to older observations at the beginning of the interval as to newer observations at the end of the interval (Menard and Daley, 1996). Derber (1989) suggested a method of correcting for a constant model error (a constant shape within the assimilation interval), see also Zupanski (1993).

A variation in the cost function when the control variable $\mathbf{x}(t_0)$ is changed by a small amount $\delta\mathbf{x}(t_0)$ is given by

$$\delta J = J[\mathbf{x}(t_0) + \delta\mathbf{x}(t_0)] - J[\mathbf{x}(t_0)] \approx \left[\frac{\partial J}{\partial \mathbf{x}(t_0)}\right]^T \cdot \delta\mathbf{x}(t_0) \tag{5.6.15}$$

where the gradient of the cost function $[\partial J/\partial\mathbf{x}(t_0)]_j = \partial J/\partial x_j(t_0)$ is a column vector. As suggested by (5.6.15), iterative minimization schemes require the estimation of the cost function gradient. In the simplest scheme, the steepest descent method, the change in the control variable after each iteration is chosen to be opposite to the gradient $\delta\mathbf{x}(t_0) = -a\nabla_{x(t_0)} J = -a\ \partial J/\partial\mathbf{x}(t_0)$. Other, more efficient methods, such as the conjugate gradient or quasi-Newton (Navon and Legler, 1987) method, also require the use of the gradient, so that in order to solve this minimization problem efficiently, we need to be able to compute the gradient of J with respect to the elements of the control variable.

As we saw in Sections 5.4 and 5.5, given a symmetric matrix $\mathbf{A}$ and a functional $J = \frac{1}{2}\mathbf{x}^T\mathbf{A}\mathbf{x}$, the gradient is given by $\partial J/\partial\mathbf{x} = \mathbf{A}\mathbf{x}$. If $J = \mathbf{y}^T\mathbf{A}\mathbf{y}$, and $\mathbf{y} = \mathbf{y}(\mathbf{x})$, then

$$\frac{\partial J}{\partial \mathbf{x}} = \left[\frac{\partial \mathbf{y}}{\partial \mathbf{x}}\right]^T \mathbf{A}\mathbf{y} \tag{5.6.16}$$

where $[\partial\mathbf{y}/\partial\mathbf{x}]_{k,l} = \partial y_k/\partial x_l$ is a matrix.

We can write (5.6.14) as $J = J_b + J_o$, and from the rules discussed above, the gradient of the background component of the cost function $J_b = \frac{1}{2}[\mathbf{x}(t_0) - \mathbf{x}^b(t_0)]^T\mathbf{B}_0^{-1}[\mathbf{x}(t_0) - \mathbf{x}^b(t_0)]$ with respect to $\mathbf{x}(t_0)$ is given by

$$\frac{\partial J_b}{\partial \mathbf{x}(t_0)} = \mathbf{B}_0^{-1}[\mathbf{x}(t_0) - \mathbf{x}^b(t_0)] \tag{5.6.17}$$

The gradient of the second term of (5.6.14) $J_o = \frac{1}{2}\sum_{i=0}^{N}[H(\mathbf{x}_i) - \mathbf{y}_i^o]^T\mathbf{R}_i^{-1}[H(\mathbf{x}_i) - \mathbf{y}_i^o]$ is more complicated because $\mathbf{x}_i = M_i[\mathbf{x}(t_0)]$. If we introduce a perturbation to the initial state, then $\delta\mathbf{x}_i = \mathbf{L}(t_0, t_i)\delta\mathbf{x}_0$, so that

$$\frac{\partial\left(H(\mathbf{x}_i) - \mathbf{y}_i^o\right)}{\partial \mathbf{x}(t_0)} = \frac{\partial H}{\partial \mathbf{x}_i}\frac{\partial M}{\partial \mathbf{x}_o} = \mathbf{H}_i\mathbf{L}(t_0, t_i) = \mathbf{H}_i\prod_{j=i-1}^{0}\mathbf{L}(t_j, t_{j+1}) \tag{5.6.18}$$

As indicated by (5.6.18), the matrices $\mathbf{H}_i$, $\mathbf{L}_i$ are the linearized Jacobians $\partial H/\partial\mathbf{x}_i$, $\partial M/\partial\mathbf{x}_o$. Therefore, from (5.6.16) and (5.6.18) the gradient of the observation cost function is given by

$$\left[\frac{\partial J_o}{\partial \mathbf{x}(t_0)}\right] = \sum_{i=0}^{N}\mathbf{L}(t_i, t_0)^T\mathbf{H}_i^T\mathbf{R}_i^{-1}\left[H(\mathbf{x}_i) - \mathbf{y}_i^o\right] \tag{5.6.19}$$

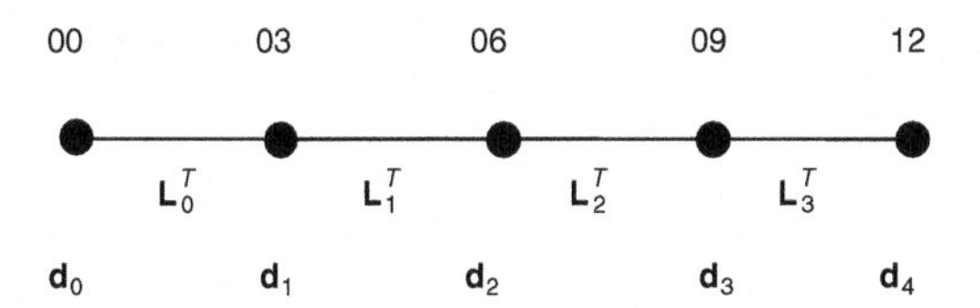

Figure 5.6.2: Schematic of the computation of the gradient of the observational cost function for a period of 12 h, observations every 3 h and the adjoint model that integrates backwards within each interval.

Equation (5.6.19) shows that every iteration of the 4D-Var minimization requires the computation of the gradient, i.e., computing the increments $[H(\mathbf{x}_i) - \mathbf{y}_i^o]$ at the observation times t_i during a forward integration, multiplying them by $\mathbf{H}_i^T \mathbf{R}_i^{-1}$ and integrating these weighted increments back to the initial time using the adjoint model. Since parts of the backward adjoint integration are common to several time intervals, the summation in (5.6.19) can be arranged more conveniently. Assume, for example that the interval of assimilation is from 00 Z to 12 Z, and that there are observations every 3 h (Fig. 5.6.2). We compute during the forward integration the weighted negative observation increments $\overline{\mathbf{d}}_i = \mathbf{H}_i^T \mathbf{R}_i^{-1}[H(\mathbf{x}_i) - \mathbf{y}_i^o] = -\mathbf{H}_i^T \mathbf{R}_i^{-1} \mathbf{d}_i$. The adjoint model $\mathbf{L}^T(t_i, t_{i-1}) = \mathbf{L}_{i-1}^T$ applied on a vector "advances" it from t_i to t_{i-1}. Then we can write (5.6.19) in the example shown in Fig. 5.6.2 as

$$\frac{\partial J_o}{\partial \mathbf{x}_o} = \overline{\mathbf{d}}_o + \mathbf{L}_0^T \left\{\overline{\mathbf{d}}_1 + \mathbf{L}_1^T \left[\overline{\mathbf{d}}_2 + \mathbf{L}_2^T \left(\overline{\mathbf{d}}_3 + \mathbf{L}_3^T \overline{\mathbf{d}}_4\right)\right]\right\} \tag{5.6.20}$$

From (5.6.17) plus (5.6.19) or (5.6.20) we obtain the gradient of the cost function, and the minimization algorithm modifies appropriately the control variable $\mathbf{x}(t_0)$. After this change, a new forward integration and new observational increments are computed and the process is repeated.

4D-Var can also be written in an incremental form with the cost function defined by

$$J(\delta\mathbf{x}_0) = \frac{1}{2}(\delta\mathbf{x}_0)^T \mathbf{B}_0^{-1}(\delta\mathbf{x}_0) \\ + \frac{1}{2}\sum_{i=0}^{N} \left[H_i \mathbf{L}(t_0, t_i)\delta\mathbf{x}_0 - \mathbf{d}_i^o\right]^T \mathbf{R}_i^{-1} \left[\mathbf{H}_i \mathbf{L}(t_0, t_i)\delta\mathbf{x}_0 - \mathbf{d}_i^0\right] \tag{5.6.21}$$

and the observational increment defined as in (5.6.10). Within the incremental formulation, it is possible to choose a *"simplification operator"* that solves the problem of minimization in a lower dimensional space $\mathbf{w}$ than that of the original model variables $\mathbf{x}$:

$$\delta\mathbf{w} = \mathbf{S}\delta\mathbf{x}$$

$\mathbf{S}$ is meant to be rank deficient (as would be the case, for example, if a lower resolution spectral truncation was used for $\mathbf{w}$ than for $\mathbf{x}$), so that its inverse doesn't exist, and we have to use a generalized inverse $\mathbf{S}^{-I} = [\mathbf{S}\mathbf{S}^T]^{-1}\mathbf{S}^T$. Then the minimum of the problem is obtained for

$$J(\delta\mathbf{w}), \mathbf{x}_0^b = \mathbf{x}_0^g + \mathbf{S}^{-I}\delta\mathbf{w}_0$$

and a new *"outer iteration"* at the full model resolution can be carried out (Lorenc, 1997).

The iteration process can also be accelerated through the use of *"pre-conditioning"*, a change of control variables that makes the cost function more "spherical", and therefore each iteration can get closer to the center (minimum) of the cost function (e.g., Parrish and Derber, 1992, Lorenc, 1997).

The most important advantage of 4D-Var is that if we assume that: (a) the model is perfect, and (b) the *a priori* error covariance at the initial time $\mathbf{B}_0$ is correct, *it can be shown that the 4D-Var analysis at the final time is identical to that of the extended Kalman filter* (Lorenc, 1986, Daley, 1991). This means that *implicitly 4D-Var is able to evolve the forecast error covariance from* $\mathbf{B}_0$ *to the final time* (Thepaut *et al.*, 1993). Unfortunately, this implicit covariance is not available at the end of the cycle, and neither is the new analysis error covariance. In other words, 4D-Var is able to find the best linear unbiased estimation but not its error covariance. To mitigate this problem, a simplified Kalman filter algorithm has been proposed to estimate the evolution of the analysis errors in the subspace of the dynamically most unstable modes (Fischer and Courtier, 1995, Cohn and Todling, 1996). 4D-Var has been successfully implemented at ECMWF and at Meteo France, and has resulted in improved forecast scores attributed principally to the fact that observations are assimilated at the time they were made, rather than at the analysis time.

5.6.4 Method of representers

Finally, we mention that Bennett (1992), Bennett *et al.* (1997) and Egbert *et al.*, (1994) developed a variational method with a weak constraint, rather than a strong constraint, thus accounting for the existence of model errors in the evolution of the forecast and the forecast error covariance. The forecast errors appear as random forcings on the dynamics, with an *a priori* forecast error covariance matrix $\mathbf{Q}$. The cost function therefore becomes

$$\begin{aligned} J[\mathbf{x}(t_0)] &= \frac{1}{2}[\mathbf{x}(t_0) - \mathbf{x}^b(t_0)]^T \mathbf{B}_0^{-1}[\mathbf{x}(t_0) - \mathbf{x}^b(t_0)] \\ &+ \frac{1}{2}\sum_{m,m'=0}^{M} [\mathbf{x}_{m+1} - M_m(\mathbf{x}_m)]^T \mathbf{Q}_{m,m'}{}^{-1}[\mathbf{x}_{m'+1} - M_{m'}(\mathbf{x}_{m'})] \\ &+ \frac{1}{2}\sum_{i=0}^{N} \left[H(\mathbf{x}_i) - \mathbf{y}_i^o\right]^T \mathbf{R}_i^{-1}\left[H(\mathbf{x}_i) - \mathbf{y}_i^0\right] \end{aligned} \tag{5.6.22}$$

where the second term accounts for the model errors. In principle this cost function makes the problem equivalent to Kalman filtering. However, the solution is made feasible using the method of representers within the observational space (Bennett, 1992, Egbert and Bennett, 1994).

Formally, the representer method can be written as a four-dimensional OI

$$\mathbf{x}(t_N) = \mathbf{x}_b(t_N) + \mathbf{W}[\mathbf{y}_o - H(\mathbf{x}_b)]$$

where the weight matrix and the observational and background vectors in $\mathbf{d}(t_i) = [\mathbf{y}_o - H(\mathbf{x}_b)]$ are defined not only in space but also in time (at the time of the observations). The details of the method of solution are described in Egbert *et al.* (1994), Bennett *et al.* (1996), and Uboldi and Kamachi (2000). Uboldi and Kamachi (2000) showed how it is applied to the nonlinear Burger equation, and how the "representers", which are the *a posteriori* error covariance functions for each observation, influence the solution beyond the time of observation.

5.7 Dynamical and physical balance in the initial conditions

We saw in Chapter 1 that Richardson's (1922) experiment resulted in a disastrous estimation of the initial surface pressure tendency (a forecast of a change of 146 hPa in 6 h, whereas the actual pressure remained almost unchanged) because of noisy data and the presence of fast inertia-gravity waves in the solution of the primitive equations. If there are fast and slow waves in the solution of a model, $\omega_{fast} >> \omega_{slow}$, and $u = U_{slow}e^{-i\omega_{slow}t} + U_{fast}e^{-i\omega_{fast}t}$, then $\partial u/\partial t = -i\omega_{slow}U_{slow}e^{-i\omega_{slow}t} - i\omega_{fast}U_{fast}e^{-i\omega_{fast}t}$. As shown schematically in Fig. 1.2.1 it is clear that, unless the amplitude of the fast waves component is made very small, the fast waves will dominate the initial tendency.

We saw in Section 2.5 that in the SWEs, the simplest example of primitive equations, there are two types of wave solutions: (a) steady or slowly evolving, quasi-geostrophically balanced "weather" modes, satisfying $\omega \approx 0$, and (b) fast inertia-gravity waves with a frequency dispersion relationship $\omega^2 \approx f^2 + gD(k^2 + l^2)$, where k, l are the horizontal wavenumbers, and D is the mean depth of the model. For the external mode, with about 10 km equivalent depth, inertia-gravity waves travel at a speed of about 300 m/s, and unless they are filtered out of the initial conditions, they can indeed produce a very noisy forecast. After a while, though, the inertia-gravity waves subside as the solution evolves towards quasi-geostrophic balance (a process known as geostrophic adjustment), so that, unless they interact nonlinearly with the slower "weather waves", the inertia-gravity waves do not necessarily ruin the forecast.

In this section we first consider the geostrophic adjustment process that takes place within a linear SWE system. This also allows us to assess what types of observations are most useful for NWP. We then consider the nonlinear case and describe the nonlinear normal mode initialization method, which was used for many years to reduce the imbalance in the initial conditions. Finally we describe a more recently introduced type of dynamical initialization, denoted digital filtering, which is simple and very effective. We should note that if the analysis is out of balance, all of the initial information that projects on inertia-gravity waves will be lost, whether the inertia-gravity waves are filtered out during the model integration through geostrophic adjustment, or through the other initialization methods. For this reason it is preferable to enforce balance within the analysis as done in 3D-Var, since this reduces the loss of information.

5.7.1 Geostrophic adjustment and the relative importance of different observations

If the initial conditions of a model are not in quasi-geostrophic balance, the balanced portion of the initial field will project on the quasi-geostrophic mode, and the unbalanced portion will project onto inertia-gravity waves. These waves have large horizontal divergence and propagate horizontally, dispersing quite fast. Because of horizontal dispersion, after a while the amplitude of the inertia-gravity waves becomes much smaller, and the leftover fields remain in quasi-geostrophic balance. Rossby (1936, 1938) first described this process of "*geostrophic adjustment*." The time scale for geostrophic adjustment is of the order of f^{-1} (about 12 h). Arakawa (1997) provided an analytic solution of the linear geostrophic adjustment problem for the SWEs.

In Section 2.4 we showed that the potential vorticity is conserved for individual parcels:

$$\frac{d}{dt}\frac{\nabla^2\psi + f}{\Phi} = 0 \tag{5.7.1}$$

Consider small perturbations on a basic state of rest. The linearized SWE potential vorticity is obtained assuming that the relative vorticity $\nabla^2\psi$ is small compared with the Coriolis parameter and that the total geopotential height $\Phi = gD + \phi$, where the perturbations to the free surface are small compared to the mean depth of the fluid. In that case, the conservation of potential vorticity becomes

$$\frac{d}{dt}\eta = \frac{d}{dt}\left(f + \nabla^2\psi - \frac{f_0}{gD}\phi\right) = 0 \tag{5.7.2}$$

Parcels will evolve conserving their *initial* potential vorticity $\eta(0) = f + \nabla^2\psi - (f_o/gD)\phi$ even as they undergo the geostrophic adjustment process. This important conservation property allows us to assess how much of the initial mass and wind increments will project on the slow modes, and be "remembered" by the model after geostrophic adjustment, and how much information will be lost through inertia-gravity waves. If we introduce a perturbation $\delta\psi(0)$, $\delta\phi(0)$ in the initial conditions through data assimilation, the perturbation in potential vorticity will remain in the solution even after geostrophic adjustment: $\delta\eta_g = \delta\eta(0)$. Recall that after geostrophic adjustment, the winds become geostrophic, so that $\psi_g = \phi_g/f_0$.

Assume that within a univariate analysis, the introduction of observations results in an *analysis increment* field associated with either mass observations $\delta\phi(x, y)$ or wind observations $\delta\psi(x, y)$. After about 12–24 h the initial unbalanced field will disperse away as inertia-gravity waves, and the remaining increments will be in geostrophic balance:

$$\delta\phi \to \delta\phi_g, \delta\psi \to \delta\psi_g \text{with } \delta\psi_g = \delta\phi_g / f_o \tag{5.7.3}$$

Assume that the analysis increment field was of the form $\delta\phi = Ae^{i(kx+ly)}$, with $k^2 + l^2 = n^2 = (2\pi/L)^2$. From (5.7.2), the final increment of potential vorticity after geostrophic adjustment is equal to the initial analysis increment $\delta\eta_g(x, y) = \delta\eta(x, y)$. Consider the effect of introducing only mass observations (performing a univariate analysis). They will result in an analysis increment $\delta\phi$, and the potential vorticity initial (and final) increments are then

$$\delta\eta = -\frac{f_0}{gD}\delta\phi = \nabla^2\delta\psi_g - \frac{f_0}{gD}\delta\phi_g = -n^2\delta\psi_g - \frac{f_0}{gD}\delta\phi_g \tag{5.7.4}$$

The *impact of the mass observation* after geostrophic adjustment is therefore

$$\delta\phi_g = \frac{1}{n^2R^2 + 1}\delta\phi \tag{5.7.5}$$

where $R = \sqrt{gD/f_0^2}$ is the *Rossby radius of deformation*, the natural quasi-geostrophic horizontal scale given a rotation rate and mean depth. Equation (5.7.5) indicates that the response depends strongly on whether the waves are short or long compared with R. For *long waves*, for which $n^2R^2 \ll 1$, the model "remembers" the mass data: $\delta\phi_g \approx \delta\phi$, i.e., it retains it after geostrophic adjustment. For *short waves*, for which $n^2R^2 \gg 1$, on the other hand, the model "forgets" the mass information: $\delta\phi_g \approx 0$.

The situation is reversed for a univariate analysis of only wind data, leading to an analysis increment $\delta\psi$. After geostrophic adjustment,

$$\delta\psi_g = \frac{n^2R^2}{n^2R^2 + 1}\delta\psi \tag{5.7.6}$$

so that for *long waves*, $n^2R^2 \ll 1$, the model "forgets" the wind data: $\delta\psi_g \approx 0$, whereas wind information is retained in *short waves*: $\delta\psi_g \approx \delta\psi$.

Now we have to determine which waves are "short" and which "long". In mid-latitudes, with $f_0 \sim 10^{-4}$, for the external or barotropic mode $D \sim$10 km, so that short waves are those for which $n^2R^2 \gg 1$ or $L \ll \sqrt{4\pi^2 gD/f_0^2} \approx 20000$ km, i.e., *all but planetary waves are very short*. This implies that the model essentially will ignore surface pressure data, which is the mass data corresponding to the external mode, and it will adjust its surface pressure to the barotropic component of the wind. For the first internal mode $D \sim 1$ km, and waves are "short" if shorter than about

6000 km. In other words, most of the energy in the mid-latitude atmosphere is actually in short waves. In the tropics, where f is an order of magnitude smaller, the statement that "most waves are short" applies even more strongly. For this reason, *winds tend to be more effective in providing initial conditions for an NWP model than mass data*, but *temperature data are more important for shallower vertical modes*, for which the wind will adjust to the temperature observations.

The "acceptance" of wind and/or mass data is enhanced by the use of multivariate analysis schemes, because the geostrophic correlation assumed in the background error covariance enforces an approximate geostrophic balance in the analysis increments. However, even with multivariate analysis, because nonlinearities do not allow for a perfect balance, wind observations still have the strongest impact on the skill of a forecast in modern data assimilation systems. The full impact of an observing system depends on the extent to which they contribute to define the potential vorticity, as discussed above, and on other factors:

(a) *Precision*

Assume we observe winds and mass, and in a multivariate scheme we combine them optimally, in which case their precisions (inverse of the error variances) are additive. This allows a simple estimation of the relative contributions to the analysis precision of mass and wind measurements. If we observe heights $\phi = f_0 \psi^{mass}$ with an error $\delta\phi_{ob} = g\delta z_{ob}$, and winds $\mathbf{v} = \mathbf{k} \times \boldsymbol{\nabla}\psi^{wind}$ with an error $|\delta\mathbf{v}_{ob}|$, they both contribute to the streamfunction. If we combine them optimally, the analysis error precision will be

$$\frac{1}{|\delta\psi|^2} = \frac{f_o^2}{g^2|\delta z_{ob}|^2} + \frac{n^2}{|\delta\mathbf{v}_{ob}|^2} \tag{5.7.7}$$

suggesting that the wind field contributes more accuracy to the analysis for short waves or in the tropics. The higher accuracy of winds for short waves is a result of the fact that they measure a gradient field, and is *independent of the geostrophic adjustment argument.*

(b) *Data coverage, both in the horizontal and in the vertical*

Obviously, the denser an observing system is, the more it will contribute to the accuracy of the analysis. This is true in the horizontal as well as in the vertical, so that vertical profiles of winds, temperature, or moisture are found to be more useful than single level observations. An observing system will also contribute more to the skill of the forecasts in the absence of other observing systems. For example, in the Northern Hemisphere, which has a relatively good network of rawinsondes, the contribution of satellite data to the improvement of NWP forecasts is much lower than in the Southern Hemisphere, where rawinsondes are much fewer.

(c) *Physical model adjustments and model spin-up*

In the same way that mass and wind initial fields undergo a dynamical adjustment towards geostrophic balance, other variables quickly evolve towards thermal and hydrological balance within the model. For example, because of the low heat capacity of the surface layer, surface air temperatures (at 2 m) adjust very rapidly towards equilibrium with the sea or land surface temperatures. As a result, it is difficult to effectively use surface air temperature observations since the model will tend to "forget" them (replace them by model adjusted values). Similarly, if the moisture analysis profiles are wetter or drier than what the model hydrological equilibrium would require (i.e., the model has a wet or dry bias), the analysis moisture profiles are quickly replaced by model adjusted profiles. Therefore, a model with a dry climatological bias will produce excessive rain during the analysis cycle, as the initial conditions bring in higher observed moisture profiles every 6 h, and the model rains out what it perceives as "excessive moisture". This adjustment process, which is also affected by other physical parameterizations such as surface fluxes and radiation, is known as the "spin-up" or "spin-down" of the model, depending on whether there is an initial increase or decrease in the precipitation. The spin-up process is strongest immediately after the analysis and takes between 12 and 36 h before reaching model balance. It can be reduced by "physical initialization" or assimilation of precipitation, in which temperature and moisture profiles are modified so that the model during the analysis cycle is forced to produce precipitation similar to the observed precipitation (e.g., Krishnamurti *et al.*, 1988, Treadon, 1996, Falkovich *et al.*, 2000).

Fig. 5.7.1 shows the results of a comparison of different 5-day forecasts performed every 12 h for February 1998 using the 2000 operational NCEP data assimilation system (courtesy of Michiko Masutani and Stephen Lord). The forecasts are based on different data assimilation experiments, in which: (a) all available data was assimilated (ALL); (b) the satellite radiances were not assimilated (No TOVS 1B); (c) the rawinsonde temperatures were not assimilated (No RAOB Temp); and (d) the rawinsondes winds were not assimilated (No RAOB Wind). The results illustrate several conclusions, in agreement with the discussion above:

- If every statistical assumption about errors was perfect, the ALL experiments should have the best forecasts. This is generally but not always true.
- In the Northern Hemisphere where the rawinsonde network is fairly abundant, the forecasts from the assimilation without satellite data No TOVS 1B, are, on the average, slightly worse than the ALL data experiments.
- Eliminating the rawinsonde winds from the data assimilation has a much larger negative impact than eliminating the rawinsonde temperatures in the Northern Hemisphere.
- In the Southern Hemisphere, where there are relatively few rawinsondes, the satellite radiances are the backbone of the information needed in the data

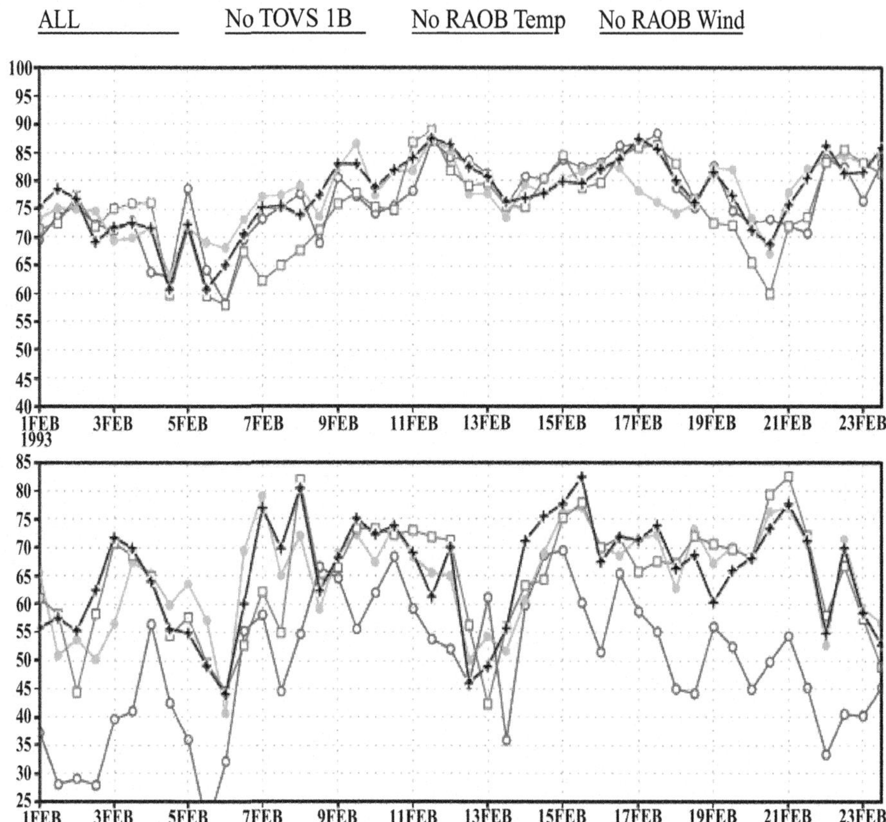

Figure 5.7.1: Verification (anomaly correlation) for the 5-day forecasts performed every 12 h with the NCEP data assimilation system with different combinations of data: crosses, all data; open circles, no TOVS radiances; open squares: no rawinsonde winds; full circles, no rawinsonde temperatures. (Courtesy of M. Masutani.)

assimilation. Without satellite radiances, the forecasts in the Southern Hemisphere substantially deteriorate.

- We should also note that there is substantial day-to-day variability in the 5-day forecast skill. This can be attributed to the changes in atmospheric predictability, i.e., on some days the atmosphere is simply easier to predict than on others. This will be studied in detail in Chapter 6.

5.7.2 Normal modes initialization

Because atmospheric motion is nonlinear, a simple linear geostrophic balance as discussed in the previous section is not enough to ensure balanced initial conditions. An approach that has been widely used to improve the initial imbalance is that of nonlinear normal modes initialization, introduced by Machenhauer (1977) and by Baer and Tribbia (1977). As indicated by its name, it requires the determination of the (linear) normal modes of a model as a first step. Daley (1991), Temperton and Williamson (1981), and others give a complete discussion about how this procedure is carried out in a three-dimensional model. Here we only illustrate how it would be applied to a simple SWE model on a periodic *f*-plane model. The SWEs are written separating them into their linear terms (on the left-hand sides) and the nonlinear terms (on the right-hand sides):

$$\left.\begin{aligned} \frac{\partial u}{\partial t} - fv + \frac{\partial \phi}{\partial x} &= R_u \\ \frac{\partial v}{\partial t} + fu + \frac{\partial \phi}{\partial y} &= R_v \\ \frac{\partial \phi}{\partial t} + gD\left(\frac{\partial u}{\partial x} + \frac{\partial v}{\partial y}\right) &= R_\phi \end{aligned}\right\} \tag{5.7.8}$$

For the SWEs the nonlinear terms $R_u,\ R_v,\ R_\phi$ are

$$\left.\begin{aligned} R_u &= -u\frac{\partial u}{\partial x} - v\frac{\partial u}{\partial y} \\ R_v &= -u\frac{\partial v}{\partial x} - v\frac{\partial v}{\partial y} \\ R_\phi &= -u\frac{\partial \phi}{\partial x} - v\frac{\partial \phi}{\partial y} \end{aligned}\right\} \tag{5.7.9}$$

Note that the left-hand sides contain linear terms about a basic state of rest.

The first step is to determine the (linear) normal modes, and for this purpose we find the eigensolutions or normal modes of (5.7.8) setting $R_u = R_v = R_\phi = 0$. In the case of a doubly periodic domain the normal modes are simply of the form

$$\begin{bmatrix} U_{kl} \\ V_{kl} \\ F_{kl} \end{bmatrix} e^{i(kx+ly)} e^{-i\omega t} = X_{kl} e^{-i\omega t} \tag{5.7.10}$$

In a three-dimensional model there would be an additional functional dependence on the vertical, which can be represented by a vertical wavenumber. If we plug (5.7.10) into (5.7.8) with zero on the right-hand side, we obtain three solutions for the frequency:

$$\left.\begin{aligned} \omega_S &= 0 \\ (\omega_F)^2 &= f^2 + gD(k^2 + l^2) \end{aligned}\right\} \tag{5.7.11}$$

where the subscripts *S* and *F* refer to the slow (quasi-geostrophic) and the fast (inertia-gravity wave) modes. Because the equations are homogeneous, $(U,\ V,\ F)$ are related to each other and the amplitude of the normal mode is arbitrary. For the slow modes, the relationship is $U_{klS} = -ilF_{klS}/f,\ V_{klS} = ikF_{klS}/f$ (geostrophic balance). For the fast modes, a similar but more complicated relationship exists which, if $f^2 \ll gD(k^2+l^2)$ reduces to

$$U_{\pm klF} = \pm \frac{kF_{\pm klF}}{\sqrt{gD(k^2+l^2)}} \qquad V_{\pm klF} = \pm \frac{lF_{\pm klF}}{\sqrt{gD(k^2+l^2)}}$$

The normal modes in (5.7.10) can therefore be written as a set of slow (Y_{kl}) and fast (Z_{kl}) modes:$X_{klm} = (Y_{kl}, Z_{kl})$ where the subscript m is S for slow modes and F for fast modes. The normal modes X_{klm} constitute a complete orthonormal basis if we normalize them by their total energy:

$$\int_x \int_y [gD({U_{klm}}^2 + {V_{klm}}^2) + {F_{klm}}^2] dx dy = 1 \tag{5.7.12}$$

Any time-dependent field of winds and heights can be expanded in terms of the normal modes:

$$\begin{bmatrix} u(x,y,t) \\ v(x,y,t) \\ \phi(x,y,t) \end{bmatrix} = \sum_k \sum_l \sum_{m=S,F} a_{klm}(t) \begin{bmatrix} U_{kl} \\ V_{kl} \\ F_{kl} \end{bmatrix}_m e^{i(kx+ly)} \tag{5.7.13}$$

where the (time-dependent) coefficients can be determined from a back Fourier transform of (5.7.13):

$$a_{klm}(t) = \int_x \int_y [gD(uU_{klm} + vV_{klm}) + \phi F_{klm}] e^{-i(kx+ly)} dy dx \tag{5.7.14}$$

Now we return to the full nonlinear SWE (5.7.8), multiply the three equations by

$$\left. \begin{array}{l} gDU_{klm} e^{-i(kx+ly)} e^{+i\omega_{klm}t} \\ gDV_{klm} e^{-i(kx+ly)} e^{+i\omega_{klm}t} \\ F_{klm} e^{-i(kx+ly)} e^{+i\omega_{klm}t} \end{array} \right\} \tag{5.7.15}$$

respectively, add them and integrate over the domain to obtain the nonlinear equations for the amplitudes of the slow and fast modes:

$$\left. \begin{array}{l} \dfrac{da_{klS}}{dt} + i\omega_{klS} a_{klS} = R_{klS} \\ \dfrac{da_{klF}}{dt} + i\omega_{klF} a_{klF} = R_{klF} \end{array} \right\} \tag{5.7.16}$$

Here we have separated slow and fast modes. $R_{klm}(Y, Z)$ is the result of applying this operation to the right-hand side nonlinear terms and depends on both the vector **Y** of slow modes coefficients and on the vector **Z** of fast modes coefficients. Recall (5.7.11) that for this simple geometry, $\omega_{klS} = 0$.

$$\frac{da_{klF}}{dt} + i\omega_{klF} a_{klF} = 0 \tag{5.7.17}$$

is the linear equation for the fast modes.

We can choose to perform a *linear normal mode initialization* by zeroing out the initial amplitude of the fast modes: $a_{klF}(t) = a_{klF}(0)e^{-i\omega_{klF}t} = 0$ in (5.7.17). This will make the *linear* time derivative of the fast modes equal to zero. But (5.7.16) shows that this initialization, which is equivalent to a perfect geostrophic balance, is not accurate enough for the realistic nonlinear case. The presence of nonlinear forcing will generate fast oscillations even if the basic state is geostrophic. Therefore Machenhauer (1977) suggested instead *to zero out the time derivative* of the fast modes in the nonlinear equation for the fast modes in (5.7.16), whose right-hand side depends on both slow and fast modes. Therefore, from

$$\frac{da_{Fkl}}{dt} + i\omega_{Fkl}a_{Fkl} = R_{Fkl}(Y(0), Z(0)) \tag{5.7.18}$$

we obtain the *nonlinear normal mode initialization* condition:

$$a_{Fkl}(0) = \frac{R_{Fkl}(Y(0), Z(0))}{i\omega_{Fkl}} \tag{5.7.19}$$

Since the coefficient $a_{klF}(0)$ appears both in the left- and right-hand sides of (5.7.19) in a component of the vector of fast coefficients Z, Machenauer (1977) suggested iterating equation (5.7.19) until convergence.

Nonlinear normal mode initialization (NLNMI) has been widely used in many operational data assimilation systems, since it is quite effective in substantially reducing the amplitude of the inertia-gravity waves from the initial conditions, much better than a simple geostrophic balance. It requires the determination of the linear normal modes of a model, but Temperton (1988) derived a formulation denoted "implicit NLNMI" without this requirement.

NLNMI has some problems, however:

(a) In the tropics, diabatic heating plays a fundamental role, essentially balancing the vertical advection of static stability. Therefore, diabatic forcing has to be included in the nonlinear terms, and this requires estimating the heating from short-term forecasts (Wergen, 1988).

(b) There is some arbitrariness in defining which "fast" modes need to be initialized. For example, inertia-gravity waves with high vertical wavenumbers are quite slow (see the discussion in the previous subsection), so that only the first few vertical modes are usually initialized. On the other hand, NLNMI eliminates the high-frequency but real atmospheric tides from the solution, since they appear as fast modes. This requires a special handling of these modes.

(c) NLNMI is only an approximation of the true slow evolution of the atmosphere: if we apply NLNMI to a model that has been running for a day or longer, ideally it should not modify it, since it has already reached slow modes equilibrium. However, NLNMI will change the initial fields significantly.

Ballish *et al.* (1992) developed a modification of the procedure denoted *incremental* NLNMI, in which the initialization is applied to the analysis increments, rather than to the full analysis field. This procedure is able to substantially solve the three problems indicated above.

In recent years the use of NLNMI after the analysis step has become less popular because of the development of three alternative approaches. The first one is the use of 3D-Var, which allows the introduction *within the cost function* of a term that penalizes the lack of balance. Parrish and Derber (1992) included a penalty term based on the global linear balance equation applied on the analysis increments. They found that this, combined with the use of a more realistic global background error covariance based on differences between 24- and 48-h forecasts verifying at the same time, yielded an analysis that was well balanced. As a result, the NLNMI step became unnecessary in the NCEP system. This is a major advantage of 3D-Var over the standard OI procedure followed by NLNMI. It eliminates the artificial separation of the analysis step, which produces fields that are close to the observations but out of balance, and the initialization step, which produces fields that are balanced but further away from the observations.

Another method that seems to achieve similar balanced results and minimize the spin-up problems is the *incremental analysis update* (Bloom *et al.*, 1996), in which the analysis increment is added in small "drips" throughout the 6-h forecast rather than once as a large change at the analysis time. Assume that there are n time steps in the 6-h forecast. In the incremental analysis update the analysis increments are computed at the analysis time. Then the forecast at the analysis time minus 3 h is integrated for 6 h adding at each time step the analysis increment divided by n, until the forecast reaches the analysis time plus 3 h. At that time a preliminary 3-h integration of the model without analysis increments is performed until the next analysis time is reached, and the cycle is repeated. The overhead of this method is only the additional preliminary integration of the model during the second half of the interval between analysis times.

A third development is the introduction of digital filter initialization, a variation of dynamic initialization that has proven to be very simple and efficient.

5.7.3 Dynamic initialization using digital filters

Some numerical schemes, like the Euler-backwards or Matsuno scheme, damp high frequencies, and this property has been used in order to reduce the accumulation of high-frequency noise within an assimilation cycle. Assume we have an equation

$$\frac{du}{dt} = -i\omega u \tag{5.7.20}$$

The Matsuno scheme is a predictor–corrector type of scheme (see Table 3.2.1) where

$$\left.\begin{aligned} \tilde{u} &= u_n - i\omega\Delta t u_n \\ u_{n+1} &= u_n - i\omega\Delta t\tilde{u} \end{aligned}\right\} \tag{5.7.21}$$

so that $u_{n+1} = \rho u_n$, where the amplification factor is $\rho = (1 - \omega^2 \Delta t^2) - i\omega\Delta t$. Therefore

$$|\rho|^2 = (1 - \omega^2 \Delta t^2 + \omega^4 \Delta t^4) \tag{5.7.22}$$

It is evident from (5.7.22) that as long as the CFL stability condition $|\omega_{max} \Delta t_{CFL}| \leq 1$ is satisfied, high frequencies are damped at every Matsuno time step. This has been found to be reasonably satisfactory for avoiding excessive accumulation of noise in the analysis cycle (e.g., Halem *et al.*, 1982), but the damping is slow except for high frequencies close to $1/\Delta t$. Using a damping time scheme does not balance the initial fields; rather the balance is achieved only after integrating the model for a while (e.g., 6 h in the analysis cycle). Several dynamical initialization methods using forward/backward integrations were suggested to balance the initial field (e.g., Nitta and Hovermale, 1969, Okamura, 1969). Grant (1975) suggested a more efficient dynamic initialization based on linear combinations of forward/backward integrations, with combinations of time steps some of which are longer than allowed by the CFL condition. Dynamic initialization never became widely used, despite its simplicity, because it is not efficient, requiring many forward/backward iterations to substantially reduce medium frequency waves.

Exercise 5.7.1: Show that the net damping at $t = 0$ after the application of a Matsuno time step followed by another Matsuno time step integrating backwards in time (changing the sign of Δt) is given by (5.7.22).

Exercise 5.7.2: Find the damping of the Okamura scheme: $u^+ = u^n(1 + i\omega\Delta t)$; $u^- = u^+(1 - i\omega\Delta t)$; $u^{n+1} = 2u^n - u^-$. The Okamura–Rivas scheme is the same except that the time step cycles over three iterations: $\Delta t = \Delta t_{CFL}$; $1.4\Delta t_{CFL}$; $2\Delta t_{CFL}$, resulting in an even faster damping (Grant, 1975).

The introduction of dynamic initialization based on digital filtering by Lynch and Huang (1992) and Lynch (1997) has changed this situation substantially, and essentially eliminated the need for NLNMI. In a digital filter, the model is integrated forward and backward in time between $-t_M$ and t_M, as in regular dynamic initialization. The difference is that the model fields are used at every time step to compute a weighted average valid at the initial time $t = 0$, and the weights are optimally chosen in order to damp high frequencies, rather than simply using a damping time scheme.

The idea of a digital filter is to choose the filtering weights in such a way that for low frequencies $\omega \ll \omega_s$ the amplitude of the solution with that frequency remains mostly unchanged, whereas for high frequencies $\omega \geq \omega_s$ the amplitude is substantially reduced. Given a time step Δt and the corresponding threshold computational frequency $\theta_s = \omega_s \Delta t$, low frequencies are characterized by $0 \leq |\theta| \ll \theta_s$, and the high frequencies that we want to filter by $\theta_s \leq |\theta| \leq \pi$.

One digital filter approach (Lynch, 1997) is based on a Dolph–Tchebychev filter, which is close to optimal, using the properties of Tchebychev polynomials:

$$T_n(x) = \begin{cases} \cos(n \cos^{-1} x) & \text{if } |x| \leq 1 \\ \cosh(n \cosh^{-1} x) & \text{if } |x| > 1 \end{cases} \tag{5.7.23}$$

from which $T_0(x) = 1$ and $T_1(x) = x$. The higher polynomials can be obtained from the recurrence relationship

$$T_n(x) = 2x T_{n-1}(x) - T_{n-2}(x) \quad n \geq 2 \tag{5.7.24}$$

The following function is ideal for such damping:

$$H(\theta) = \frac{T_{2M}[x_0 \cos(\theta/2)]}{T_{2M}(x_0)} \tag{5.7.25}$$

where $x_0 = 1/\cos(\theta_S/2)$, since for the low-frequency range, $H(\theta)$ falls from 1 to $r = 1/T_{2M}(x_0)$ as $|\theta|$ goes from 0 to θ_s, and for the high frequency range $\theta_S \leq |\theta| \leq \pi$, $H(\theta)$ oscillates within $\pm r$ (e.g., see Fig. 5.7.2).

From the definition of the Tchebychev polynomials it can be shown that

$$H(\theta) = \sum_{n=-M}^{+M} h_n e^{in\theta} \tag{5.7.26}$$

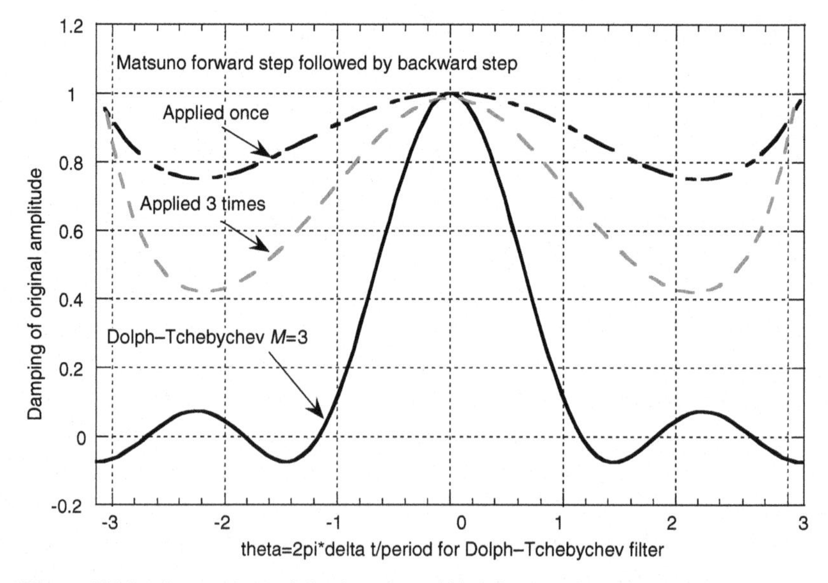

Figure 5.7.2: Comparison of the damping of high frequencies using a Dolph–Tchebychev filter with $\theta_s = \pi/3$, $M = 3$, and a forward/backward Matsuno dynamical initialization.

where the coefficients are

$$h_n = \frac{1}{2M+1}\left[1 + 2r\sum_{m=1}^{M} T_{2M}\left(x_0 \cos\frac{\theta_m}{2}\right)\cos m\theta_n\right] \tag{5.7.27}$$

The solution of the model, integrated from $-t_M$ to t_M, is weighted averaged:

$$\overline{u}(0) = \sum_{n=-M}^{M} h_n u_n \tag{5.7.28}$$

so that at the end of this procedure the amplitude of each frequency in $\overline{u}$ is modulated by $H(\theta)$. The parameters are chosen in the following way: Choose a period τ_S such that waves with periods shorter than this are to be filtered. The cut-off frequency is then given by $\theta_s = 2\pi\,\Delta t/\tau_s$. The time span of the integration $T_s = 2t_M = 2M\Delta t$ gives a filter of order $N = 2M + 1$.

In practice, Lynch (1997) recommended to first perform a backward integration with just the *dry adiabatic dynamics* (since they are reversible), from $t = 0$ to $t = -T_S$. An application of the weighted average (5.7.28) gives a filtered field centered at $t = -T_S/2$. Then a forward integration from $t = -T_S/2$ to $t = +T_S/2$ using the *full model with physics* results in a field centered at $t = 0$, *filtered for the second time including filtering of the effects of irreversible diabatic processes*. In the high resolution-limited area model, with a time step of half an hour, and a filtering period of 3 h, the time span is also half an hour, with $N = 2M = 7$. The double filter gives a reduction in energy of the high frequencies of more than 99%.

Figure 5.7.2 shows the response of dynamic initialization for this case, using the Dolph–Tchebychev filter with $\Delta t = 30$ min, $\theta_s = \pi/3$ (filtering periods shorter than 3 h) and $M = 3$. It compares this with the result of using one and three iterations of the forward/backward Matsuno time step.

Another digital filter is based on the Lanczos filter, which is widely used to filter out high frequencies from time series (Duchon, 1979, Lynch and Huang, 1992). The filter is similar to the Dolph–Tchebychev one, but now the Lanczos weights in (5.7.28) are given by

$$h_n = \frac{\sin(n\theta_{crit}\Delta t)}{\pi n}\frac{\sin(n\pi/M)}{n\pi/M} \tag{5.7.29}$$

Fig. 5.7.3 compares the responses of the Dolph–Tchebychev and Lanczos filters requiring the same number of time steps (6), since for the Lanczos filter the coefficients for $n = 4$ and 5 are zero. The response for the Lanczos filter with $\theta_{crit} = \pi/4$, $M = 5$ is competitive with the Dolph–Tchebychev filter.

In summary, initialization using digital filtering is a very simple process that avoids the determination of the model normal modes and the need of NLNMI. Most importantly, it dampens the high-frequency solutions according to their actual model-determined frequency rather than from an arbitrary separation into inertia-gravity waves and quasi-geostrophic modes. It does not make any additional approximation

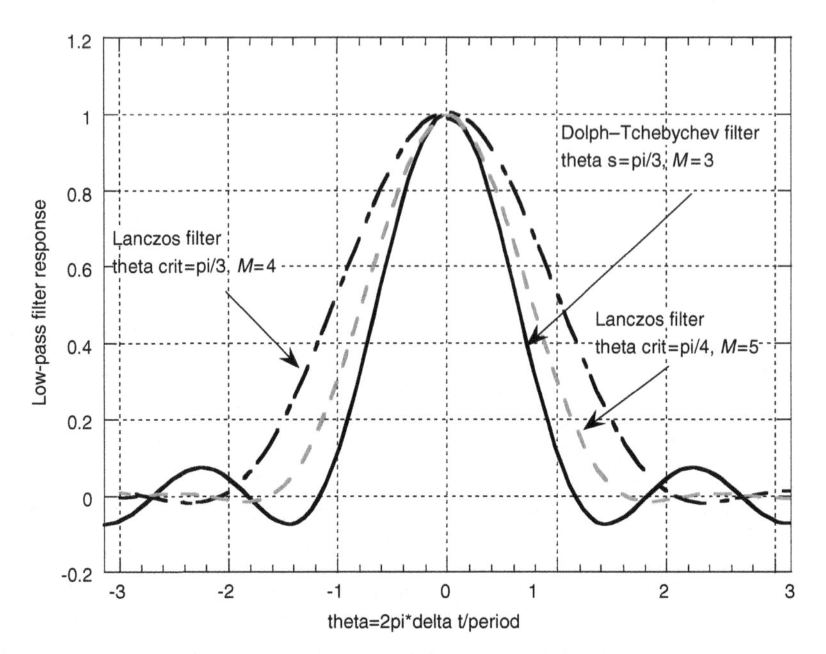

Figure 5.7.3: Comparison of the responses of the Dolph–Tchebychev filter, with $\theta_s = \pi/3$, $M = 3$, and the Lanczos filter, with $\theta_{crit} = \pi/3$, $M = 4$ and $\theta_{crit} = \pi/4$, $M = 5$. All the choices require six time integrations and are therefore computationally comparable.

and it can include the full diabatic and nonreversible effects to determine the classification of slow and fast modes.

Exercise 5.7.3: Is it better to achieve balance within 3D-Var or to apply a digital filter afterwards?

5.8 Quality control of observations

The reported atmospheric observations used in data assimilation are not perfect; they contain several kinds of errors, including instrumental errors and errors of human origin. The reported observations may also contain "errors of representativeness", i.e., actually correct observations may reflect the presence of a subgrid-scale atmospheric phenomenon that cannot be resolved by the model or the analysis. The representativeness error indicates the observation is not representative of the areally averaged measurement required by the model grid. The instrumental and representativeness errors can be systematic or random. Systematic errors and biases should be determined by calibration or other means such as time averages. Random errors are generally assumed to be normally distributed.

In addition to randomly distributed errors, the reported observations may contain errors that are so large that the observations have no useful information content and should be tossed out. Frequently, these rough or gross errors are of human origin, and take place during the computation or the transmission of the observation. There are other sources of observation errors (wrong date, time or location, uncalibrated instruments, etc.). The use of an observation with a rough error can cause a disproportionally big error in the analysis, so there has been a tendency to use observations conservatively ("when in doubt, throw it out"). In recent years, however, quality control systems have become more sophisticated and many observations that would have been thrown out in the past are now corrected, resulting in an improvement in the initial conditions, and hence in the forecasts. Newer quality control systems allow for a continuous weighting of the observation suspected of having a gross error, rather than a "yes or no" decision to toss it out.

Quality control is based on a comparison between observations and some kind of expected value (which could be based on climatology, an average of nearby observations, or the first guess). The difference between the expected value and the reported observation is denoted the "residual". If the residual is very large (measured in standard deviations of the estimate), the observation may be considered to be erroneous. The more sophisticated the estimate of the expected value (i.e., the smaller its expected standard deviation with respect to the true value), the more discriminating the quality control algorithm will be, i.e., the better it will be able to distinguish between observations with large errors, which should not be given credence, and correct observations reporting unusual states of the atmosphere, such as very low pressures or unusually high winds reported within an area affected by a tropical cyclone. In the latter case it is important to keep the observation in order to improve the initial conditions of the forecast.

Earlier quality control systems were based on several checks performed in series (one after another) before the analysis. For example DiMego *et al.* (1985) compared each observation with a climatological distribution to see whether it was within a reasonable range ("gross check"). If the reported value was outside a prescribed range, or differed from the climatological mean by more than, for example, five standard deviations, the observation was tossed out. If the observation survived this test, it was then compared to the average of nearby observations, and again tossed out if outside a reasonable range. This check (called a "buddy check", comparing the observations with their "buddies") could also salvage an observation previously tossed out even if it was quite different from the expected climatological value.

An analysis is, in principle, more accurate than either the first guess or the observation. This led to the development of an "OI" quality control: each observation is compared with a simple OI value that would be obtained at the observation location using the first guess (background) field and nearby observations, but without including the observation being checked (Lorenc, 1981, Woollen, 1991). When the residual (difference between the observed and analyzed value) is larger than a certain number of analysis error standard deviations, it is tossed out. The analysis is iterated so some observations may be salvaged after first being tossed out.

Gandin (1988) introduced the idea of *complex quality control* (where the word "complex" means that it uses several tests simultaneously rather than in series). The basic idea is to estimate several independent residuals and then apply a decision making algorithm based on the information provided by all the independent residuals, rather than performing decisions either in a sequential order, or as a single OI quality control check using all the information at once. Collins and Gandin (1990) and Collins (1998) applied this approach to rawinsonde heights and temperatures with great success. The power of this method lies in the fact that several independent checks can support each other and reduce the level of uncertainty. Furthermore, if the residuals are large and agree reasonably well with each other, they provide the basis for a correction of the observation.

The independent residuals obtained from the different checks used at NCEP for the complex quality control of rawinsondes temperature and heights are: (1) Incremental check (the residual is the increment between the reported observation and the 6-h forecast) (2) Horizontal check based on a simple OI horizontal analysis of the increments, using one observation per quadrant, and only observations within 1000 km of the reported observation. The horizontal residual is the difference between the increment at the observation location and the horizontally interpolated value. (3) Vertical check: the vertical residual is the difference between the observed increment and the increment interpolated vertically from the nearest data points for the same station, one above and one below. (4) Hydrostatic check, the most powerful of the checks since it takes advantage of the redundancy between temperature and height information reported in rawinsonde observations. The hydrostatic residual is the difference between the values of the thickness of a layer between mandatory level heights, calculated using the reported heights, and the thickness calculated independently from the reported virtual temperatures. (5) Baseline check computed by making a hydrostatic computation downward, from the first mandatory level above the surface with complete heights and temperatures, to the reported surface pressure. The baseline residual is the difference between the station elevation, given by the report, and the hydrostatically determined height at the surface pressure. Another possible check used at NCEP is based on a temporal interpolation of observations at the same station 12 or 24 h before and after the observation time. This check is particularly useful for isolated stations within a "reanalysis" mode. It is also possible to perform a check of the stability of the lapse rate.

Collins and Gandin (1990) and Collins (1998) developed a sophisticated decision making algorithm that makes generally confident decisions correcting computation or communication errors of human origin. They assumed that human errors have a simple structure: a single digit or a sign is wrong or missing. The following example of such error detected at a single level in the heights is typical:

Example of CQC and the decision making algorithm correction of gross errors (Collins and Gandin, 1998)

Reported 1000 hPa height: 8 m

Computed residuals:

Incremental residual: −72 m;

Vertical residual: −66 m;

Hydrostatic residual (using significant levels): 60 m;

Hydrostatic (using only mandatory levels): 65 m;

Baseline residual: −58 m.

With this information, the decision making algorithm concluded that a simple correction changing one digit (adding 60 m to make the observation 68 m instead of the reported 8 m) could be confidently made.

The NCEP global system has also an OI-based complex quality control for all other data (winds, moisture, satellite retrievals) that makes the first three checks discussed above and rejects observations with large residuals but does not attempt to make corrections, since the strong redundancy of the hydrostatic check is not available for variables other than rawinsonde temperatures and heights (Woollen, 1991).

The effect of modern quality control systems is difficult to gauge, but Kistler *et al.* (2001) showed an impressive example of the positive impact from the modern approach compared with the quality control that was operational at NCEP in the 1970s. They pointed out that in 1974 NMC (now NCEP) introduced a modern observation formatting system (known as Office Note 29, ON29), which later became the basis of the official World Meteorological Organization (WMO) Binary Universal Format Representation system for the encoding of observations. ON29 included more information about the observation than previously used encoders. This change in formatting required a complete overhaul of the NMC decoding system, and errors must have been introduced during this complex reprogramming process. The NMC operational forecast skill actually went down and it took a few years before it recovered to the pre-1974 error levels (Kalnay *et al.*, 1998). During the production of the NCEP/NCAR reanalysis (Kalnay *et al.*, 1996), both the complex quality control for rawinsonde heights and temperatures and the OI quality control for other observations were used to screen observations. The complex quality control found and corrected an unusually large number of rawinsondes errors starting in 1974, and, presumably as a result of this correction and the OI quality control screening of the other information, the benefit of using the more advanced formatting system ON29 became realized. In the reanalysis, the forecast skill increased substantially in 1974, rather than deteriorating as in the operational forecasts.

Another approach that has also become popular is variational quality control performed within 3D-Var or 4D-Var, rather than before the analysis, like OI and complex quality controls (Purser, 1984, Lorenc and Hammon, 1988, Ingleby and Lorenc, 1993, Andersson and Jarvinen, 1999, Collins, 2001a,b). It has the advantage that it is performed as part of the analysis itself, rather than as a preprocessing step like OI quality control, but because it computes a single (iterative) residual for each observation, it is not able to correct observations like complex quality control.

The variational quality control approach is based on modifying the observational component J_o of the variational cost function $J = J_b + J_o$ to take into account the possibility of gross errors. Note that in the variational analysis approach, the gradient $\nabla_{y_o} J_o$ of the cost function with respect to an observation y_o determines how quickly the analysis estimate $\mathbf{x}$ will shift towards that observation (Fig. 5.8.1(b)).

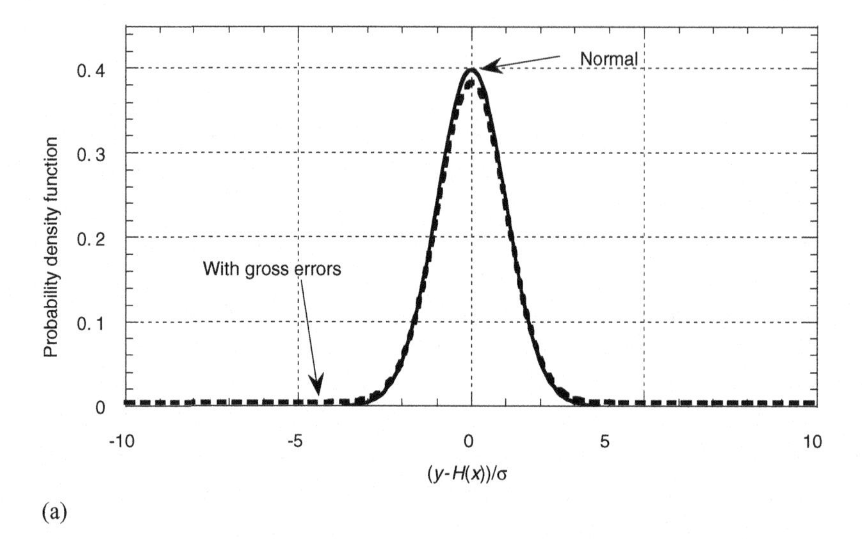

(a)

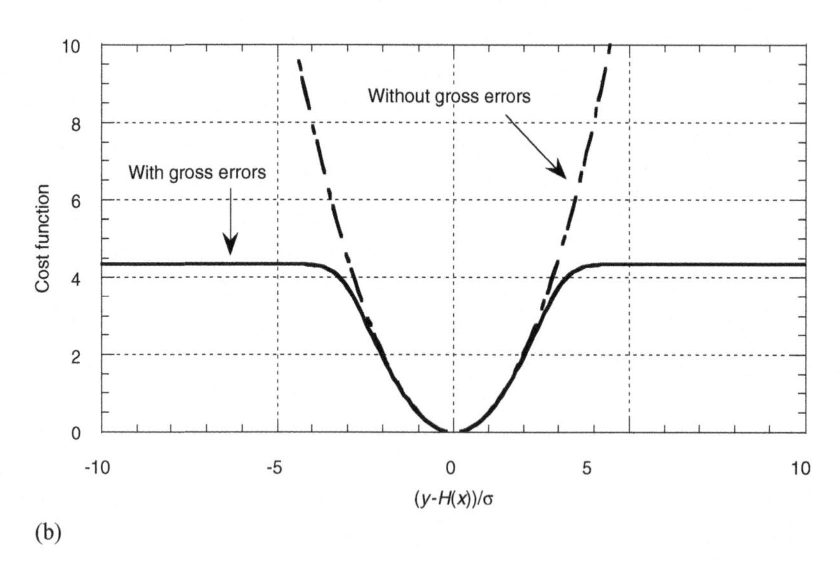

(b)

Figure 5.8.1: (a) Probability density function for an observation with a given scaled residual, without gross errors (normal distributions) and with gross errors. $A = 0.05$, $d = 5$. (b) Observational cost function derived from the logarithm of the error distributions in (a). (c) Weight factor applied to the gradient of the cost function for different values of the *a priori* probability of gross errors A and the width of the flat distribution.

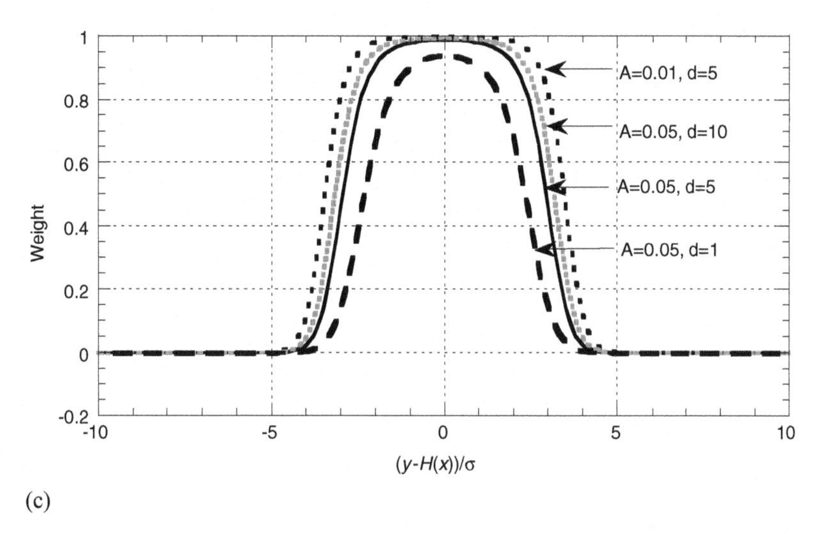

(c)

Figure 5.8.1: *(cont.)*

Consider the cost function term for a single uncorrelated observation y_o without allowing for gross errors:

$$J_o^N = \frac{1}{2}\left[\frac{y_o - H(x)}{\sigma_o}\right]^2 \text{ with } \nabla J_o^N = \left[\frac{y_o - H(x)}{\sigma_o^2}\right] \tag{5.8.1}$$

where x is a close approximation of the true value (analysis). In variational quality control one assumes that there is an *a priori* probability A of having a gross error, estimated from past statistics. $1 - A$ is then the *a priori* probability of not having a gross error, in which case the observation is assumed to have random errors with a Gaussian distribution. Without gross errors ($A = 0$), the probability that an observation y_o, if $H(x)$ is the true value, is given by the normal distribution is

$$p(y_o) = N = \frac{1}{\sqrt{2\pi}\sigma_o} e^{-J_o^N} \tag{5.8.2}$$

where (as discussed in Section 5.3) $J_o^N = -\ln N + const.$ (Fig. 5.8.1(a), full line) and the constant is chosen arbitrarily to make $J^N(0) = 0$.

If there are gross errors with an *a priori* probability A (let's say, 0.05), then the probability of an observation y_o (Fig. 5.8.1(a), bold dashes) is modified:

$$p^{QC}(y_o) = (1 - A)N + AF \tag{5.8.3}$$

where F is a flat (uniform) distribution for the gross errors:

$$F = \begin{cases} \dfrac{1}{2d\sigma_o} & \text{if } |y_o - H(x)| < d\sigma_o \\ 0 & \text{otherwise} \end{cases} \tag{5.8.4}$$

Here d is the maximum number of standard deviations allowed for gross errors (e.g., $d = 5$). If $|y_o - H(x)| > d\sigma_o$, it is assumed that the observation was so obviously wrong that it was eliminated in a preliminary check against climatology or the background. The integral of F is therefore equal to 1.

We can then modify the contribution to the cost function made by the observation by including the probability of gross errors: $J_o^{QC} = -\ln p^{QC}(y_o) + const.$ (Fig. 5.8.1(a), full line). Since the flat probability distribution of gross errors does not depend on the value y_o, the modified gradient of the cost function is:

$$\begin{aligned}\nabla_{y_o} J_o^{QC} &= \nabla[-\ln p^{QC}(y_o)] = \frac{1-A}{p^{QC}(y_o)} N \nabla J_o^N \\ &= \frac{(1-A)N}{(1-A)N + AF} \nabla J_o^N = (1-P)\nabla J_o^N \end{aligned} \tag{5.8.5}$$

where

$$P = \frac{AF}{(1-A)N + AF} \tag{5.8.6}$$

is the *a posteriori* probability of having a gross error (after making the observation y_o). In other words, the gradient of the cost function including variational quality control is the gradient without variational quality control multiplied by a weight

$$W^{QC} = \frac{(1-A)N}{(1-A)N + AF} \tag{5.8.7}$$

(the probability of NOT having a gross error), which is close to 1 for $|y_o - H(x)|/\sigma_o < d$ and goes to zero for $|y_o - H(x)|/\sigma_o \geq d$. Figure 5.8.1(c) shows that the weights are not very sensitive to the choice of parameters. A small value for the *a priori* probability of gross errors A will result in a steeper reduction of weights. Because of the shape of the normal probability function, the weights are not too sensitive to d unless the value chosen is rather small (e.g., $d = 1$ in the graph).

Because $H(x)$ is assumed to be close to the truth, at ECMWF variational quality control is not turned on (i.e., the weight multiplying the gradient during the minimization is $W = 1$) during the first 40 iterations of the 4D-Var algorithm, while the solution starts to converge towards the analysis. It is then turned on ($W = W^{QC}$) for the last 30 iterations of the 4D-Var, thus giving less weight to observations that are likely to contain gross errors (Andersson and Jarvinen, 1999).

6

Atmospheric predictability and ensemble forecasting

6.1 Introduction to atmospheric predictability

In his 1951 paper on NWP, Charney indicated that he expected that even as models improved there would still be a limited range to skillful atmospheric predictions, but he attributed this to inevitable model deficiencies and finite errors in the initial conditions. Lorenz (1963a,b) discovered the fact that the atmosphere, like any dynamical system with instabilities, has a *finite limit of predictability* (which he estimated to be about two weeks) *even if the model is perfect, and even if the initial conditions are known almost perfectly*. He did so by performing what is now denoted an "identical twin" experiment: he compared two runs made with the same model but with initial conditions that differed only very slightly. Just from round-off errors, he found that after a few weeks the two solutions were as different from each other as two random trajectories of the model.

Lorenz (1993) described how this fundamental discovery took place: His original goal had been to show that statistical prediction could not match the accuracy attainable with a nonlinear dynamical model, and therefore that NWP had a potential for predictive skill beyond that attainable purely through statistical methods. He had acquired a Royal-McBee LGP-30 computer, with a memory of 4K words and a speed of 60 multiplications per second, which for the late 1950s was very powerful. He developed and programmed in machine language a "low-order" atmospheric model (i.e., a model whose evolution was described by only 12 variables) driven by external heating and damped by dissipation. During 1959 he changed parameters in the model for several months trying to find a nonperiodic solution (since a periodic solution would be perfectly predictable from past statistics, and that would have defeated his purpose). He submitted a preliminary title, "The statistical prediction of solutions of dynamical equations", to the NWP conference that was going to take place during 1960 in Tokyo, gambling that he would indeed be able to find, for the first time in history, a nonperiodic numerical solution. After making the external heating a function of both latitude and longitude, he finally found the nonperiodic behavior that he was seeking. He rounded off and printed the evolution of the variables with three significant digits, which seemed sufficient to define the state of the model with plenty of accuracy. After running the model for several simulated years and satisfying himself that the solution had no periodicities, he decided to repeat part of an integration in more detail. When he came back from a coffee break, Lorenz found that the new solution was completely different from the original run. Before calling service for the computer, he checked the results and found that at the beginning the new run did coincide with the original printed numbers, but that after a few days the last digit became different, and then the next one, and after about two months any resemblance with the original integration disappeared (Lorenz, 1993):

> The initial round-off errors were the culprits; they were steadily amplifying until they dominated the solution. In today's terminology, there was chaos.... It soon struck me that, if the real atmosphere behaved like the simple model, long-range forecasting would be impossible.... In due time I convinced myself that the amplification of small differences was *the cause* of the lack of periodicity. Later, when I presented my results at the Tokyo meeting, I added a brief description of the unexpected response of the equations to the round-off errors.

Lorenz (1963a,b) thus discovered the fundamental theorem of predictability: Unstable systems have a finite limit of predictability, and conversely, stable systems are infinitely predictable (since they are either stationary or periodic), as suggested by the schematic Fig. 6.1.1.

In his 1972 talk "Predictability: does the flap of a butterfly's wings in Brazil set off a tornado in Texas?" Lorenz further reviewed basic ideas on atmospheric predictability (Lorenz, 1993):

> ...I am proposing that over the years minuscule disturbances neither increase nor decrease the frequency of occurrence of various weather events such as tornadoes; the most that they can do is to modify the sequence in which these events occur. The question which really interests us is whether they can even do this – whether, for example, two particular weather situations differing by as little as the immediate influence of a single butterfly will generally after sufficient time evolve into two situations differing by as much as the presence of a tornado. In more technical language, is the behavior of the atmosphere *unstable* with respect to perturbations of small amplitude?[1]

1 Gleick (1987) pointed out that the concept of a "butterfly effect" existed in some form from literary sources long before Lorenz's work. There is a short story by Ray Bradbury which deals with this nearly as well as Lorenz does, right down to the butterfly. In this story, *A Sound of*

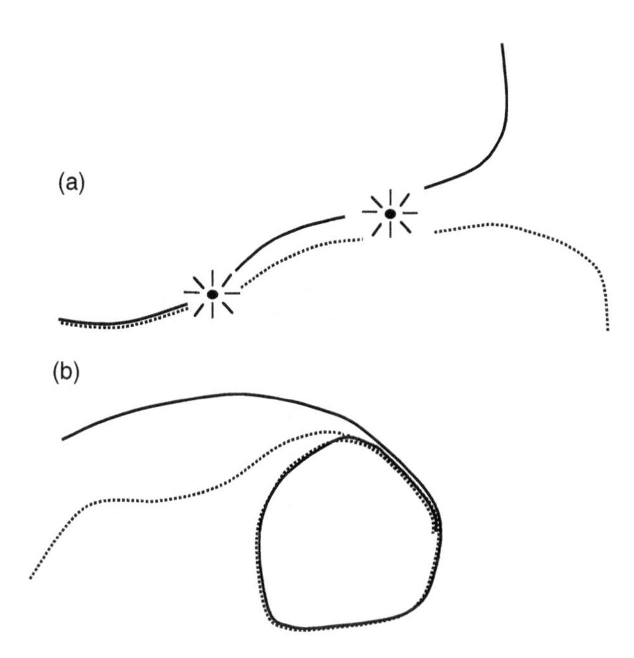

Figure 6.1.1: Schematic illustrating trajectories of: (a) a dynamical system with instabilities, which no matter how close they initially are, inevitably drift apart, and (b) a stable system with stationary or periodic orbits: after a possible transient stage, the trajectories stay close to each other, i.e., they become infinitely predictable.

The connection between this question and our ability to predict the weather is evident. Since we do not know how many butterflies there are, nor whether they are all located, let alone which ones are flapping their wings at any instant, we cannot, if the answer to our question is affirmative, accurately predict the occurrence of tornadoes at a sufficient distant future time. More significantly, our general ability to detect systems as large as thunderstorms when they slip between weather stations may impair our ability to predict the general weather pattern even in the near future....

...The evidence [that the answer to the question whether the atmosphere is unstable is affirmative] is overwhelming. The most significant results are the following:

1. Small errors in the coarser structure of the weather pattern – those features which are readily resolved by conventional observing networks – tend to double in about three days.[2] As the errors become larger the growth rate subsides. This limitation alone would allow us to extend the range of acceptable prediction by three days every time we cut the observation error in

Thunder, some time-travelers go back to the prehistoric era and are very careful not to touch anything lest they alter the future to which they wish to return. When they return to the present, they find everything altered (for the worse it seems to them). It turns out one of them had accidentally stepped on a butterfly. ' "Not a little thing like that! Not a butterfly!" cried Eckels. It fell to the floor, an exquisite thing, a small thing that could upset balances and knock down a line of small dominoes and then big dominoes and then gigantic dominoes, all down the years across Time. Eckels' mind whirled. It couldn't change things. Killing one butterfly couldn't be that important! could it?'. This story was first published in 1952. (Courtesy of Bill Martin, pers. comm. 1998.)

2 Current estimates are that small errors in synoptic (coarser) scales double even faster, in about 2 days (e.g., Simmons *et al.*, 1995, Dalcher and Kalnay, 1987, Toth and Kalnay, 1993).

half, and would offer the hope of eventually making good forecasts several weeks in advance.

2. Small errors in the finer structure – e.g. the positions of individual clouds – tend to grow much more rapidly, doubling in hours or less. This limitation alone would not seriously reduce our hopes for extended-range forecasting, since ordinarily we do not forecast the finer structure at all.
3. Errors in the finer structure, having attained appreciable size, tend to produce errors in the coarser structure. This result, which is less firmly established than the previous ones, implies that after a day or so there will be appreciable errors in the coarser structure, which will thereafter grow just as if they had been present initially. Cutting the observations in the finer structure in half – a formidable task – would extend the range of acceptable prediction of even the coarser structure only by hours or less. The hopes for predicting two weeks or more in advance are thus greatly diminished.
4. Certain special quantities such as weekly average temperatures and weekly total rainfall may be predictable at a range at which entire weather patterns are not.

Since the early days of Lorenz's momentous discovery, which gave impetus to the new science of chaos,[3] additional progress has been made, but his findings have not been changed in any fundamental way. In NWP, substantial progress has been made through the realization that the chaotic behavior of the atmosphere requires the replacement of single "deterministic" forecasts by "ensembles" of forecasts with differences in the initial conditions and in the model characteristics that realistically reflect the uncertainties in our knowledge of the atmosphere. This realization led to the introduction of operational ensemble forecasting at both NCEP and ECMWF in December 1992. It also led to work on extending the usefulness of NWP forecasts through a systematic exploitation of the chaotic nature of the atmosphere.

6.2 Brief review of fundamental concepts about chaotic systems

Lorenz (1963a) introduced a three-variable model that is a prototypical example of chaos theory. These equations were derived as a simplification of Saltzman's (1962) nonperiodic model for convection. Like Lorenz's (1962) original 12-variable model, the three-variable model is a **dissipative** system. This is in contrast to **Hamiltonian**

3 It should be noted that Poincaré (1897, see Alligood *et al.*, 1997) had already discovered that the planetary system is chaotic, i.e., that the orbits of the planets cannot be predicted well beyond a certain number of (millions of) years. He showed this for the simplest three-body problem of two stars with circular orbits moving on a plane around their center of mass, and a third "asteroid" with negligible mass in comparison with the first two, moving in the same plane. He found that the motion of the third body was *sensitively dependent on the initial conditions*, the hallmark of chaos (Alligood *et al.*, 1997).

systems, which **conserve total energy** or some other similar property of the flow. The system is **nonlinear** (it contains products of the dependent variables) but **autonomous** (the coefficients are time-independent). Sparrow (1982) wrote a whole book on the Lorenz three-variable model that provides a nice introduction to the subject of chaos, bifurcations and strange attractors. Lorenz (1993) is a superbly clear introduction to chaos with a very useful glossary of the nomenclature used in today's literature. Alligood *et al.* (1997) is also a very clear introduction to dynamical systems and chaos. In this section we use bold type to introduce some of the words used in the dynamical system vocabulary.

The Lorenz (1963a) equations are

$$\left.\begin{aligned}\frac{dx}{dt} &= \sigma(y-x)\\ \frac{dy}{dt} &= rx - y - xz\\ \frac{dz}{dt} &= xy - bz\end{aligned}\right\} \tag{6.2.1}$$

The solution obtained by integrating the differential equations in time is called a **flow**. The **parameters** σ, b, r are kept constant within an integration, but they can be changed to create a **family of solutions** of the **dynamical system** defined by the differential equations. The particular parameter values chosen by Lorenz (1963a), $\sigma = 10,\ b = 8/3,\ r = 28$, result in **chaotic** solutions (sensitively dependent on the initial conditions), and since this publication they have been widely used in many papers. The solution of a time integration from a given **initial condition** defines a **trajectory** or **orbit** in **phase space**. The coordinates of a **point in phase space** are defined by the simultaneous values of the independent variables of the model, $x(t), y(t), z(t)$. The **dimension of the phase space** is equal to the number of independent variables (in this case three). The dimension of the subspace actually visited by the solution after an initial transient period (i.e., the **dimension of the attractor**) can be much smaller than the dimension of the phase space. A **volume** in phase space can be defined by a set of points in phase space such as a hypercube $V = \delta x \delta y \delta z$, a hypersphere $V = \{\delta\mathbf{r};\ |\delta\mathbf{r}| \le \varepsilon\}$, etc.

The fact that the Lorenz system (6.2.1) is dissipative can be seen from the **divergence** of the **flow:**

$$\frac{\partial \dot{x}}{\partial x} + \frac{\partial \dot{y}}{\partial y} + \frac{\partial \dot{z}}{\partial z} = -(\sigma + b + 1) \tag{6.2.2}$$

which shows that an original volume V contracts with time to $Ve^{-(\sigma+b+1)t}$. This proves the existence of a **bounded globally attracting set of zero volume** (i.e., an attractor of **dimension** smaller than n, the dimension of the phase space). A solution may start from a point away from the attracting set but it will eventually settle on the **attractor**. This initial portion of the trajectory is known as a **transient**. The **attracting set** (the set of points approached again and again by the trajectories after the transients are over) is called the **attractor** of the system. The attractor can have several components: **stationary** points (equilibrium or steady state solutions of the dynamical equations), **periodic orbits**, and more complicated structures known as **strange attractors** (which can also include periodic orbits). The different components of the attractor have corresponding **basins of attraction** in the phase space, which are all the initial conditions that will evolve to the same attractor. The fact that any initial volume in phase space contracts to zero with time is a general property of dissipative bounded systems, including atmospheric models with friction. Hamiltonian systems, on the other hand, are **volume-conserving**.

If we change the **parameters** of a dynamical system (in this example σ, b, r) and obtain families of solutions, we find that there is a point at which the behavior of the flow changes abruptly. The point at which this sudden change in the characteristics of the flow occurs is called a **bifurcation point**. For example in Lorenz's equations the origin is a stable, stationary point for $r < 1$, as can be seen by investigating the local stability at the origin. The **local stability** of a point can be studied by linearizing the flow about the point and computing the eigenvalues of the linear flow. For $r < 1$ the stationary point is stable: all three eigenvalues are negative. This means that all orbits near the origin tend to get closer to it. At $r = 1$ there is a bifurcation, and for $r > 1$ two new additional stationary points $C_\pm$ are born, with coordinates $(x, y, z)_\pm = (\pm\sqrt{b(r-1)}, \pm\sqrt{b(r-1)}, r-1)$. For $r > 1$ the origin becomes **nonstable:** one of the three eigenvalues becomes positive (while the other two remain negative), indicating that the flow diverges locally from the origin in one direction. For $1 < r < 24.74\ldots$, C_+ and C_- are stable, and at $r = 24.74\ldots$ there is another bifurcation so that above that critical value C_+ and C_- also become unstable. As discussed by Lorenz (1993), a ubiquitous phenomenon is the occurrence of bifurcations of periodic motion leading to **period doubling**, and **sequences of period doubling bifurcations** leading to chaotic behavior (see Sauer *et al.*, 1991).

A solution of a dynamical system can be defined to be **stable** if it is bounded, and if any other solution once sufficiently close to it remains close to it for all times. This indicates that a bounded stable solution must be **periodic** (repeat itself exactly) or at least **almost periodic**, since once the trajectory approaches a point in its past history, the trajectories will remain close forever (Fig. 6.1.1(b)). A solution that is not periodic or almost periodic is therefore **unstable**: two trajectories that start very close will eventually diverge completely (Fig. 6.1.1(a)).

The long-term stability of a dynamical system of n-variables is characterized by the **Lyapunov exponents**. Consider a point in a trajectory, and introduce a (hyper)sphere of small perturbations about that point. If we apply the model to evolve each of those perturbations, we find that after a short time the sphere will be deformed into a (hyper)ellipsoid. In an unstable system, at least one of the axes of the ellipsoid will become larger with time, and once nonlinear effects start to be significant the ellipsoid will be deformed into a "banana" (Fig. 6.2.1). Consider the linear phase,

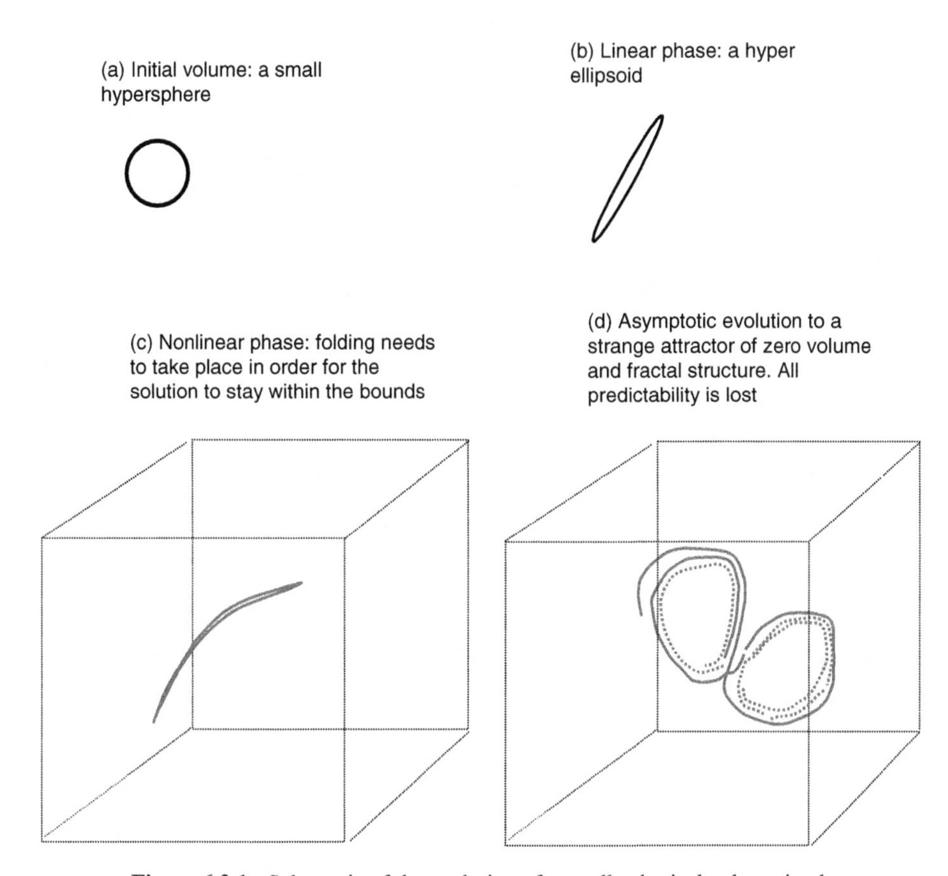

(a) Initial volume: a small hypersphere
(b) Linear phase: a hyper ellipsoid
(c) Nonlinear phase: folding needs to take place in order for the solution to stay within the bounds
(d) Asymptotic evolution to a strange attractor of zero volume and fractal structure. All predictability is lost

Figure 6.2.1: Schematic of the evolution of a small spherical volume in phase space in a bounded dissipative system. Initially (during the linear phase) the volume is stretched into an ellipsoid while the volume decreases. The solution space is bounded, and a bound is schematically indicated in the figure by the hypercube. The ellipsoid continues to be stretched in the unstable directions, until (because the solution phase space is bounded) it has to fold through nonlinear effects. This stretching and folding continues again and again, evolving into an infinitely foliated (fractal) structure. This structure, of zero volume and fractal dimension, is called a "strange attractor." The attractor is the set of states whose vicinity the system will visit again and again (the "climate" of the system). Note that in phases (a), (b), and (c), there is predictive knowledge: we know where the original perturbations generally are. In (d), when the original sphere has evolved into the attractor, all predictability is lost: we only know that each original perturbation is within the climatology of possible solutions, but we don't know where, or even in which region of the attractor it may be.

during which the sphere evolves into an ellipsoid. We can maintain the linear phase for an infinitely long period by taking an infinitely small initial sphere, or, alternatively, by periodically scaling down the ellipsoid dimensions dividing all its dimensions by the same scalar. Each axis j of the ellipsoid grows or decays over the long term by amounts given by $e^{\lambda_j t}$, where the λ_js are the Lyapunov exponents ordered by size $\lambda_1 \geq \lambda_2 \geq \ldots \geq \lambda_n$. The total volume of the ellipsoid will evolve like $V_0 e^{-(\lambda_1+\lambda_2+\cdots\lambda_n)t}$. Therefore, a Hamiltonian (volume-conserving) system is characterized by a sum of Lyapunov exponents equal to zero, whereas for a dissipative system the sum is negative.

Because the attractor of a dissipative system is bounded (the trajectories are enclosed within some hyperbox), if the first Lyapunov exponent is greater than zero, at least one of the axes of the ellipsoid keeps getting longer with time. The ellipsoid will eventually be distorted into a banana shape: it has to be folded in order to continue fitting into the box. The banana will be further stretched along the unstable axis and then necessarily folded again and again onto itself in order to continue fitting into the box. Since the volume of the ellipsoid eventually goes to zero for a dissipative system, the repeated stretching and folding of the ellipsoid of a chaotic system eventually converges to a zero-volume attractor with an infinitely foliated structure (a process similar to the stretching and folding used to make "phyllo" dough!). This structure is known as "**strange attractor**" (Ruelle, 1989). It has a **fractal** structure: a dimension which in general is not an integer and is smaller than the original space dimension n, estimated by Kaplan and Yorke (1979) to be

$$d = k + (\lambda_1 + \cdots + \lambda_k)/|\lambda_{k+1}| \tag{6.2.3}$$

where the sum of the first k Lyapunov exponents is positive, and the sum of the first $k + 1$ exponents is negative. If the system is Hamiltonian, its invariant manifold has the same dimension as the phase space.

In summary, a **stable** system has all Lyapunov exponents less than or equal to zero. A **chaotic** system has *at least one Lyapunov exponent greater than zero:* if at least $\lambda_1 > 0$ chaotic behavior will take place because at least one axis of the ellipsoid will be continuously stretched, leading to the separation of orbits originally started closely along that axis. Note that a **chaotic bounded flow** must also have *a Lyapunov exponent equal to zero*, with the corresponding local Lyapunov vector parallel to an orbit. This can be understood by considering two initial conditions such that the second is equal to the first after applying the model for one time step. The solutions corresponding to these initial conditions will remain close together, since the second orbit will always be the same as the first orbit shifted by one time step, and on the average, the distance between the solutions will remain constant. If we add a tiny perturbation, though, the second solution will diverge from the first one because there is a positive Lyapunov exponent.

6.3 Tangent linear model, adjoint model, singular vectors, and Lyapunov vectors

In 1965 Lorenz published another paper based on a low-order model that behaved like the atmosphere. It was a quasi-geostrophic two-level model in a periodic channel, with a "Lorenz" vertical grid (velocity and temperature variables defined at the same two levels, see Section 3.3), and a spectral (Fourier) discretization in longitude and latitude. By keeping only two Fourier components in latitude and three in longitude, and choosing appropriate values for the model parameters, he was able to find a model able to reproduce baroclinic instability and nonlinear wave interactions with just 28 variables. In this fundamental paper, Lorenz introduced for the first time (without using their current names) the concepts of the tangent linear model, adjoint model, singular vectors, and Lyapunov vectors for the low-order atmospheric model, and their consequences for ensemble forecasting. He also pointed out that the predictability of the model is not constant with time: it depends on the stability of the evolving atmospheric flow (the basic trajectory or reference state). In the following introduction to these subjects we follow Lorenz (1965), Szunyogh *et al.* (1997) and Pu *et al.* (1997b).

6.3.1 Tangent linear model and adjoint model

Consider a nonlinear model. Once it has been *discretized in space* using, for example, finite differences or a spectral expansion leading to n independent variables (or degrees of freedom), the model can be written as a set of n nonlinear coupled ordinary differential equations:

$$\frac{d\mathbf{x}}{dt} = \mathbf{F}(\mathbf{x}) \qquad \mathbf{x} = \begin{bmatrix} x_1 \\ \vdots \\ x_n \end{bmatrix} \qquad \mathbf{F} = \begin{bmatrix} F_1 \\ \vdots \\ F_n \end{bmatrix} \tag{6.3.1}$$

This is the model in differential form. Once we choose a time-difference scheme (e.g., Crank–Nicholson, see Table 3.2.1), it becomes a set of nonlinear-coupled *difference* equations. Typically, an atmospheric model consists of one such system of difference equations which, for example, using a two-time level Crank–Nicholson scheme would be of the form

$$\mathbf{x}^{n+1} = \mathbf{x}^n + \Delta t\mathbf{F}\left(\frac{\mathbf{x}^n + \mathbf{x}^{n+1}}{2}\right) \tag{6.3.2}$$

A numerical solution of (6.3.1) starting from an initial time t_0 can be readily obtained by integrating the model numerically using (6.3.2) between t_0 and a final time t (i.e., "running the model"). This gives us a *nonlinear model solution* that depends only on the initial conditions:

$$\mathbf{x}(t) = M[\mathbf{x}(t_0)] \tag{6.3.3}$$

where M is the time integration of the numerical scheme from the initial condition to time t. A small perturbation $\mathbf{y}(t)$ can be added to the basic model integration $\mathbf{x}(t)$:

$$\begin{aligned} M[\mathbf{x}(t_0) + \mathbf{y}(t_0)] &= M[\mathbf{x}(t_0)] + \frac{\partial M}{\partial \mathbf{x}}\mathbf{y}(t_0) + O[\mathbf{y}(t_0)^2] \\ &= \mathbf{x}(t) + \mathbf{y}(t) + O[\mathbf{y}(t_0)^2] \end{aligned} \tag{6.3.4}$$

At any given time, the linear evolution of the small perturbation $\mathbf{y}(t)$ will be given by

$$\frac{d\mathbf{y}}{dt} = \mathbf{J}\mathbf{y} \tag{6.3.5}$$

where $\mathbf{J} = \partial \mathbf{F}/\partial \mathbf{x}$ is the Jacobian of $\mathbf{F}$.

This system of linear ordinary differential equations is the tangent linear model in differential form. Its solution between t_0 and t can be obtained by integrating (6.3.5) in time using the same time difference scheme used in the nonlinear model (6.3.3):

$$\mathbf{y}(t) = \mathbf{L}(t_0, t)\mathbf{y}(t_0) \tag{6.3.6}$$

Here $\mathbf{L}(t_0, t) = \partial M/\partial \mathbf{x}$ is an $(n \times n)$ matrix known as the *resolvent* or *propagator* of the tangent linear model: it propagates an initial perturbation at time t_0 into the final perturbation at time t. Because it is linearized over the flow from t_0 to t, $\mathbf{L}$ depends on the *basic trajectory* $\mathbf{x}(t)$ (the solution of the nonlinear model), but it does not depend on the perturbation $\mathbf{y}$. (The original nonlinear model is autonomous since $\mathbf{F}(\mathbf{x})$ depends on $x(t)$ but not explicitly on time, but the linear tangent model is nonautonomous). Lorenz (1965) introduced the concept of the tangent linear model of an atmospheric model, but he actually obtained it directly from (6.3.4), neglecting terms quadratic or higher order in the perturbation $\mathbf{y}$:

$$M[\mathbf{x}(t_0)] + \mathbf{L}(t_0, t)\mathbf{y}(t_0) = \mathbf{x}(t) + \mathbf{y}(t) \approx M[\mathbf{x}(t_0) + \mathbf{y}(t_0)] \tag{6.3.7}$$

He did so by creating as initial perturbations a "sphere" of small perturbations of size ε along the n unit basis vectors $\mathbf{y}_i(t_0) = \varepsilon \mathbf{e}_i$ and applying (6.3.7) to each of these perturbations. With this choice of initial perturbations, subtracting (6.3.3) he obtained the matrix that defines the tangent linear model:

$$\mathbf{L}(t_0, t)[\varepsilon\mathbf{e}_1, \ldots, \varepsilon\mathbf{e}_n] = \varepsilon\mathbf{L}(t_0, t) = [\mathbf{y}_1(t), \ldots, \mathbf{y}_n(t)] \tag{6.3.8}$$

The Euclidean *norm* of a vector is the inner product of the vector with itself:

$$\|\mathbf{y}\|^2 = \mathbf{y}^T\mathbf{y} = \langle \mathbf{y}, \mathbf{y} \rangle \tag{6.3.9}$$

The Euclidean norm of $\mathbf{y}(t)$ is therefore related to the initial perturbation by

$$\|\mathbf{y}(t)\|^2 = (\mathbf{L}\mathbf{y}(t_0))^T\mathbf{L}\mathbf{y}(t_0) = \langle \mathbf{L}\mathbf{y}(t_0), \mathbf{L}\mathbf{y}(t_0) \rangle = \langle \mathbf{L}^T\mathbf{L}\mathbf{y}(t_0), \mathbf{y}(t_0) \rangle \tag{6.3.10}$$

The *adjoint* of an operator $\mathbf{K}$ is defined by the property $\langle \mathbf{x}, \mathbf{K}\mathbf{y} \rangle \equiv \langle \mathbf{K}^T\mathbf{x}, \mathbf{y} \rangle$. In this case of a model with real variables, the *adjoint* of the tangent linear model $\mathbf{L}(t_0, t)$ is simply the *transpose* of the tangent linear model.

Now assume that we separate the interval (t_0, t) into two successive time intervals. For example, if $t_0 < t_1 < t$,

$$\mathbf{L}(t_0, t) = \mathbf{L}(t_1, t)\mathbf{L}(t_0, t_1) \tag{6.3.11}$$

Since the **adjoint** of the tangent linear model is the transpose of the TLM, the property of the transpose of a product is also valid:

$$\mathbf{L}^T(t_0, t) = \mathbf{L}^T(t_0, t_1)\mathbf{L}^T(t_1, t) \tag{6.3.12}$$

Equation (6.3.11) shows that the tangent linear model can be cast as a product of the tangent linear model matrices corresponding to short integrations, or even single time steps. Equation (6.3.12) shows that the adjoint of the model can also be separated into single time steps, but they are executed backwards in time, starting from the last time step at t, and ending with the first time step at t_0. For low-order models the tangent linear model and its adjoint can be constructed by repeated integrations of the nonlinear model for small perturbations, as done by Lorenz (1965), equation (3.7), and by Molteni and Palmer (1993) with a global quasi-geostrophic model.

For large NWP models this approach is too time consuming, and instead it is customary to develop the linear tangent and adjoint codes from the nonlinear model code following some rules discussed in Appendix B. An example of a FORTRAN code for a nonlinear model, and the corresponding tangent linear model and adjoint models are also given in Appendix B.

6.3.2 Singular vectors

Recall that for a given basic trajectory and an interval (t_0, t_1) the tangent linear model is a matrix that when applied to a small initial perturbation $\mathbf{y}(t_0)$ produces the final perturbation $\mathbf{y}(t_1)$:

$$\mathbf{y}(t_1) = \mathbf{L}(t_0, t_1)\mathbf{y}(t_0) \tag{6.3.13}$$

Singular value decomposition theory (e.g., Golub and Van Loan, 1996) indicates that for any matrix $\mathbf{L}$ there exist two orthogonal matrices $\mathbf{U}, \mathbf{V}$ such that

$$\mathbf{U}^\mathbf{T}\mathbf{L}\mathbf{V} = \mathbf{S} \tag{6.3.14}$$

where

$$\mathbf{S} = \begin{bmatrix} \sigma_1 & 0 \cdots & 0 \\ 0 & \sigma_2 \cdots & 0 \\ \vdots & \vdots & \vdots \\ 0 & 0 \cdots & \sigma_n \end{bmatrix}$$

and

$$\mathbf{U}\mathbf{U}^T = \mathbf{I} \qquad \mathbf{V}\mathbf{V}^T = \mathbf{I} \tag{6.3.15}$$

$\mathbf{S}$ is a diagonal matrix whose elements are the *singular values* of $\mathbf{L}$.

If we left multiply (6.3.14) by $\mathbf{U}$, we obtain

$$\mathbf{L}\mathbf{V} = \mathbf{U}\mathbf{S} \quad \text{i.e., } \mathbf{L}(\mathbf{v}_1, \ldots, \mathbf{v}_n) = (\sigma_1\mathbf{u}_1, \ldots, \sigma_n\mathbf{u}_n) \tag{6.3.16}$$

where $\mathbf{v}_i$ are the columns of $\mathbf{V}$ and $\mathbf{u}_i$ the columns of $\mathbf{U}$. This implies that

$$\mathbf{L}\mathbf{v}_i = \sigma_i \mathbf{u}_i \tag{6.3.17}$$

Equation (6.3.17) defines the $\mathbf{v}_i$s as the *right singular vectors of* $\mathbf{L}$, hereafter referred to as *initial singular vectors*, since they are indeed valid at the beginning of the optimization interval over which $\mathbf{L}$ is defined.

We now right multiply (6.3.14) by $\mathbf{V}^T$ and obtain:

$$\mathbf{U}^T\mathbf{L} = \mathbf{S}\mathbf{V}^T \tag{6.3.18}$$

Transposing (6.3.18), we obtain

$$\mathbf{L}^T\mathbf{U} = \mathbf{V}\mathbf{S} \quad \text{i.e.,} \quad \mathbf{L}^T(\mathbf{u}_1, \ldots, \mathbf{u}_n) = (\sigma_1\mathbf{v}_1, \ldots, \sigma_n\mathbf{v}_n) \tag{6.3.19}$$

so that

$$\mathbf{L}^T\mathbf{u}_i = \sigma_i \mathbf{v}_i \tag{6.3.20}$$

The $\mathbf{u}_i$s are the *left singular vectors* of $\mathbf{L}$ and will be referred to as *final* (or *evolved*) *singular vectors*, since they correspond to the end of the interval of optimization.

From (6.3.17) and (6.3.20) we obtain

$$\mathbf{L}^T\mathbf{L}\mathbf{v}_i = \sigma_i\mathbf{L}^T\mathbf{u}_i = \sigma_i^2\mathbf{v}_i \tag{6.3.21}$$

Therefore the initial singular vectors can be obtained as the eigenvectors of $\mathbf{L}^T\mathbf{L}$, a normal matrix whose eigenvalues are the squares of the singular values. Since $\mathbf{U}, \mathbf{V}$ are orthogonal matrices, the vectors $\mathbf{v}_i$ and $\mathbf{u}_i$ that form them constitute orthonormal bases, and any vector can be written in the following form:

$$\mathbf{y}(t_0) = \sum_{i=1}^{n} \langle \mathbf{y}_0, \mathbf{v}_i \rangle \mathbf{v}_i \tag{6.3.22a}$$

$$\mathbf{y}(t_1) = \sum_{i=1}^{n} \langle \mathbf{y}_1, \mathbf{u}_i \rangle \mathbf{u}_i \tag{6.3.22b}$$

where $\langle \mathbf{x}, \mathbf{y} \rangle$ is the inner product of two vectors $\mathbf{x}, \mathbf{y}$. Therefore, using (6.3.22a) and (6.3.17)

$$\mathbf{y}(t_1) = \mathbf{L}(t_0, t_1)\mathbf{y}(t_0) = \sum_{i=1}^{n} \langle \mathbf{y}_0, \mathbf{v}_i \rangle \sigma_i \mathbf{u}_i \tag{6.3.23}$$

If we now take the inner product of (6.3.23) with $\mathbf{u}_i$ we obtain

$$\langle \mathbf{y}(t_1), \mathbf{u}_i \rangle = \sigma_i \langle \mathbf{y}(t_0), \mathbf{v}_i \rangle \tag{6.3.24}$$

This indicates that by applying the tangent linear model $\mathbf{L}$ each initial vector $\mathbf{v}_i$ component will be stretched by an amount equal to the singular value σ_i (or contracted if $\sigma_i < 1$), and the direction will be rotated to that of the evolved vector $\mathbf{u}_i$. Similarly applying the adjoint of the tangent linear model, $\mathbf{L}^T$, each final vector $\mathbf{u}_i$ will be stretched by an amount equal to the singular value σ_i and rotated to the initial vector $\mathbf{v}_i$.

Exercise 6.3.1: Use (6.3.20) and (6.3.22b) to show that $\langle \mathbf{y}(t_0), \mathbf{v}_i \rangle = \sigma_i \langle \mathbf{y}(t_1), \mathbf{u}_i \rangle$.

If we consider all the perturbations $\mathbf{y}(t_0)$ of size 1, from (6.3.24) we obtain that for each of them

$$\sum_{i=1}^{n} \left(\frac{\langle \mathbf{y}(t), \mathbf{u}_i \rangle}{\sigma_i} \right)^2 = \sum_{i=1}^{n} \langle \mathbf{y}(t_0), \mathbf{v}_i \rangle^2 = \|\mathbf{y}(t_0)\|^2 = 1 \tag{6.3.25}$$

so that an initial sphere of radius 1 becomes a hyperellipsoid of semiaxes σ_i. The first initial singular vector $\mathbf{v}_1$ is also called an "optimal vector" since it gives the direction in phase space (i.e., the shape in physical space) of the perturbation that will attain maximum growth σ_1 in the interval (t_0, t_1) (Fig. 6.3.1).

Note that applying $\mathbf{L}$ is the same as running the tangent linear model forward in time, from t_0 to t_1. Applying $\mathbf{L}^T$ is like running the adjoint model backwards, from t_1 to t_0. From (6.3.21) we see that if we apply the adjoint model to a sphere of final perturbations of size 1 (expanded on the basis formed by the evolved or left singular vectors), they also become stretched and rotated into a hyperellipsoid of semiaxes in the directions of the $\mathbf{v}_i$ with length σ_i (Fig. 6.3.2).

Therefore, if we apply $\mathbf{L}^T\mathbf{L}$ (i.e., run the tangent linear model forward in time, and then the adjoint backwards in time, the first initial singular vector will grow by a factor σ_1^2 (see Fig. 6.3.3), and the other initial singular vectors will grow or decay by their corresponding singular value squared σ_i^2. In other words, the (initial) singular vectors $\mathbf{v}_i$ are the eigenvectors of $\mathbf{L}^T\mathbf{L}$ with singular values σ_i^2. Conversely, if we apply the adjoint model first (integrate the adjoint model backwards from the final to the initial time), followed by the tangent linear model (integrate forward to the final time), the final singular vectors $\mathbf{u}_i$ will grow both backward and forward, by a

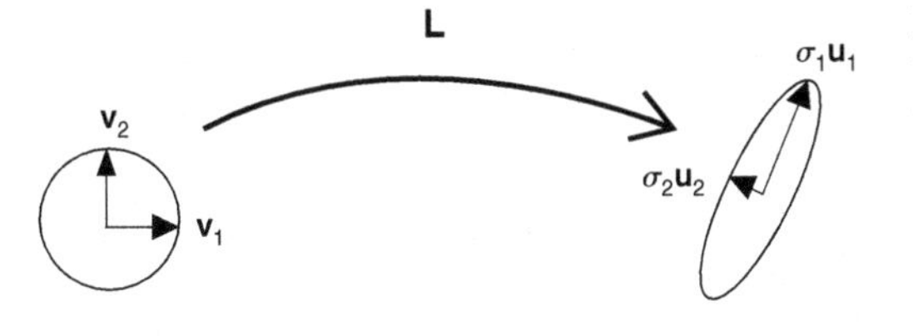

Figure 6.3.1: Schematic of the application of the tangent linear model to a sphere of perturbations of size 1 for a given interval (t_0, t_1).

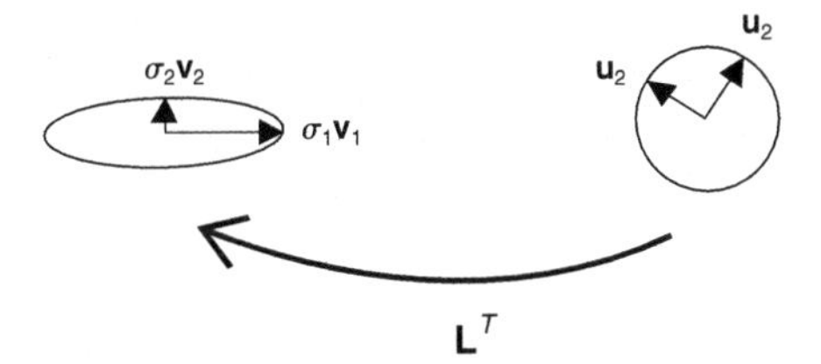

Figure 6.3.2: Schematic of the application of the adjoint of the tangent linear model to a sphere of perturbations of size 1 at the final time.

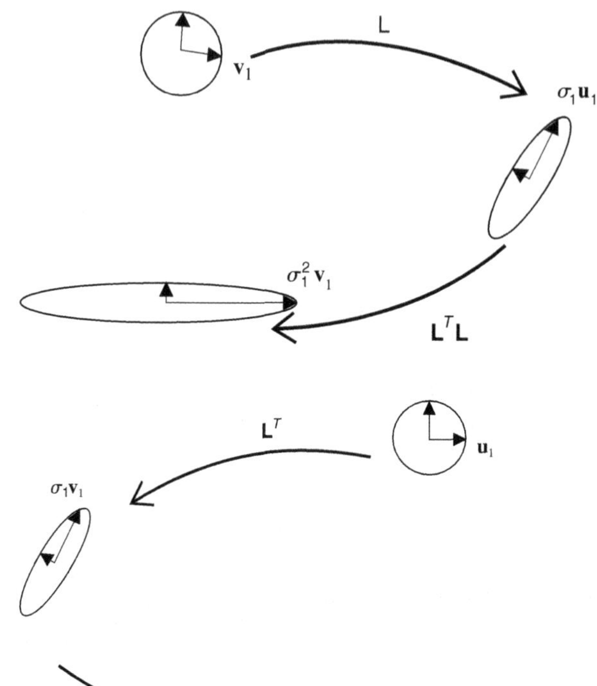

Figure 6.3.3: Schematic of the application of the tangent linear model forward in time followed by the adjoint of the tangent linear model to a sphere of perturbations of size 1 at the initial time.

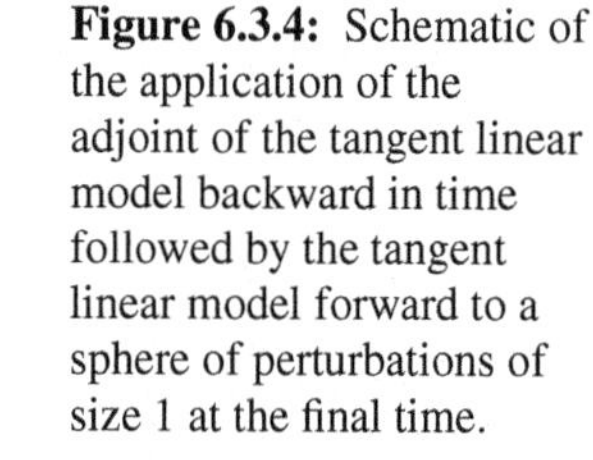

Figure 6.3.4: Schematic of the application of the adjoint of the tangent linear model backward in time followed by the tangent linear model forward to a sphere of perturbations of size 1 at the final time.

total factor also equal to σ_i^2 (Figure 6.3.4). In other words, the final singular vectors are the eigenvectors of $\mathbf{L}\mathbf{L}^T$, and again they have eigenvalues equal to the square of the singular values of $\mathbf{L}$. Alternatively, once the initial singular vectors are obtained using, for example, the Lanczos algorithm, the final singular vectors can be derived by integrating the tangent linear model ((6.3.17)).

If we apply $\mathbf{L}^T\mathbf{L}$ repeatedly over the same interval (t_0, t), we obtain the leading initial singular vector, or first optimal vector. Additional leading singular vectors can be obtained by a generalization of the power method (Lanczos algorithm, Golub and Van Loan, 1996), which requires running the tangent linear model and its adjoint about three times the number of singular vectors required. For example, to get the leading 30 singular vectors optimized for $t_1 - t_0 = 36$ h, the ECMWF performed 100 iterations, equivalent to running the tangent linear model for about 300 days (Molteni *et al.*, 1996).

It is important to note that the adjoint model and the singular vectors are defined with respect to a given norm. So far we have used an Euclidean norm in which the weight matrix that defines the inner product is the identity matrix:

$$\|\mathbf{y}\|^2 = \mathbf{y}^T\mathbf{y} = \langle \mathbf{y}, \mathbf{y} \rangle \tag{6.3.26}$$

The leading (initial) singular vectors are the vectors of equal size (initial norm equal to one $\|\mathbf{y}(t_0)\|^2 = \mathbf{y}(t_0)^T\mathbf{y}(t_0) = \langle \mathbf{y}(t_0), \mathbf{y}(t_0) \rangle = 1$), that grow fastest during the optimization period (t_0, t_1), i.e., the initial vectors that maximize the norm at the final time:

$$J(\mathbf{y}(t_0)) \equiv \|\mathbf{y}(t_1)\|^2 = [\mathbf{L}\mathbf{y}(t_0)]^T\mathbf{L}\mathbf{y}(t_0) = \langle \mathbf{L}^T\mathbf{L}\mathbf{y}(t_0), \mathbf{y}(t_0)\rangle \tag{6.3.27}$$

If we define a norm using any other weight matrix $\mathbf{W}$ applied to $\mathbf{y}$, then the requirement that the initial perturbations be of equal size implies:

$$\|\mathbf{y}(t_0)\|^2 = (\mathbf{W}\mathbf{y}(t_0))^T\mathbf{W}\mathbf{y}(t_0) = \mathbf{y}(t_0)^T\mathbf{W}^T\mathbf{W}\mathbf{y}(t_0) = 1 \tag{6.3.28}$$

We can use a different norm to define the size of the perturbation to be maximized at the final time than the norm $\mathbf{W}$ used for the initial time (6.3.28). For example the final norm could be a projection operator $\mathbf{P}$ at the end of the interval. Then the function that we want to maximize is, instead of (6.3.27):

$$J(\mathbf{y}(t_0)) = [\mathbf{PL}\mathbf{y}(t_0)]^T\mathbf{PL}\mathbf{y}(t_0) = \mathbf{y}(t_0)^T\mathbf{L}^T\mathbf{P}^T\mathbf{PL}\mathbf{y}(t_0) \tag{6.3.29}$$

subject to the *strong constraint* (6.3.28).

From the calculus of variations, the maximum of (6.3.29) subject to the strong constraint (6.3.28) can be obtained by the unconstrained maximum of another function:

$$\begin{aligned} K(\mathbf{y}(t_0)) &= J(\mathbf{y}(t_0)) + \lambda[1 - \mathbf{y}(t_0)^T\mathbf{W}^T\mathbf{W}\mathbf{y}(t_0)] \\ &= \mathbf{y}(t_0)^T\mathbf{L}^T\mathbf{P}^T\mathbf{PL}\mathbf{y}(t_0) + \lambda[1 - \mathbf{y}(t_0)^T\mathbf{W}^T\mathbf{W}\mathbf{y}(t_0)] \end{aligned} \tag{6.3.30}$$

where the λ are the Lagrange multipliers multiplying the square brackets (equal to zero due to the constraint (6.3.28)).

The unconstrained minimization of K is obtained by computing its gradient with respect to the control variable $\mathbf{y}(t_o)$ and making it equal to zero. From Remark 5.4.1(d), we can compute this gradient as:

$$\nabla_{y(t_0)}K = \mathbf{L}^T\mathbf{P}^T\mathbf{PL}\mathbf{y}(t_0) - \lambda\mathbf{W}^T\mathbf{W}\mathbf{y}(t_0) = 0 \tag{6.3.31}$$

It is convenient, given the constraint (6.3.28), to change variables:

$$\mathbf{W}\mathbf{y}(t_0) = \hat{\mathbf{y}}(t_0) \qquad \text{or} \qquad \mathbf{y}(t_0) = \mathbf{W}^{-1}\hat{\mathbf{y}}(t_0) \tag{6.3.32}$$

Then, (6.3.31) becomes

$$(\mathbf{W}^{-1})^T\mathbf{L}^T\mathbf{P}^T\mathbf{PLW}^{-1}\hat{\mathbf{y}}(t_0) = \lambda\hat{\mathbf{y}}(t_0) \tag{6.3.33}$$

subject to the constraint

$$\hat{\mathbf{y}}^T(t_0)\hat{\mathbf{y}}(t_0) = 1 \tag{6.3.34}$$

Therefore, the transformed vectors $\hat{\mathbf{y}}(t_0)$ are the eigenvectors of the matrix $(\mathbf{W}^{-1})^T\mathbf{L}^T\mathbf{P}^T\mathbf{PLW}^{-1}$ in (6.3.33), with eigenvalues equal to the Lagrange multipliers λ_i. After the leading eigenvectors $\hat{\mathbf{y}}(t_0)$ are obtained (using, for example, the Lanczos algorithm), the variables are transformed back to $\mathbf{y}(t_0)$ using (6.3.32). The eigenvalues of this problem are the square of the singular values of the tangent linear model: $\lambda_i = \sigma_i^2$.

This allows great generality (as well as arbitrariness[4]) in the choice of initial norm and final projection operator. Errico and Vukicevic (1992), showed that the singular vectors are very sensitive to both the choice of norm and the length of the optimization interval (the interval from t_0 to t_1). In another example, Palmer *et al.* (1998) tested different weight matrices **W** defining the initial norm. They used "streamfunction," "enstrophy," "kinetic energy," and "total energy" norms, which measured, as the "initial size" the square of the perturbation streamfunction, vorticity, wind speed and weighted temperature, wind and surface pressure, respectively. They found that the use of different initial norms resulted in extremely different initial singular vectors, and concluded that the total energy was the norm of choice for ensemble forecasting. In 1995, ECMWF included in their ensemble system a projection operator **P** that measures only the growth of perturbations north of 30° N, i.e., a matrix that multiplies variables that correspond to latitudes greater than or equal to 30° N by the number 1, and by 0 otherwise) (Buizza and Palmer, 1995). One could use any other pair of initial **W** and final **P** weights (norms) to answer the related question of *forecast sensitivity*. An example of a forecast sensitivity problem is: "What is the optimal (minimum size) initial perturbation (measured by the square of the change in surface pressure over the states of Oklahoma and Texas) that produces the maximum final change after a 1-day forecast (measured by the change in vorticity between surface and 500 hPa over the eastern USA)?" ECMWF has been routinely carrying out experiments to find out "What is the change in the initial conditions from 3 days ago that would lead to the best verification of today's analysis?" (see Errico (1997), Rabier *et al.* (1996), Pu *et al.* (1997a,b) for more details).

6.3.3 Lyapunov vectors

As we saw in Section 6.2, if we start a set of perturbations on a sphere of very small size, it will evolve into an ellipsoid. The growth of the axis of the hyperellipsoid after a finite interval s is given by the singular values $\sigma_i(t_0 + s)$. The (global) Lyapunov exponents describe the linear *long-term* growth of the hyperellipsoid:

$$\lambda_i = \lim_{s \to \infty} \frac{1}{s} \ln[\sigma_i(t_0 + s)] \tag{6.3.35}$$

4 Jon Ahlquist (2000, pers. communication) showed, given a linear operator **L**, a set of arbitrary vectors $\mathbf{x}_i$, and a set of arbitrary nonnegative numbers σ_i arranged in decreasing order, how to construct an inner product and a norm such that the σ_i and the $\mathbf{x}_i$ are ith singular values and singular vectors of **L**. He pointed out that "Because anything not in the null space can be a singular vector, even the leading singular vector, one cannot assign a physical meaning to a singular vector simply because it is a singular vector. Any physical meaning must come from an additional aspect of the problem. Said in another way, nature evolves from initial conditions without knowing which inner products and norms the user wants to use."

In other words, the Lyapunov exponents describe the long-term average exponential rate of stretching or contraction in the attractor. (We call the Lyapunov exponents "global" to distinguish them from the finite time or "local" Lyapunov exponents which are useful in predictability applications.) There are as many Lyapunov exponents as the dimension of the model (number of independent variables or degrees of freedom). If the model has at least one λ_i greater than zero, then the system can be called chaotic, i.e., there is exponential separation of trajectories. In other words, there is at least one direction of the ellipsoid that continues to be stretched, and therefore two trajectories will diverge in time and eventually become completely different. Conversely, a system with all negative Lyapunov exponents is stable, and will remain predictable at all times. The first Lyapunov exponent can be estimated by running the tangent linear model for a long time starting from any randomly chosen initial perturbation $\mathbf{y}(t_0)$. During a long integration the growth rate of any random perturbation will converge to the first Lyapunov exponent:

$$\lambda_1 = \lim_{s \to \infty} \frac{1}{s} \ln \left[\frac{\|\mathbf{y}(t_0 + s)\|}{\|\mathbf{y}(t_0)\|} \right] \tag{6.3.36}$$

which is independent of the norm. In practice, the first Lyapunov exponent is obtained by running the tangent linear model for a long period from random initial conditions, and renormalizing the perturbation vector periodically in order to avoid computational overflow.

When we are dealing with atmospheric predictions, we are not really interested in the *global* growth properties, which correspond to the atmosphere's attractor (climatology), i.e., relevant average properties over many decades. Instead, in predictability problems we are interested in the growth rate of perturbations at a given time and space: we need to know the *local* stability properties in space and time, which are related to our ability to make skillful forecasts. We can define the leading local Lyapunov vector (LLV) at a certain time t, as the vector towards which *all* random perturbations $\mathbf{y}(t - s)$ started a long time s before t will converge (Fig. 6.3.5).

$$\mathbf{l}_1(t) = \lim_{s \to \infty} \mathbf{L}(t - s, t)\mathbf{y}(t - s) \tag{6.3.37}$$

Once a perturbation has converged to the leading LLV $\mathbf{l}_1(t)$, the leading local Lyapunov exponent can be computed from the rate of change of its norm. In practice, the local leading Lyapunov exponent, also known as finite time Lyapunov exponent, can be estimated over a finite period τ:

$$l_1 \approx \frac{1}{\tau} \ln \left[\frac{\|\mathbf{l}_1(t + \tau)\|}{\|\mathbf{l}_1(t)\|} \right] \tag{6.3.38}$$

The argument of the logarithm is defined as the amplification rate $A(t, \tau)$.

The ***first LLV is independent of the definition of norm***, and represents the direction in which maximum sustainable growth (or minimum decay) can occur in a system without external forcing. In fact, after a finite transition period T takes place, every

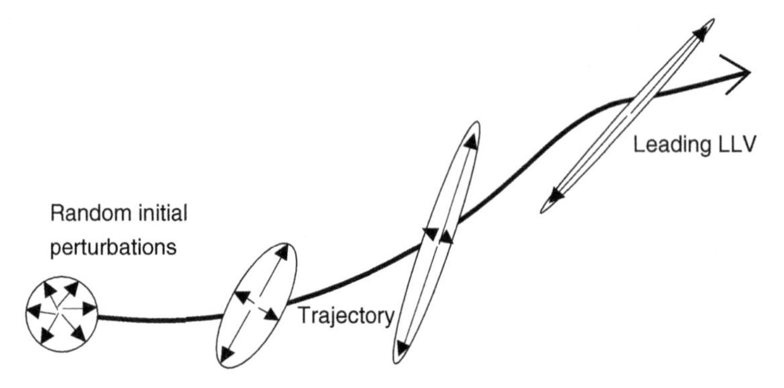

Figure 6.3.5: Schematic of how all perturbations will converge towards the leading LLV.

initial perturbation will turn in the direction of the LLV at every point of the trajectory. This also includes the final singular vectors $\mathbf{u}_i$ for a sufficiently long optimization interval.

Trevisan and Legnani (1995) introduced the notion of the leading LLV. Additional LLVs can be obtained by Gramm–Schmidt orthogonalization, and this would seem to indicate that they are norm-dependent. However, Trevisan and Pancotti (1998) showed that it is also possible, at least in theory, to define additional LLVs (denoted *characteristic* vectors by Legras and Vautard, 1996) *without the use of norms*. The LLVs are therefore a fundamental characteristic of dynamical systems. It should be noted that regrettably, at this time, there is not a universally accepted nomenclature for LLVs. Legras and Vautard (1996) call the LLVs "backward Lyapunov vectors", since they were started an infinitely long time in the past. Unfortunately, this name is extremely confusing, since they represent forward evolution rather than backward evolution as this name would imply. The LLVs are also the final singular vectors optimized for an infinitely long time, i.e., the eigenvectors (valid at time t) of $\mathbf{L}(t-T,t)\mathbf{L}^T(t-T,t)$ for $T \to \infty$. Similarly, Legras and Vautard define as "forward Lyapunov vectors" the *initial* singular vectors obtained from a very long *backward* integration with the adjoint of the model, i.e., they are the eigenvectors (valid at time t) of $\mathbf{L}^T(t,t+T)\mathbf{L}(t,t+T)$ for very large T.

Legras and Vautard (1996) showed (as did Trevisan and Pancotti (1998)) that a complete set of LLVs (which they denote *characteristic* Lyapunov vectors) can be defined from the intersection of the subspaces spanned by the "forward" and "backward" Lyapunov vectors. The (characteristic) LLVs are therefore independent of the norm, and grow in time with a rate given by the local Lyapunov exponents. As such, they are a fundamental characteristic of dynamical systems.

Several authors have shown that the leading (first few) LLVs of low-dimensional dynamical systems span the attractor, i.e., they are parallel to the hypersurface in phase space that the dynamical system visits again and again ("realistic solutions").

Leading singular vectors, on the other hand, have very different properties. They can grow much faster than the leading LLVs, but are initially *off* the attractor: they point to areas in the phase space where solutions do not naturally occur (e.g., Legras and Vautard, 1996, Trevisan and Legnani, 1995, Trevisan and Pancotti, 1998, Pires *et al.*, 1996), see also next section.

For ensemble forecasting, Ehrendorfer and Tribbia (1997) showed that if $\mathbf{V}$ is the initial analysis error covariance (which unfortunately we don't know and can only estimate), then the initial singular vectors defined with the norm $\mathbf{W} = \mathbf{V}^{-1/2}$ evolve into the eigenvectors of the evolved error covariance matrix. This implies that the leading singular vectors, defined using the initial error covariance, are optimal in describing the forecast errors at the end of the optimization period. The initial error covariance norm yields singular vectors quite different from those derived using the energy norm. Barkmeijer *et al.* (1998) used the ECMWF estimated 3D-Var error covariance as the initial norm (instead of the total energy norm) and obtained initial perturbations with structures closer to the bred vectors (i.e., leading LLVs) used at NCEP (see Section 6.5.1).

6.3.4 Simple examples of singular vectors and eigenvectors

In order to get a more intuitive feeling of the relationship between singular vectors and Lyapunov vectors, we consider a simple linear model in two dimensions:

$$\begin{bmatrix} x_1(t+T) \\ x_2(t+T) \end{bmatrix} = \mathbf{M}_T[x(t)] = \begin{bmatrix} 2x_1(t) + 3x_2(t) + 7 \\ 0.5x_2(t) - 4 \end{bmatrix} \tag{6.3.39}$$

We compute the two-dimensional tangent linear model, constant in time:

$$\mathbf{L} = \begin{bmatrix} \dfrac{\partial M_1}{\partial x_1} & \dfrac{\partial M_1}{\partial x_2} \\ \dfrac{\partial M_2}{\partial x_1} & \dfrac{\partial M_2}{\partial x_2} \end{bmatrix} = \begin{bmatrix} 2 & 3 \\ 0 & 0.5 \end{bmatrix} \tag{6.3.40}$$

The propagation or evolution of any perturbation (difference between two solutions) over a time interval $(t, t+T)$ is given by

$$\delta\mathbf{x}(t+T) = \mathbf{L}\delta\mathbf{x}(t) \tag{6.3.41}$$

Note that the translation terms in (6.3.39) do not affect the perturbations. The eigenvectors of L (which for this simple constant tangent linear model are also the Lyapunov vectors) are proportional to

$$\mathbf{l}_1 = \begin{pmatrix} 1 \\ 0 \end{pmatrix} \qquad \mathbf{l}_2 = \begin{pmatrix} -2 \\ 1 \end{pmatrix}$$

corresponding to the eigenvalues $\lambda_1 = 2,\ \lambda_2 = 0.5$, respectively, which in this case are the two Lyapunov numbers (their logarithms are the Lyapunov exponents). If we normalize them, so that they have unit length, the Lyapunov vectors are

$$\mathbf{l}_1 = \begin{pmatrix} 1 \\ 0 \end{pmatrix} \qquad \mathbf{l}_2 = \begin{pmatrix} -0.89 \\ 0.45 \end{pmatrix} \tag{6.3.42}$$

The Lyapunov vectors are not orthogonal, they are separated by an angle of 153.4° (Fig. 6.3.6(a)). We will see that because they are not orthogonal it is possible to find linear combinations of the Lyapunov vectors that grow faster than the leading Lyapunov vector. We will also see that the leading Lyapunov vector is the attractor of the system, since repeated applications of $\mathbf{L}$ to any perturbation makes it evolve towards $\mathbf{l}_1$.

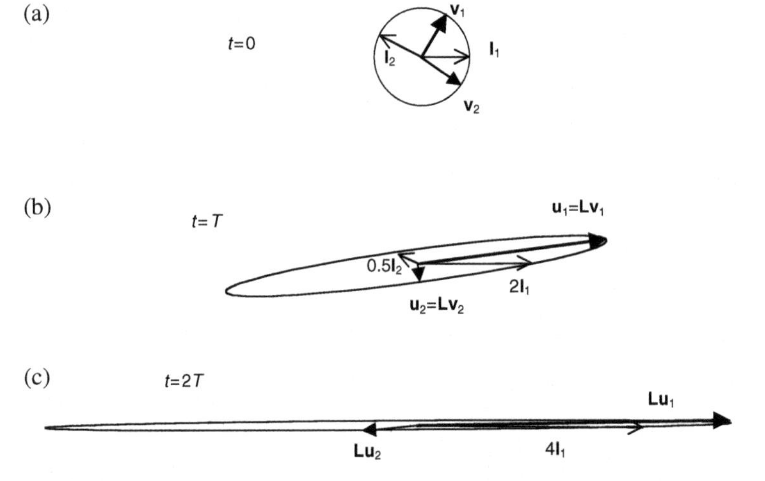

Figure 6.3.6: Schematic of the evolution of the two nonorthogonal Lyapunov vectors (thin arrows $\mathbf{l}_1$ and $\mathbf{l}_2$), and the corresponding two initial singular vectors (thick arrows $\mathbf{v}_1(0)$and $\mathbf{v}_2(0)$), optimized for the interval (0, T), for the tangent linear model

$$\mathbf{L} = \begin{bmatrix} 2 & 3 \\ 0 & 0.5 \end{bmatrix}$$

with eigenvalues 2 and 0.5. (a) Time $t = 0$, showing the initial singular vectors $\mathbf{v}_1(0)$ and $\mathbf{v}_2(0)$, as well as the Lyapunov vectors $\mathbf{l}_1$ and $\mathbf{l}_2$. (b) Time $t = T$, evolved singular vectors, $\mathbf{u}_1(T) = \mathbf{L}\mathbf{v}_1(0),\ \mathbf{u}_2(T) = \mathbf{L}\mathbf{v}_2(0)$ at the end of the optimization period; the Lyapunov vectors have grown by factors of 2 and 0.5 respectively, whereas the leading singular vector has grown by 3.63. The second evolved singular vector has grown by 0.275, and is still orthogonal to the first singular vector. (c) Time $t = 2T$. Beyond the optimization period T, the evolved singular vectors $\mathbf{u}_1(t + 2T) = \mathbf{L}\mathbf{u}_1(t + T),\ \mathbf{u}_2(2T) = \mathbf{L}\mathbf{u}_2(T)$ are not orthogonal and they approach the leading Lyapunov vector with similar growth rates.

Applying first $\mathbf{L}$ and then its transpose $\mathbf{L}^T$ we obtain the symmetric matrix

$$\mathbf{L}^T\mathbf{L} = \begin{bmatrix} 4 & 6 \\ 6 & 9.25 \end{bmatrix} \tag{6.3.43}$$

whose eigenvectors are the *initial singular vectors*, and whose eigenvalues are the squares of the singular values. The initial singular vectors (eigenvectors of $\mathbf{L}^T\mathbf{L}$) are

$$\mathbf{v}_1 = \begin{pmatrix} 0.55 \\ 0.84 \end{pmatrix} \mathbf{v}_2 = \begin{pmatrix} 0.84 \\ -0.55 \end{pmatrix} \tag{6.3.44}$$

with eigenvalues $\sigma_1^2 = 13.17,\ \sigma_2^2 = 0.076$. As indicated before, the *singular values* of $\mathbf{L}$ are the square roots of the eigenvalues of $\mathbf{L}^T\mathbf{L}$, i.e., $\sigma_1 = 3.63,\ \sigma_2 = 0.275$. Note that this implies that during the optimization period $(0, T)$ the leading singular vector grows almost twice as fast as the leading Lyapunov vector (3.63 vs. 2). The angle that the leading *initial singular vector* has with respect to the leading Lyapunov vector is 56.82°, whereas the second initial singular vector is perpendicular to the first one (Fig.6.3.6(a)).

The final or *evolved SVs at the end of the optimization period* $(0, T)$ are the eigenvectors of

$$\mathbf{L}\mathbf{L}^T = \begin{bmatrix} 13 & 1.5 \\ 1.5 & 0.25 \end{bmatrix} \tag{6.3.45}$$

and after normalization, they are

$$\mathbf{u}_1 = \begin{pmatrix} 0.99 \\ 0.12 \end{pmatrix} \qquad \mathbf{u}_2 = \begin{pmatrix} 0.12 \\ -0.99 \end{pmatrix} \tag{6.3.46}$$

Note again that the operators $\mathbf{L}^T\mathbf{L}$ and $\mathbf{L}\mathbf{L}^T$ are quite different, and the final singular vectors are different from the initial singular vectors, but they have the same singular values $\sigma_1^2 = 13.17,\ \sigma_2^2 = 0.076$.

Alternatively, the evolved singular vectors at the end of the optimization period can also be obtained by applying $\mathbf{L}$ to the initial singular vectors, which is computationally inexpensive. In this case,

$$\mathbf{u}_1(T) = \mathbf{L}\mathbf{v}_1(0) = \begin{bmatrix} 3.6 \\ 0.42 \end{bmatrix} \qquad \mathbf{u}_2(T) = \mathbf{L}\mathbf{v}_2(0) = \begin{bmatrix} 0.03 \\ -0.27 \end{bmatrix}$$

which is the same as (6.3.46) but without normalization.

The final leading singular vector has strongly rotated towards the leading Lyapunov vector: at the end of the optimization period the angle between the leading singular vector and the leading Lyapunov vector is only 6.6° (Fig. 6.3.6(b)), and because the singular vectors have been optimized for this period, the final singular vectors are still orthogonal.

To obtain the evolution of the singular vectors *beyond the optimization period* (0, T) we apply $\mathbf{L}$ again to the evolved singular vector valid at $t = T$ and obtain

$$\mathbf{u}_1(t+2T) = \mathbf{L}\mathbf{u}_1(t+T) = \begin{bmatrix} 8.47 \\ 0.21 \end{bmatrix} \qquad \mathbf{u}_2(t+2T) = \mathbf{L}\mathbf{u}_2(t+T) = \begin{bmatrix} -0.76 \\ -0.14 \end{bmatrix}$$

During the interval $(T, 2T)$ the leading singular vector grows by a factor of just 2.33, which is not very different from the growth rate of the leading Lyapunov vector. At the end of this second period (Fig. 6.3.6(c)) the angle with the leading Lyapunov vector is only 1.41°. The angle of the second evolved singular vector at time T, after applying the linear tangent model $\mathbf{L}$ and the leading Lyapunov vector is also quite small (10.24^o), and because it was further away from the attractor, the second singular vector (whose original, transient, singular value was 0.5), grows by a factor of 2.79. This example shows how quickly *all perturbations, including all singular vectors, evolve towards the leading Lyapunov vector, which is the attractor of the system.* It is particularly noteworthy that during the optimization period (0, T), the first singular vector grows very fast as it rotates towards the attractor, but once it gets close to the leading Lyapunov vector, its growth returns to the normal leading Lyapunov vector's growth.

Let us now choose as the tangent linear model another matrix

$$\mathbf{L} = \begin{bmatrix} 2 & 30 \\ 0 & 0.5 \end{bmatrix}$$

with the same eigenvalues 2 and 0.5, i.e., with eigenvectors (Lyapunov vectors) that still grow at a rate of $2/T$ and $0.5/T$ respectively. However, now the angle between the first and the second Lyapunov vector is 177°, i.e. the Lyapunov vectors are almost antiparallel. In this case, the first singular vector grows by a factor of over 30 during the optimization period, but beyond the optimization period it essentially continues evolving like the leading Lyapunov vector.

These results do not depend on the fact that one Lyapunov vector grows and the other decays. As a third example, we choose

$$\mathbf{L} = \begin{bmatrix} 2 & 3 \\ 0 & 1.5 \end{bmatrix}$$

with two Lyapunov vectors growing with rates $2/T$ and $1.5/T$. The Lyapunov vectors are almost parallel, with an angle of 170°, and the leading singular vector grows during the optimization period by a factor of 3.83. Applying the tangent linear model again to the evolved singular vectors we obtain that at time $2T$ the leading singular vector has grown by a factor of 2.9 and its angle with respect to the leading Lyapunov vector is 1°. Because it is not decaying, the second Lyapunov vector is also part of the attractor, but only those perturbations that are exactly parallel to it will remain parallel, all others will move towards the first Lyapunov vector.

These examples illustrate the fact that the fast growth of the singular vectors during the optimization period depends on the lack of orthogonality between Lyapunov vectors. *A very fast "supergrowth" of singular vectors is associated with the presence of almost parallel Lyapunov vectors*, and it takes place when the initial singular vector, which is not in the attractor, rotates back towards the attractor. At the end of the optimization period, the leading singular vector tends to be much closer to the attractor, more parallel to the leading Lyapunov vector. The second (trailing) singular vector is also moving towards the leading Lyapunov vector.

Finally, we point out that this introductory discussion is appropriate for relatively low-dimensional systems. For extremely high-dimensional systems like the atmosphere, there may be multiple sets of Lyapunov exponents corresponding to different types of instabilities. For example, as pointed out by Toth and Kalnay (1993), convective instabilities have very fast growth but small amplitudes, whereas baroclinic instabilities have slower growth but much larger amplitudes, and each of these can lead to different types of Lyapunov vectors. If we are interested in the predictability characteristics associated with baroclinic instabilities, then the analysis of growth rates of infinitesimally small Lyapunov vectors over infinitely long times may not be appropriate for the problem (Lorenz, 1996). In that case, it may be better to consider the finite amplitude, finite time extension of Lyapunov vectors introduced by Toth and Kalnay (1993, 1997) as bred vectors. Bred vectors are discussed in Section 6.5.1, and their relationship to Lyapunov vectors in Kalnay *et al.* (2002).

6.4 Ensemble forecasting: early studies

We saw in previous sections that Lorenz (1963a,b,1965) showed that the forecast skill of atmospheric models depends not only on the accuracy of the initial conditions and on the realism of the model (as it was generally believed at the time), but also on the instabilities of the flow itself. He demonstrated that any nonlinear dynamical system with instabilities, like the atmosphere, has a finite limit of predictability. The growth of errors due to instabilities implies that the smallest imperfection in the forecast model or the tiniest error in the initial conditions, will *inevitably* lead to a total loss of skill in the weather forecasts after a *finite* forecast length. Lorenz estimated this *limit of weather predictability* as about two weeks. With his simple model he also pointed out that predictability is strongly dependent on the evolution of the atmosphere itself: some days the forecasts can remain accurate for a week or longer, and on other days the forecast skill may break down after only 3 days. This discovery made inevitable the realization that NWP needs to account for the stochastic nature of the evolution of the atmosphere (Fig. 6.4.1). As previously discussed, Lorenz (1965) studied the error growth of a complete "ensemble" of perturbed forecasts, with the ensemble size equal to the dimension of the phase space (one perturbation for each of the 28 model variables). In this paper he introduced for the first time concepts related

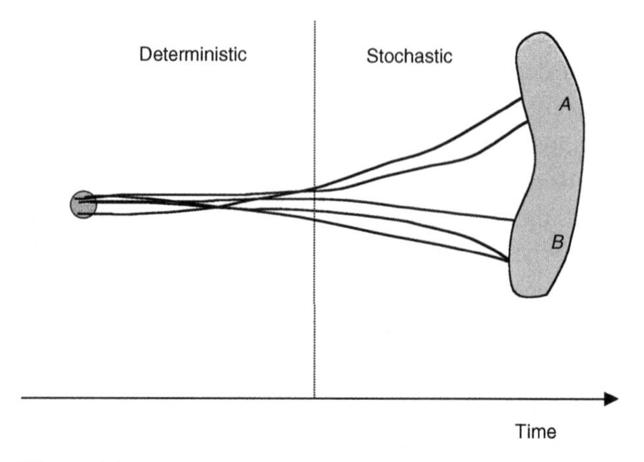

Figure 6.4.1: Schematic of ensemble prediction, with individual trajectories drawn for forecasts starting from a representative set of perturbed initial conditions within a circle representing the uncertainty of the initial conditions (ideally the analysis error covariance) and ending within the range of possible solutions. For the shorter range, the forecasts are close to each other, and they may be considered deterministic, but beyond a certain time, the equally probable forecasts are so different that they must be considered stochastic. The transition time is of the order of 2–3 days for the prediction of large-scale flow, but can be as short as a few hours for mesoscale phenomena like the prediction of individual storms. The transition time is shorter for strongly nonlinear parameters: even for large-scale flow, precipitation forecasts show significant divergence faster than the 500-hPa fields. The forecasts may be clustered into subsets A and B. (Adapted from Tracton and Kalnay, 1993.)

to singular vectors and LLVs discussed in the previous section. This was followed by several early approaches to the problem of accounting for the variable predictability of the atmosphere reviewed in this section.

6.4.1 Stochastic-dynamic forecasting

Historically, the first forecasting method to explicitly acknowledge the uncertainty of atmospheric model predictions was developed by Epstein (1969), who introduced the idea of ***stochastic-dynamic forecasting***. He derived a continuity equation for the probability density $\varphi(X;t)$ of a model solution X of a dynamical model $\dot{X} = G(X(t))$, where the model has dimension D:

$$\frac{\partial \varphi}{\partial t} + \nabla_D \cdot (\dot{X}\varphi) = 0 \qquad (6.4.1)$$

This equation indicates that in an ensemble of forecast solutions, "no member of the ensemble may be created or destroyed". An ensemble starting from an infinite number of perturbed integrations spanning the analysis uncertainty gives the "true" probability distribution (with all its moments), but even for a simple low-order model, the integration of (6.4.1) is far too expensive. Therefore Epstein introduced an approximation to predict only the first and second moments of the probability distribution (expected means and covariances) rather than the full probability distribution.

Epstein assumed that the model equations are of the form

$$\dot{x}_i = \sum_{j,k} a_{ijk} x_j x_k - \sum_j b_{ij} x_j + c_i \tag{6.4.2}$$

The forecast equations for the expected first and second moments are

$$\left.\begin{aligned} \dot{\mu}_i &= E(\dot{x}_i) \\ \dot{\rho}_{ij} &= E(x_i \dot{x}_j + \dot{x}_i x_j) \end{aligned}\right\} \tag{6.4.3}$$

The covariances ρ_{ij} are related to the second order moments by $\sigma_{ij} = E[(x_i - \mu_i)(x_j - \mu_j)]$. Substituting (6.4.2) into (6.4.3) gives rise to forecast equations for $\dot{\mu}$ and $\dot{\sigma}$ that contain triple moments $(x_i x_j x_k)$. As done in turbulence models with a second order closure for the triple products (Chapter 4), Epstein introduced a closure assumption for the third order moments around the mean $\tau_{ijk} = E[(x_i - \mu_i)(x_j - \mu_j)(x_k - \mu_k)]$. He assumed that $\Sigma_{kl} a_{jkl} \tau_{ikl} + a_{ikl} \tau_{jkl} = 0$, which then gives a closed set of equations for the means and covariances:

$$\left.\begin{aligned} \dot{\mu}_i &= \sum_{jk} a_{ijk}(\sigma_{jk} + \mu_j \mu_k) - \sum_j b_{ij} \mu_j + c_i \\ \dot{\sigma}_{ij} &= \sum_{kl} a_{jkl}(\mu_k \sigma_{il} + \mu_l \sigma_{ik}) + a_{ikl}(\mu_k \sigma_{jl} + \mu_l \sigma_{jk}) \\ &\quad - \sum_k (b_{ik} \sigma_{jk} + b_{jk} \sigma_{ik}) \end{aligned}\right\} \tag{6.4.4}$$

Epstein tested these "approximate" stochastic equations for a Lorenz three-variable model. The "true" probability distribution was computed from a Monte Carlo ensemble of 500 members, and the comparison indicated good agreement, at least for several simulated days. Note that in his case, the number of ensemble members was much larger than the number of degrees of freedom of the model, a situation that would be impossible to replicate with current models with millions of degrees of freedom. In his paper, Epstein also introduced the idea of using stochastic-dynamic forecasting in the *analysis cycle*, with the background forecast and error covariance provided by stochastic-dynamic forecasts combined with observations that also contain errors (cf. Sections 5.3–5.5).

Unfortunately, although the stochastic-dynamic forecasting method was introduced as a shortcut to an "infinite" Monte Carlo ensemble, in a model with N degrees of freedom, it requires $N(N+1)/2 + N$ forecast equations, equivalent to making about $(N+3)/2$ model forecasts. Although this was practical with a three-variable model, it is completely unfeasible for a modern model, with millions of degrees of freedom.

6.4.2 Monte Carlo forecasting

In 1974, Leith proposed the idea of performing ensemble forecasting with a limited number m of ensemble members instead of the conventional single (deterministic) control forecast. He also proposed performing an "optimal estimation" of the verification using linear regression on the dynamical forecasts, with optimal weights determined from forecast error covariances (cf. Sections 5.3–5.5). Since forecasts lose their skill at longer lead times, and individual forecasts eventually are further away from the verification than the climatology (cf. eqs. (6.4.5) and (6.4.6)), optimal estimation of the verification is equivalent to *tempering* (i.e., *hedging* the forecast towards climatology).

He cast his analysis using, instead of model variables, their deviation $\mathbf{u}$ with respect to climatology (also known as forecast *anomalies*). The true state of the atmosphere is denoted $\mathbf{u}_0$, and $\hat{\mathbf{u}}$ then denotes an unbiased estimate of $\mathbf{u}_0$, whose expected value (average over many forecasts, represented by the angle brackets) is equal to zero: $\langle\hat{\mathbf{u}}\rangle = 0$.

We can compute the expected error covariance of a climatological forecast (i.e., a forecast of zero anomaly):

$$\langle(0 - \mathbf{u}_0)(0 - \mathbf{u}_0)^T\rangle = \langle\mathbf{u}_0\mathbf{u}_0^T\rangle = \mathbf{U} \tag{6.4.5}$$

A single (deterministic) forecast $\hat{\mathbf{u}}$, on the other hand, has, on average, an error covariance given by

$$\langle(\hat{\mathbf{u}} - \mathbf{u}_0)(\hat{\mathbf{u}} - \mathbf{u}_0)^T\rangle = \langle\hat{\mathbf{u}}\hat{\mathbf{u}}^T + \mathbf{u}_0\mathbf{u}_0^T - \hat{\mathbf{u}}\mathbf{u}_0^T - \mathbf{u}_0\hat{\mathbf{u}}^T\rangle \underset{t\to\infty}{\longrightarrow} 2\mathbf{U} \tag{6.4.6}$$

This limit occurs because the last two terms in the second angle brackets go to zero as the forecasts become decorrelated with the true atmosphere at long lead times, and we assume that the model covariance is also unbiased. This indicates that for long lead times an individual deterministic forecast has twice the error covariance of a climatological forecast. Therefore, a "regressed" forecast, tempered towards climatology, must be better than a single deterministic forecast (in a least square error sense), with an error covariance that asymptotes to $\mathbf{U}$, and not $2\mathbf{U}$.

A regressed forecast $\hat{\mathbf{u}}_0 = \hat{\mathbf{u}}\mathbf{A}$ is obtained by linear regression, minimizing the square of the regressed error $\varepsilon^T\varepsilon = \langle(\mathbf{u}_0 - \hat{\mathbf{u}}\mathbf{A})^T(\mathbf{u}_0 - \hat{\mathbf{u}}\mathbf{A})\rangle$ with respect to the elements of the matrix of constant regression coefficients $\mathbf{A}$. As we did in the derivation of the optimal weight matrix for the observational increments in Section 5.4, we make use of the linear regression formulas: if the linear prediction equation is $\hat{\mathbf{y}} = \mathbf{x}\mathbf{A}$, then the error is given by $\varepsilon = \mathbf{y} - \mathbf{x}\mathbf{A}$. The matrix of the derivatives of the (scalar) squared error $\varepsilon^T\varepsilon$ with respect to each element of $\mathbf{A}$ is given by $\partial\varepsilon^T\varepsilon/\partial\mathbf{A} = -2\mathbf{x}^T(\mathbf{y} - \mathbf{x}\mathbf{A}) = 0$, which gives the normal equation $\mathbf{x}^T\mathbf{y} = \mathbf{x}^T\mathbf{x}\mathbf{A}$, or $\mathbf{A} = (\mathbf{x}^T\mathbf{x})^{-1}(\mathbf{x}^T\mathbf{y})$. Applying this to the regressed forecast we obtain $\langle\hat{\mathbf{u}}^T(\mathbf{u}_0 - \hat{\mathbf{u}}\mathbf{A})\rangle = 0$, or

$$\mathbf{A} = \langle\hat{\mathbf{u}}^T\hat{\mathbf{u}}\rangle^{-1}\langle\hat{\mathbf{u}}^T\mathbf{u}_0\rangle \tag{6.4.7}$$

Estimating the required forecast statistics in (6.4.7) involves considerable work. The size of the regression matrix is usually large compared to the size of the sample available to estimate it, and in order to reduce the number of parameters to be estimated additional approximations are needed (e.g., by parameterizing error growth, Hoffman and Kalnay, 1983).

Now, instead of regression let's consider an ensemble of m forecasts computed from perturbations $\mathbf{r}_i$ to the initial best estimate (analysis) $\hat{\mathbf{u}}$. Ideally, the perturbations should be chosen so that their outer product is a good estimate of the initial error covariance (i.e., the analysis error covariance $\langle \mathbf{r}\mathbf{r}^T \rangle = \mathbf{P}_a$, as suggested in Fig. 6.4.1). In practice, however, the analysis error covariance can only be approximately estimated (e.g., Barkmeijer *et al.*, 1998).

If $\overline{\mathbf{u}} = (1/m)\Sigma_{i=1}^{m}\mathbf{u}_i$ is the average of an ensemble of m forecasts, then its error covariance evolves like

$$\langle (\overline{\mathbf{u}} - \mathbf{u}_0)(\overline{\mathbf{u}} - \mathbf{u}_0)^T \rangle = \langle \overline{\mathbf{u}}\overline{\mathbf{u}}^T + \mathbf{u}_0\mathbf{u}_0^T - \overline{\mathbf{u}}\mathbf{u}_0^T - \mathbf{u}_0\overline{\mathbf{u}}^T \rangle \underset{t\to\infty}{\longrightarrow} \left(1 + \frac{1}{m}\right)\mathbf{U} \tag{6.4.8}$$

since the last two terms in the second angle brackets go to zero at long time leads, and the first one evolves like

$$\langle \overline{\mathbf{u}}\overline{\mathbf{u}}^T \rangle = \frac{1}{m}\sum_{i=1}^{m}\mathbf{u}_i \frac{1}{m}\sum_{j=1}^{m}\mathbf{u}_j^T \underset{t\to\infty}{\longrightarrow} \frac{m}{m^2}\mathbf{U} \tag{6.4.9}$$

Equation (6.4.8) shows that averaging a Monte Carlo ensemble of forecasts approximates the tempering of the forecasts towards climatology, *without the need to perform regression*. It suggests that such tempering may be substantially achieved with a relatively small number of ensemble members (compare (6.4.8) with (6.4.5) and (6.4.6)). Leith (1974) used an analytical turbulence model to test this hypothesis, and concluded that a Monte Carlo forecasting procedure represents a practical, computable approximation to the stochastic-dynamic forecasts proposed by Epstein (1969). He suggested that adequate accuracy would be obtained for the best estimate of the forecast (i.e., *the ensemble mean*) with sample sizes as small as 8, but that the estimation of *forecast errors* may require a larger number of ensemble members. Monte Carlo forecasting is thus a feasible approach for ensemble forecasting, requiring only a definition of the initial perturbations and m forecasts.

6.4.3 Lagged average forecasting

In 1983, Hoffman and Kalnay proposed *lagged average forecasting* (LAF) as an alternative to Monte Carlo forecasting, in which the forecasts initialized at the current initial time, $t = 0$, as well as at previous times, $t = -\tau, -2\tau, \ldots, -(N-1)\tau$ are combined to form an ensemble (see Fig. 6.4.2). In an operational set up, τ is typically 6, 12 or 24 hours, so that the forecasts are already available, and the perturbations are generated automatically from the forecast errors. Since the ensemble comprises forecasts of different "age", Hoffman and Kalnay (1983) weighted them according to their expected error, which they estimated by parameterizing the observed error

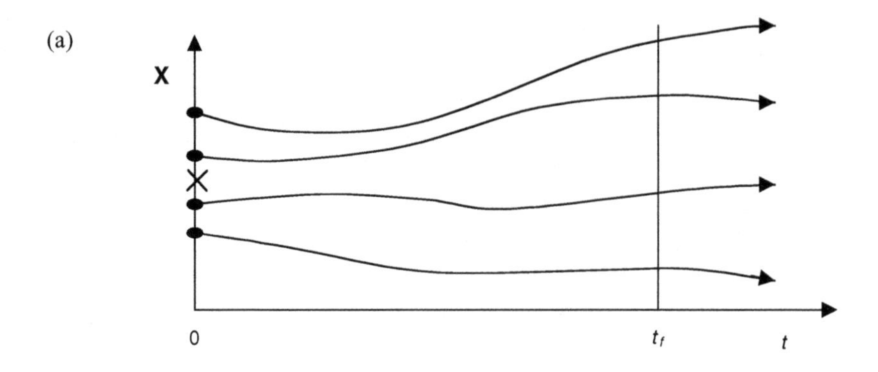

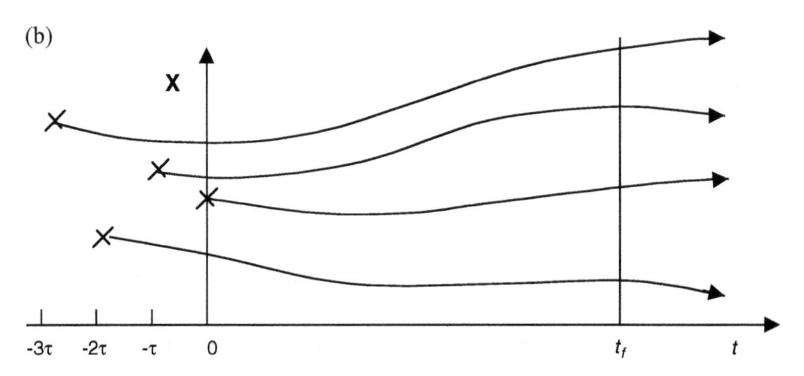

Figure 6.4.2: Schematic time evolutions of Monte Carlo forecasts (a) and lagged average forecasts (b). The abscissa is forecast time t, and the ordinate is the value of a forecast variable **X**. The crosses represent analyses obtained at time intervals τ, and the dots, randomly perturbed initial conditions; t_f is a particular forecast time. The initial "perturbation" for the lagged average forecast is the previous forecasts' error at the initial time. (Adapted from Hoffman and Kalnay, 1983.)

covariance growth. They compared the lagged average forecasting and Monte Carlo forecasting methods within a simulation system, using a primitive equations model as "nature", and a quasi-geostrophic model to perform the "forecasts". In this way they allowed for model errors, unlike the previous "identical twin" experiments that assumed a perfect model. They "observed" the required variables and introduced random "observation errors" every 6 hours and performed many ensemble forecast experiments separated by 50 days of integration. They compared the results of single forecasts (ordinary dynamical forecasts), Monte Carlo forecasting, lagged average forecasting and tempered ordinary dynamical forecasts, as well as persistence-climatology forecast (the most skillful baseline forecast).

Hoffman and Kalnay looked at the error growth of individual forecasts (Fig. 6.4.3). Note in this figure that the individual forecast errors grow slowly and then at a certain time there is a rapid error growth until nonlinear saturation takes place (only the period

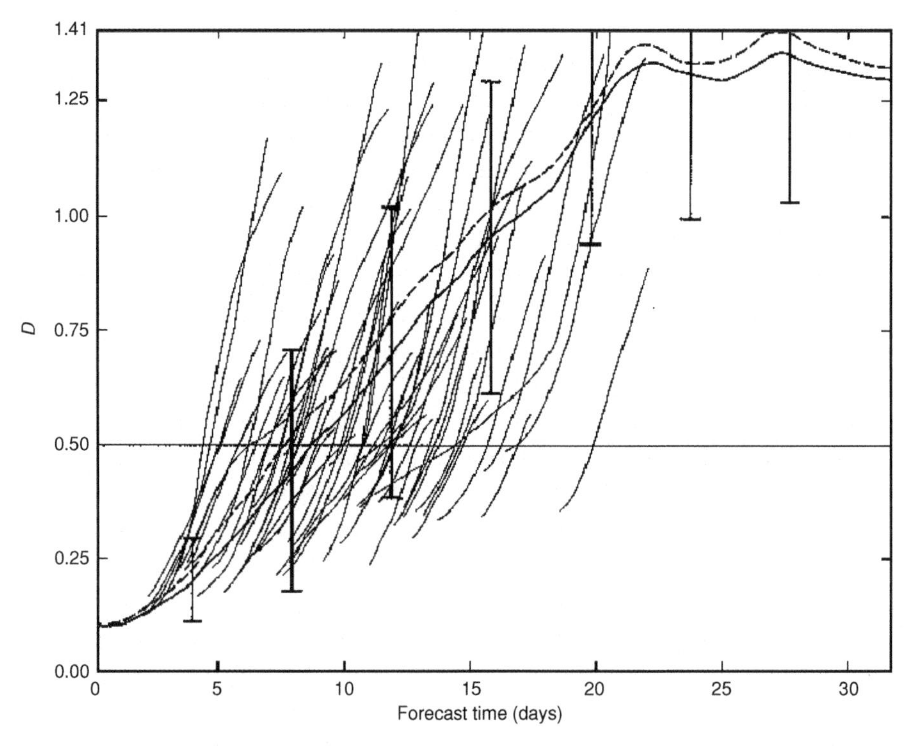

Figure 6.4.3: Time evolution of D, the individual forecast errors scaled by the climatological forecast error, plotted only during the period the forecast error crossed $D = 0.5$. Also plotted are two measures of average forecast error. (Adapted from Hoffman and Kalnay, 1983.)

of rapid growth is plotted). Note also that the forecast errors saturate around $\sqrt{2}$ of the climatological variability, as indicated by (6.4.6).

In these simulated forecasts, like in real weather forecasts, the forecast skill exhibits a lot of day-to-day variability. The rapid growth takes place at a time that varies from a minimum of 5 days to a maximum of 20 days. Hoffman and Kalnay tested the ability of the ensemble to predict the time at which the forecast error crossed 50% of the climatological standard deviation. They used as the predictor the spread of the ensembles (standard deviation with respect to their mean). They found that the lagged average forecasting ensemble average forecast was only slightly better than the Monte Carlo forecasting, but the advantage of lagged average forecasting in predicting forecast skill was much more apparent, with the correlation between predicted and observed time of crossing the 50% level being 0.68 for Monte Carlo forecasting and 0.79 for lagged average forecasting.

The advantages of lagged average forecasting over Monte Carlo forecasting are probably due to the fact that lagged average forecasting perturbations in the initial conditions were not *randomly chosen* errors like in Monte Carlo forecasting but included dynamical influences and therefore contained "*errors of the day*". This is because the perturbations are generated from actual forecast errors and therefore they are influenced by the evolution of the underlying background large-scale flow.

Lagged average forecasting has been frequently used for experimental ensemble forecasting, both for medium-range and climate prediction. However, the statistics required to estimate the weights of the members of the lagged average forecasting ensemble according to their "age" are very hard to obtain, so that except for the study by Dalcher *et al.* (1988), all the lagged average forecasting members have been generally given equal weight. The advantages of lagged average forecasting are: (a) some of the forecasts are already available in operational centers; (b) it is very simple to perform and does not require special generation of perturbations; and (c) the perturbations contain "errors of the day" (Lyapunov vectors). Lagged average forecasting has also major disadvantages: (a) a large LAF ensemble would have to include excessively "old" forecasts; (b) without the use of optimal weights, the lagged average forecasting ensemble average may be tainted by the older forecasts.

Ebisuzaki and Kalnay (1991) introduced a variant of lagged average forecasting denoted *scaled lagged average forecasting* (SLAF) that reduces these two disadvantages. The perturbations are obtained by computing the forecast error of forecasts started at $t = -j\tau,\ j = 1, \ldots, N-1$, and multiplying these errors by $\pm 1/j$. This assumes that the errors grow approximately linearly with time during the first 2–3 days, and that the perturbations can be subtracted from and not just added to the analysis. The advantages of scaled lagged average forecasting are: (a) the initial perturbations of the ensemble members are all of approximately the same size (this can be enforced using a more sophisticated rescaling than linear growth), and (b) their number is doubled with respect to lagged average forecasting, so that only shorter-range forecasts are needed to create scaled lagged average forecasting. In practice, it has been observed that pairs of initial perturbations with opposite sign, as used in scaled lagged average forecasting, yield better ensemble forecasts, presumably because the Lyapunov vectors within the analysis errors can have either sign, whereas lagged average forecasting tends to maintain a single sign in the error. Experiments with the NCEP global model showed that scaled lagged average forecasting ensembles were better than lagged average forecasting ensembles (Ebisuzaki and Kalnay, 1991). This method is also easier to implement in regional ensemble forecasts, since it generates boundary condition perturbations consistent with the interior perturbations (Hou *et al.*, 2001).

6.5 Operational ensemble forecasting methods

Figure 6.5.1(a) shows the elements of a typical ensemble forecasting system: (1) the *control* forecast (labeled C) starts from the analysis (denoted by a cross), i.e., from the best estimate of the initial state of the atmosphere; (2) two *perturbed* ensemble forecasts (labeled P^+ and P^-) with initial perturbations added and subtracted from

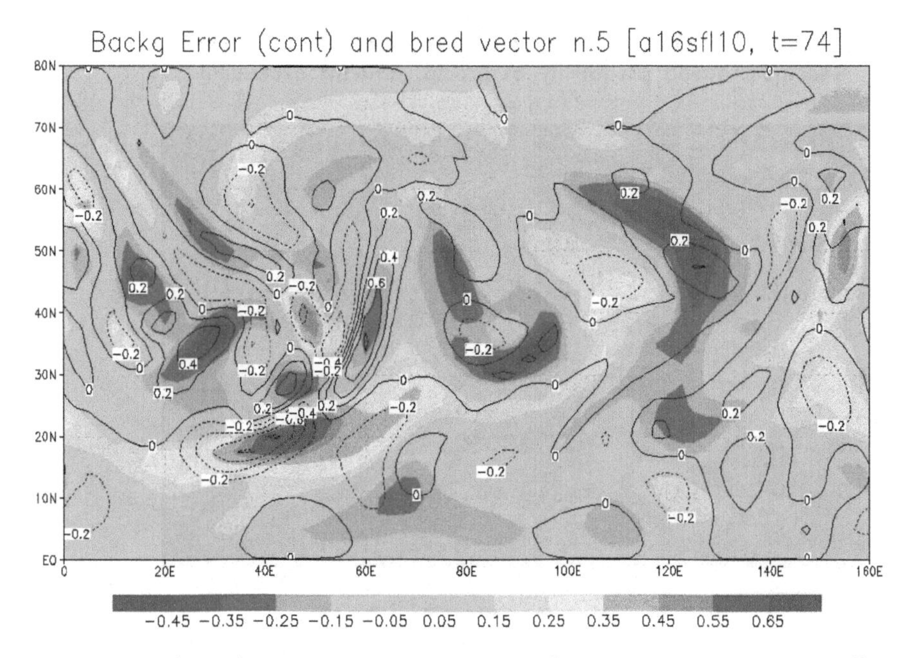

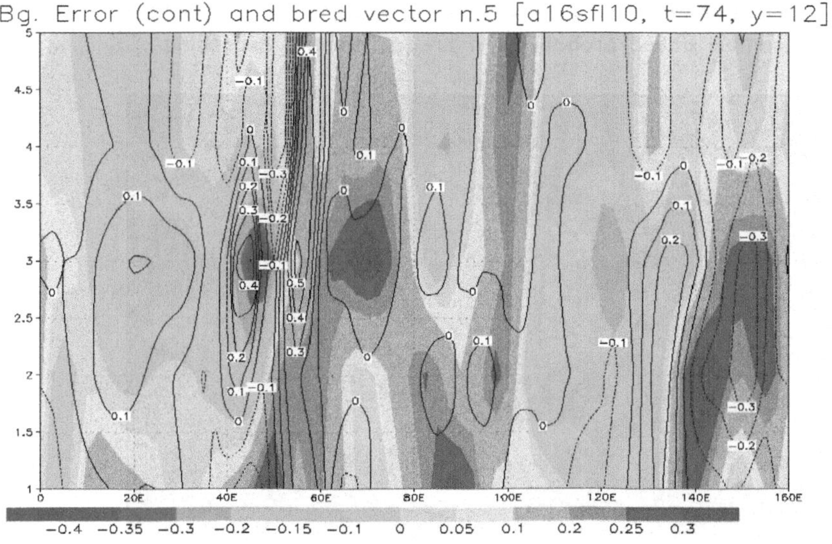

Plate 1 Comparison between the 12-h forecast error used as background (contours) and a randomly chosen bred vector for a data assimilation simulation system. The first image at the center level of the model. The second is a vertical cross-section (from Corazza et al, 2002).

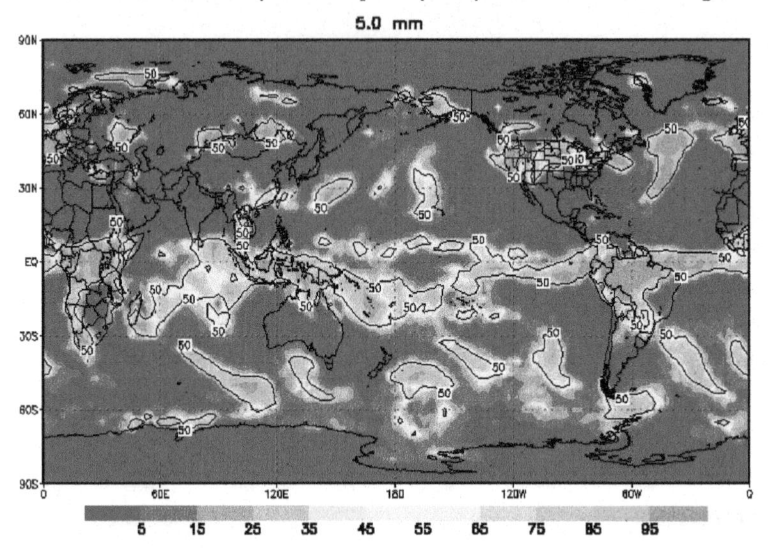

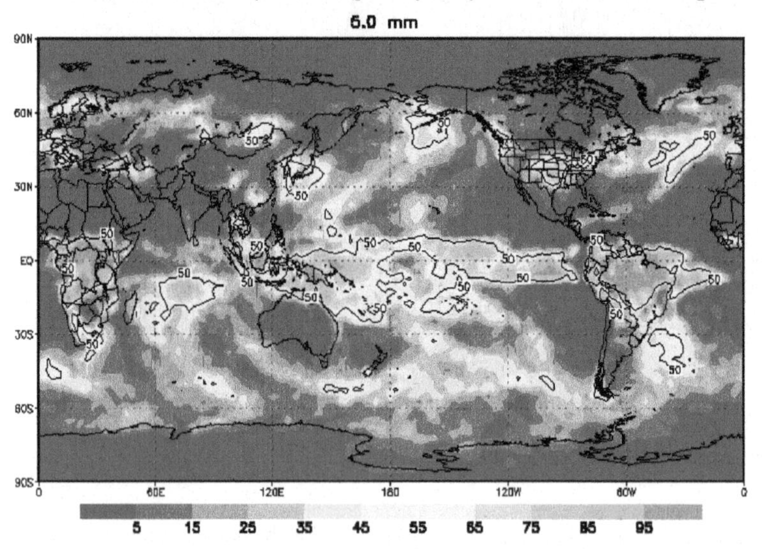

Plate 2 Example of a probabilistic forecast of accumulated precipitation greater than 5 mm. The probabilities are computed simply as the number of ensemble members with at least the indicated threshold of accumulated precipitation divided by the total number of ensemble forecasts. Both the 24-h and the 7-day forecast verify on 6 April 2001. Courtesy of NCEP/NWS.

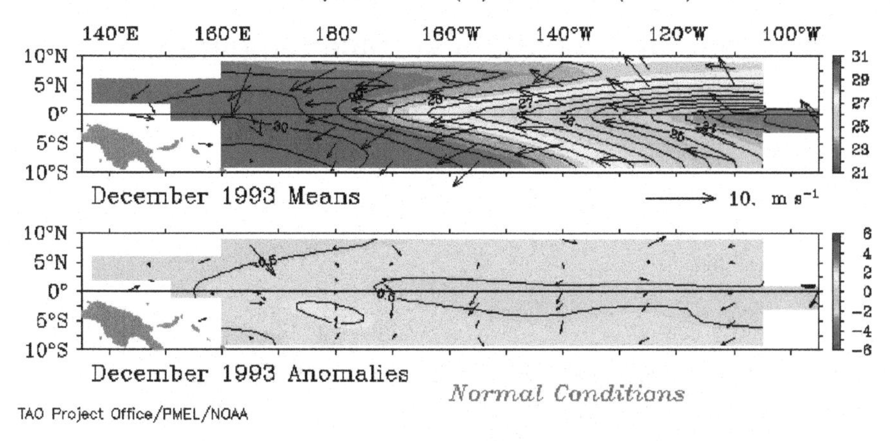

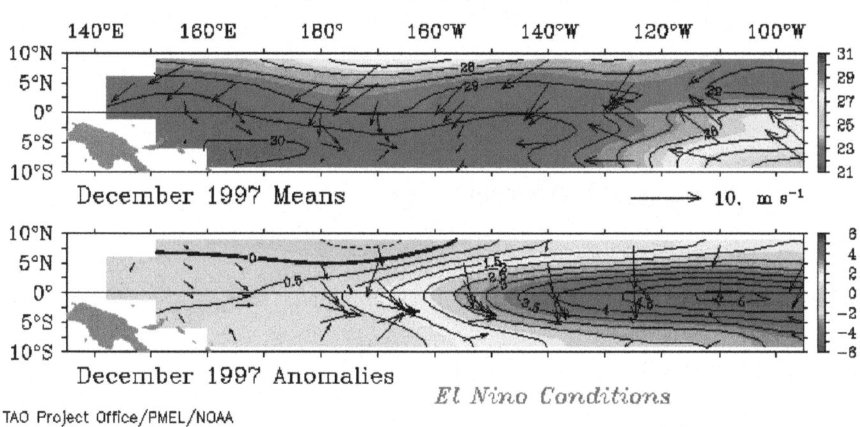

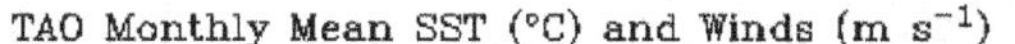

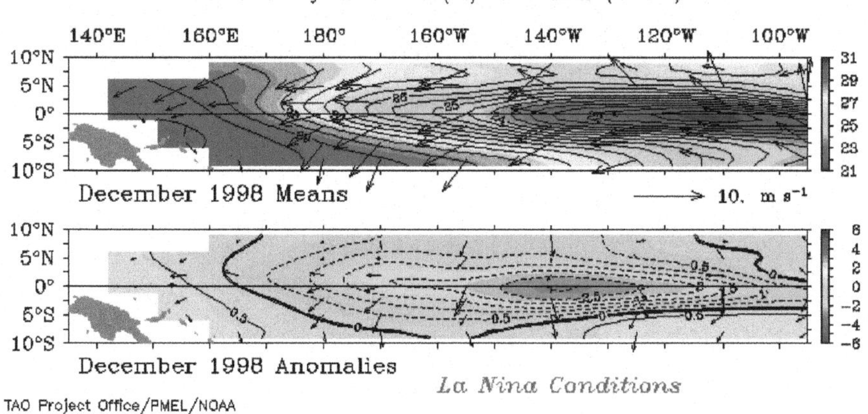

Plate 3 Equatorial Pacific SST and surface winds from TOGA TAO buoys during December 1993 (considered to be normal), December 1997 (during El Niño) and December 1998 (during La Niña). The top panels shows the fields and the bottom panels the anomalies. Courtesy of NOAA/PMEL.

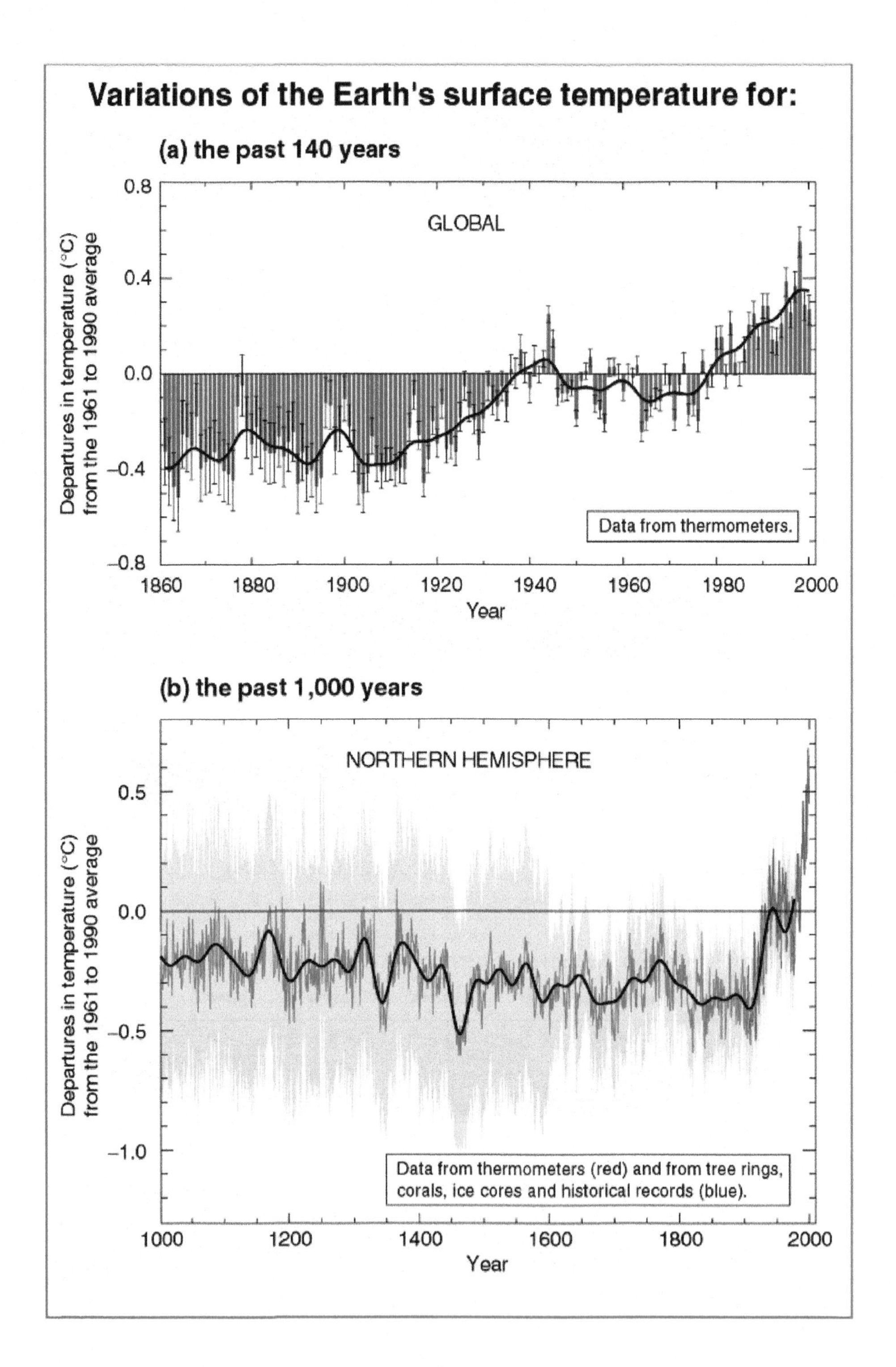

Plate 4 Variation of the surface temperature over: (a) the last 140 years and (b) the last millennium. In (a) the red bars represent the annual average for the globe based on thermometer data, and the whiskers the 95% confidence range, including uncertainties due to coverage, biases and urbanization. The black line is a 10-year moving average. In (b) the blue line represents proxy data. Adapted from IPCC (2002).

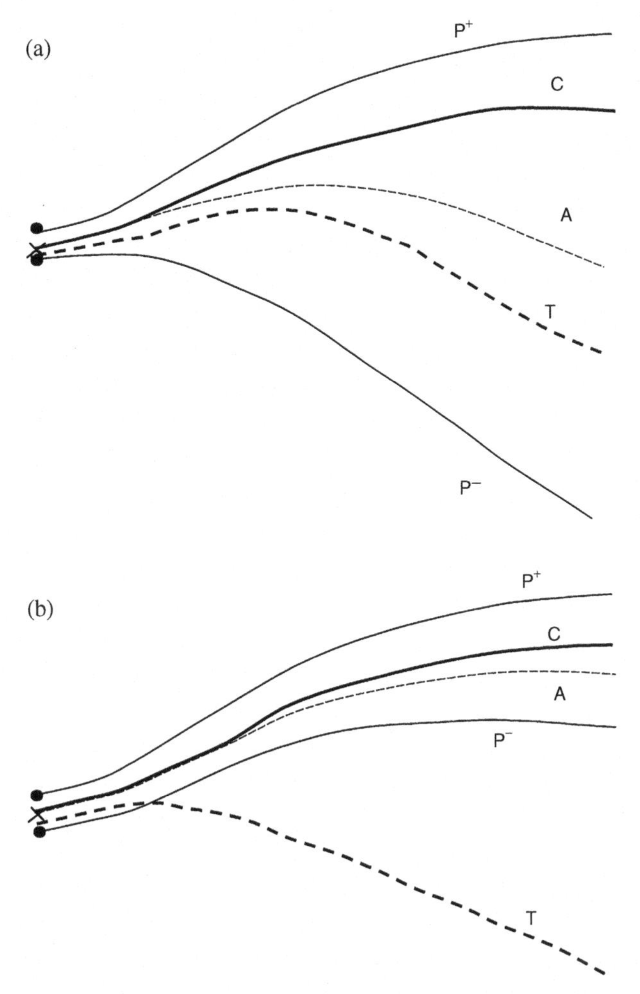

Figure 6.5.1: (a) Schematic of the components of a typical ensemble: (1) the control forecast (labeled C) which starts from the analysis (denoted by a cross), which is the best estimate of the true initial state of the atmosphere; (2) two perturbed ensemble forecasts (labeled P^+ and P^-) with initial perturbations added and subtracted from the control; (3) the ensemble average denoted A; and (4) the "true" evolution of the atmosphere labeled T. This is a "good" ensemble since the "truth" appears as a plausible member of the ensemble. Note that because of nonlinear saturation, the error of the ensemble member initially further away from the truth (in this case P^+) tends to grow more slowly than the error of the member initially closer to the truth. This results in a nonlinear filtering of the errors: the average of the ensemble members tends to be closer to the truth than the control forecast (Toth and Kalnay, 1997, also compare with Fig. 1.7.1). (b) Schematic of a "bad" ensemble in which the forecast errors are dominated by system errors (such as model deficiencies). In this case, the ensemble is not useful for forecasting, but it helps to identify the fact that forecast errors are probably due to the presence of systematic errors, rather than to the chaotic growth of errors in the initial conditions.

the control; (3) the *ensemble average* labeled A, and (4) the *true evolution* of the atmosphere (not known in real time), labeled T. This is an example of a "good" ensemble since the true evolution appears to be a plausible member of the ensemble. Figure 6.5.1(b) shows an example of a "bad" ensemble, in which the forecast errors are dominated by problems in the forecasting system (such as model deficiencies) rather than the chaotic growth of initial errors. In this case, the true evolution is quite different from the members of the ensemble, but the ensemble is still useful in identifying the presence of a deficiency in the forecasting system, which, with a single forecast, could not be distinguished from the growth of errors in the initial conditions.

Ensemble forecasting has essentially three basic goals. The first is *to improve the forecast by ensemble averaging*. The improvement is a result of the tendency of the ensemble average to filter out the components of the forecast that are uncertain (where the members of the ensemble differ from each other) and to retain those components that show agreement among the members of the ensemble. The filtering can take place only during the nonlinear evolution of the perturbations: if the perturbations are added to and subtracted from the analysis, the ensemble average forecast is equal to the control while the perturbations remain linear. The improvement of the ensemble average with respect to the control, shown schematically in Fig 6.5.1(a), is noticeable in Fig. 1.7.1 after a few days of forecasts with the NCEP global ensemble. The second goal is to *provide an indication of the reliability of the forecast*: if the ensemble forecasts are quite different from each other, it is clear that at least some of them are wrong, whereas if there is good agreement among the forecasts, there is more reason to be confident about the forecast (cf. e.g., Fig. 1.7.2(a) and (b)). The quantitative relationship between the ensemble spread and the forecast error (or conversely, between the forecast agreement and the forecast skill) has yet to be firmly established, but is now routinely taken into consideration by human forecasters. The third goal of ensemble forecasting is to *provide a quantitative basis for probabilistic forecasting*. In the example in Fig. 6.4.1, one could claim that the ensemble indicates a 40% probability of cluster A and 60% for cluster B.

An ensemble forecasting system requires the definition of the initial amplitude and the horizontal and vertical structure of the perturbations. Typically, the initial amplitude is chosen to be close to the estimated analysis error. The amplitude of the analysis uncertainty depends on the distribution of the observations. Its statistical distribution can be estimated from the analysis error covariance (Chapter 5), which depends on the accuracy of the statistical assumptions, or empirically, from the rms differences between independent analysis cycles (Fig. 6.5.2).

Ensemble forecasting methods differ mostly in the way the initial perturbations are generated, and can be classified into essentially two classes: those that have random initial perturbations, and those where the perturbations depend on the dynamics of the underlying flow. In the first class, which we can denote Monte Carlo forecasting, the initial perturbations are chosen to be "realistic", i.e., they have horizontal and vertical structures *statistically* similar to forecast errors, and amplitudes compatible

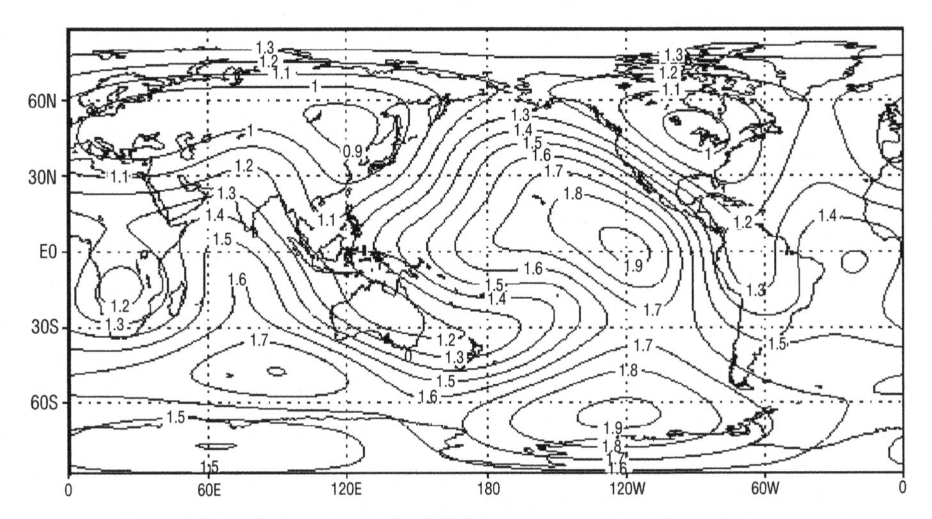

Figure 6.5.2: Estimation of the 500-hPa geopotential height analysis uncertainty obtained from running two independent analysis cycles, computing their rms difference, and using a filter to retain the planetary scales. The units are arbitrary. Note the minima over and downstream of rawinsonde-rich land regions and the maxima over the oceans (Courtesy I. Szunyogh, University of Maryland.)

with the estimated analysis uncertainty. In the Monte Carlo ensembles, the amplitudes are realistic but the perturbations themselves are chosen randomly, without regard to the "dynamics of the day". For example, Errico and Baumhefner (1987) and Mullen and Baumhefner (1994) developed a Monte Carlo method that results in realistic perturbations compatible with the average estimated analysis error. However, by construction, this type of Monte Carlo forecast does not include finite-size "growing errors of the day" which are almost certainly present in the analysis. The experiments of Hollingsworth (1980), Hoffman and Kalnay (1983), and Kalnay and Toth (1996) suggest that random initial perturbations do not grow as fast as the real analysis errors, even if they are in quasi-geostrophic balance. A second class of methods which includes errors of the day has been developed, tested, and implemented at several operational centers. The first two methods of this class implemented operationally are known as "breeding" and "singular vector" (or optimal perturbations) methods. In contrast to Monte Carlo forecasting, they are characterized by including in the initial perturbations growing errors that depend on the evolving underlying atmospheric flow. Two other methods in this class that are also very promising are based on ensembles of data assimilations, and ensembles based on operational systems from different centers, combining different models and data assimilations.

6.5.1 Breeding

Ensemble experiments performed at NCEP during 1991 showed that initial ensemble perturbations based on lagged average forecasting, scaled lagged average forecasting and on the forecast differences between forecasts verifying at the same initial time, grew much faster than Monte Carlo perturbations with the same overall size and statistical distribution (Kalnay and Toth, 1996). It was apparent that the differences in the growth rate were due to the fact that the first group included perturbations that, by construction, "knew" about the evolving underlying dynamics. Toth and Kalnay (1993, 1996, 1997) created a special operational cycle designed to "breed" fast growing "errors of the day" (Fig. 6.5.3(a)). Given an evolving atmospheric flow (either a series of atmospheric analyses, or a long model run), a breeding cycle is started by introducing a random initial perturbation ("random seed") with a given initial size (measured with any norm, such as the rms of the geopotential height or the kinetic energy). It should be noted that the random seed is introduced only once. The same nonlinear model is integrated from the control and from the perturbed initial conditions. From then on, at fixed time intervals (e.g., every 6 hours or every 24 hours), the control forecast is subtracted from the perturbed forecast. The difference is scaled down so that it has the same amplitude (defined using the same arbitrary norm) as the initial perturbation, and then added to the corresponding new analysis or model state. It was found that beyond an initial transient period of 3–4 days after random perturbations were introduced, the perturbations generated in the breeding cycle (denoted bred vectors), acquired a large growth rate, faster than the growth rate for Monte Carlo forecasting or even scaled lagged average forecasting and forecast differences.

Toth and Kalnay (1993, 1997) also found that (after the transient period of 3–4 days) the shape or structure of the perturbation bred vectors did not depend on either the norm used for the rescaling or the length of the scaling period. The bred vectors did depend on the initial random seed in the sense that regional bred vector perturbations would have the same shape but different signs, and that in many areas two or more "competing bred vectors" appeared in cycles originated from different random seeds. The breeding method is a *nonlinear generalization of the method used to construct Lyapunov vectors* (performing two nonlinear integrations and obtaining the approximately linear perturbation from their difference). Since the bred vectors are related to Lyapunov vectors localized in both space and time, it is not surprising that they share their lack of dependence on the norm or on the scaling period. Toth and Kalnay (1993) have argued that breeding is similar to the analysis cycle. In the analysis cycle (represented schematically in Fig. 6.5.3(b)) errors are evolved in time through the forecast used as background, and they are only partially corrected through the use of noisy data. Therefore, Toth and Kalnay argued that the analysis errors should project strongly on bred vectors. Corazza *et al.* (2002) compared bred vectors and background errors for a quasi-geostrophic model data assimilation system developed by Morss *et al.* (2001) Plate 1 shows a typical comparison, depicting that in fact there is a strong resemblance between the structure of the errors of the forecast used as a first guess and the bred vectors valid at the same time (Corazza *et al.*, 2002). Since the analysis errors are dominated by the background errors, especially

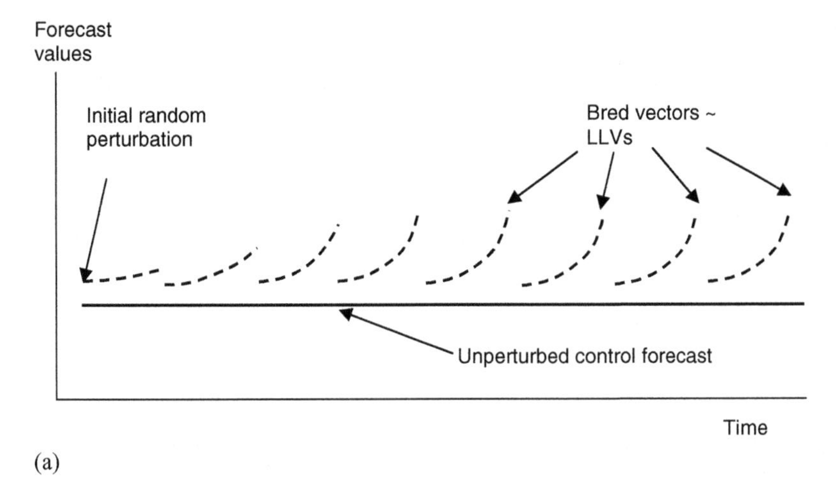

(a)

Analysis and forecasts
Analysis
Observations with errors
True atmosphere
Time

(b)

Figure 6.5.3: (a) Schematic of a breeding cycle run on an unperturbed (control) model integration. The initial growth after introducing a random initial perturbation is usually very small, but with time, the perturbation is more dominated by growing errors. The initial transient with slow growth lasts about 3–5 days. The difference of the complete perturbed (dashed line) and control (full line) forecasts is scaled back periodically (e.g., every 6 or every 24 hours) to the initial amplitude. The rescaling is done by dividing all the forecast differences by the same observed growth (typically about 1.5/day for mid-latitudes). In operational NWP, the unperturbed model integration is substituted by short-range control forecasts started from consecutive analysis fields. The breeding cycle is a nonlinear, finite-time, finite-amplitude generalization of the method used to obtain the leading Lyapunov vector. (Adapted from Kalnay and Toth, 1996.) (b) Schematic of the 6-h analysis cycle. Indicated on the vertical axis are differences between the true state of the atmosphere (or its observational measurements, burdened with observational errors). The difference between the forecast and the true atmosphere (or the observations) increases with time in the 6-h forecast because of the presence of growing errors in the analysis. (Adapted from Kalnay and Toth, 1996.)

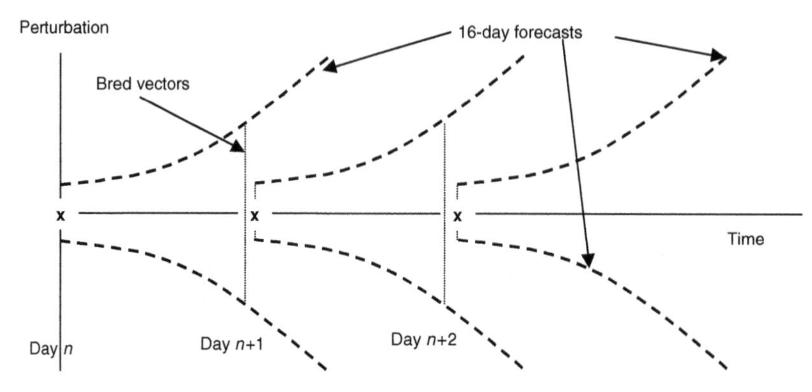

Figure 6.5.4: Schematic of a self-breeding pair of ensemble forecasts used at NCEP. Every day, the 1-day forecast from the negative perturbation is subtracted from the 1-day forecast from the positive perturbation. This difference is divided by 2, and then scaled down (by dividing all variables by the 1-day growth), so that difference is of the same size as the initial perturbation. The scaled difference is then added and subtracted from the new analysis, generating the initial conditions for the new pair of forecasts. This self-breeding is part of the extended ensemble forecast system, and does not require computer resources to generate initial perturbations beyond running the ensemble forecasts. (Adapted from Toth and Kalnay, 1997.)

when they are large, this resemblance indicates that the forecast and analysis errors do indeed project strongly on the bred vectors.

Figure 6.5.4 shows a schematic of how breeding cycles are self-propagated from the ensemble forecasts. The bred perturbations are defined every day from the difference between the one-day positive and the negative perturbation forecasts divided by 2, they are scaled down by their growth during that day, and added and subtracted to the new analysis valid at the time. This provides the initial positive and negative perturbations for the ensemble forecasts at no additional cost beyond that of computing the ensemble forecasts. Separate breeding cycles differ only in the choice of random initial perturbations (performed only once). It has been found empirically that for the atmosphere the finite amplitude bred vectors do not converge to a single "leading bred vector" (Kalnay *et al.*, 2002).

Figures 6.5.5(a) and (b) show two out of five operational bred perturbations corresponding to 5 March 2000 at 00UTC. Figure 6.5.5(c) presents an estimate of the effective local dimension of the subspace of the five perturbations using the bred vector dimension defined in Patil *et al.* (2001).[5] Only the areas where the local dimension has collapsed from the original five independent directions (shapes or structure of the

5 The local bred vector dimension is obtained as $\psi(\sigma_1, \ldots, \sigma_k) = \left(\Sigma_{i=1}^{k}\sigma_i\right)^2 / \Sigma_{i=1}^{k}\sigma_i^2$, where σ_i are the singular values corresponding to the k bred vectors within a region of about 10^6 m by 10^6 m, and it defines the effective local dimensionality. For example, if four out of five bred vectors lie along one direction, and one lies along a second direction, the bred vector dimension would be $\psi(\sqrt{4}, 1, 0, 0, 0) = 1.8$, less than 2 because one direction is more dominant than the other in representing the original data (Patil *et al.*, 2001).

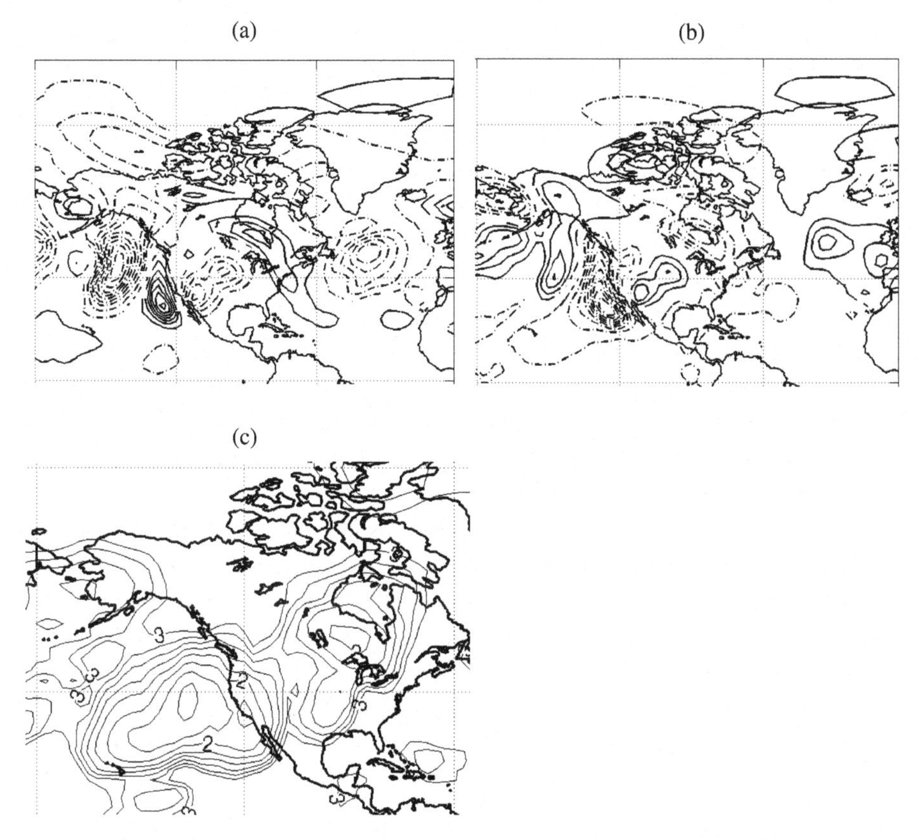

Figure 6.5.5: Examples of bred vectors (500-hPa geopotential height field differences, without plotting the zero contour) from the NCEP operational ensemble system valid at 5 March 2000: (a) bred vector 1; (b) bred vector 5. Note that over large parts of the eastern Pacific Ocean and western North America, the two perturbations have shapes that are very similar but of opposite signs and/or different amplitudes. In other areas the shape of the perturbations is quite different. (c) The bred-vector-local dimension of the five perturbations subspace (Patil *et al.*, 2001). Only dimensions less than or equal to 3 are contoured with a contour interval 0.25. In these areas the five independent bred vectors have aligned themselves into a locally low-dimensional subspace with an effective dimension less than or equal to 3. (Courtesy of D. J. Patil.)

bred vectors) to three or less are contoured. Note that these are the areas where the independently bred vectors aligned themselves into a smaller subspace. The collapse of the perturbations into fewer dimensions is what one could expect if there are locally growing dominant Lyapunov vectors expressing the regional dominant instability of the underlying atmospheric flow. These low-dimensional areas are organized into horizontal and vertical structures and have a lifetime of 4–7 days, similar to those of baroclinic developments (Patil *et al.*, 2002).

The breeding ensemble forecasting system was introduced operationally in December 1992 at NCEP, with two pairs of bred vectors (Tracton and Kalnay, 1993).

In 1994, seven pairs of self-breeding cycles replaced the original four perturbed forecasts. In addition, a regional rescaling was introduced that allowed larger perturbation amplitudes over ocean than over land proportionally to the estimate of the analysis uncertainty, Fig. 6.5.2 (Toth and Kalnay, 1997).

Toth and Kalnay (1993) found that when the initial amplitude was chosen to be within the range of estimated analysis errors (i.e., between 1 m and 15 m for the 500-hPa geopotential height) the bred vectors developed faster in strong baroclinic areas. Their horizontal scale was that of short baroclinic waves, and their hemispherical average growth rate was about 1.5/day (similar to the estimated growth of analysis errors). However, if the initial amplitude was chosen to be much smaller than the estimated analysis errors (10 cm or less), then a different type of bred vector appeared, associated with convective instabilities, which grew much faster than baroclinic instabilities (at a rate of more than 5/day). The faster instabilities saturated at amplitudes much smaller than the analysis error range (Fig. 6.5.6). Toth and Kalnay (1993) suggested that the use of nonlinear perturbations in breeding has the advantage of filtering Lyapunov vectors associated with fast growing but energetically irrelevant instabilities, like convection. This was confirmed by Lorenz (1996), who performed experiments with a low-order model containing large amplitude but slowly growing modes coupled with fast growing modes with very small amplitudes. Lorenz found that the use of breeding using finite amplitudes yielded the Lyapunov vectors of the large amplitude, slowly growing vectors, as desired, whereas for very small

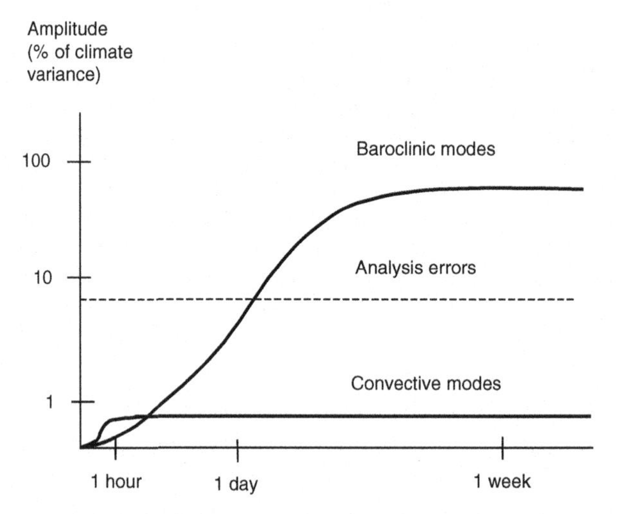

Figure 6.5.6: Schematic of the time evolution of the rms amplitude of high-energy baroclinic modes and low-energy convective modes. Note that although initially growing much faster than the baroclinic modes, convective modes saturate at a substantially lower level. These modes are therefore insignificant in the analysis/ensemble perturbation problem, since the errors in the analysis (dashed line) are much larger than the convective saturation level. (Adapted from Toth and Kalnay, 1993.)

amplitudes the Lyapunov vectors of the fast system were recovered. In a complex system like the atmosphere, with multiple-scale instabilities, using breeding may thus be more appropriate than using Lyapunov vectors, which, in a model including *all* atmospheric instabilities, would yield vectors associated with Brownian motion, which are the fastest, though clearly irrelevant, instabilities present in the system (Kalnay *et al.*, 2002). The nonlinear saturation of irrelevant fast growing modes is an advantage that suggests the use of breeding for other problems. For example, for seasonal and interannual forecasting using a coupled ocean–atmosphere system, the slower growing (but very large energy amplitude) coupled ENSO instabilities could perhaps be captured, while eliminating through nonlinear saturation the irrelevant details of weather perturbations (Cai *et al.*, 2002).

Figure 1.7.2 shows two examples of one of the ways information on ensemble forecasts are presented to the users, the "spaghetti plots", or plots showing one contour line for each forecast. In one case, a 5-day forecast verifying on 15 November 1995, the agreement in intensity and location of the contours indicated to the forecasters that this was a very predictable snowstorm (see also the book cover). In the second case, a 2.5-day forecast verifying on 21 October 1995, the ensemble members show unusually strong divergence in the location of a winter storm, warning the human forecasters that this situation is intrinsically unpredictable. Note that although the ensemble forecasts show a wide divergence in the location of the storm, this is also a case in which, in perturbation space, there is very low dimensionality, since the perturbations align themselves along the same basic shape (the perturbations for the winter storm are a one-parameter family so that the local dimension is about 1).

The second example shows the potential value of ensembles in a new area of research: *targeted observations*. In cases like this in which the ensemble indicates a region of large uncertainty in the short-term forecasts, it should be possible to find the area that originated this region of uncertainty in time to launch new observations for the next analysis cycle, and thus decrease significantly the forecast error. Finding the area where the observations should be launched can be done through several approaches. They are the adjoint sensitivity approach, the use of singular vectors (Rabier *et al.*, 1996, Langland *et al.*, 1999, Pu *et al.*, 1998, and others), the use of the quasiinverse of the tangent linear model (Pu *et al.*, 1998), and ensemble-based singular value decomposition (Bishop and Toth, 1999). These methods were tested during FASTEX (Jan–Feb 1997 in the Atlantic) and NORPEX (winter of 1997–1998 in the North Atlantic (Langland *et al.*, 1999, Pu and Kalnay, 1999)). The experience in the North Pacific has been so successful that targeted observations are now performed operationally over the Gulf of Alaska every winter (Szunyogh *et al.*, 2000).

Plate 2 shows another example of how the massive amount of information contained in the ensemble forecast can be conveyed to the forecasters. It shows a probabilistic presentation of a 1-day and a 7-day forecast of precipitation above a threshold of 5 mm in 24 h. The probabilities are simply computed as a percentage of the ensemble members with accumulated precipitation at least as large as the indicated threshold. They both verify on 6 April 2001. Note that the short-range forecast has many areas with probabilities equal to zero or above 95%, indicating that *all* the ensemble members agree that there will be no precipitation or at least 5 mm accumulated precipitation respectively. In the 7-day forecast, the areas with maximum probability of precipitation are generally in agreement with the short-range forecast, indicating the presence of skill. However, by this time there are few areas of consensus on either rain or no rain among the forecasts, since their solutions have dispersed significantly over a week.

6.5.2 Singular vectors

ECMWF developed and implemented operationally in December 1992 an ensemble forecasting system based on initial perturbations that are linear combinations of the singular vectors of the 36-h tangent linear model (Molteni *et al.*, 1996, Molteni and Palmer, 1993, Buizza, 1997, Buizza *et al.*, 1997).

As discussed in Section 6.3, the singular vectors $\mathbf{y}_i(t_0)$ used to create the initial perturbations at the time t_0 are obtained as the leading eigenvectors $\hat{\mathbf{y}}_i(t_0)$ of

$$(\mathbf{W}^{-1})^T\mathbf{L}^T\mathbf{P}^T\mathbf{P}\mathbf{L}\mathbf{W}^{-1}\hat{\mathbf{y}}(t_0) = \sigma^2\hat{\mathbf{y}}(t_0) \tag{6.5.1}$$

subject to

$$\hat{\mathbf{y}}^T(t_o)\hat{\mathbf{y}}(t_o) = 1 \quad \text{and} \quad \mathbf{y}(t_o) = \mathbf{W}^{-1}\hat{\mathbf{y}}(t_o) \tag{6.5.2}$$

ECMWF used as the projection operator $\mathbf{P}$, a symmetric projector operator that includes only forecast perturbations north of 30° N, and as the initial norm the total energy norm $\mathbf{W}^{-1}$. The linear tangent model $\mathbf{L}$ and its adjoint $\mathbf{L}^T$ are computed for the 36-h forecasts (and more recently for the 48-h forecast), which determine the length of the "optimization interval" (Section 6.3). Barkmeijer *et al.* (1998) tested the use of the analysis error covariance as the initial norm instead of the energy norm with good results. The singular vectors obtained with this norm were closer to bred vectors than those obtained with the total energy norm. They also found that the use of evolved vectors (also closer to Lyapunov or bred vectors) resulted in improved results. More recent experiments with a simplified Kalman filter also resulted in promising results (Fischer *et al.*, 1998).

From (6.5.1) and (6.5.2), the initial singular vectors $\mathbf{y}_i(t_0)$ are the perturbations with maximum energy growth north of 30° N, for the time interval 0–36 h (Buizza, 1994), or more recently, 0–48 h. The method used to obtain the singular vectors is the Lanczos algorithm (Golub and Van Loan, 1996), which requires integrating forward with $\mathbf{L}$ for a period t (36 or 48 h), and backward with $\mathbf{L}^T$. This forward–backward integration has to be performed about three times the number of singular vectors desired. Figure 6.5.7 shows an example of the horizontal structure corresponding to the initial and final singular vectors #1, 3 and 6. Figure 6.5.8 shows the corresponding initial and evolved vertical energy structure (Buizza, 1997). Singular vectors defined

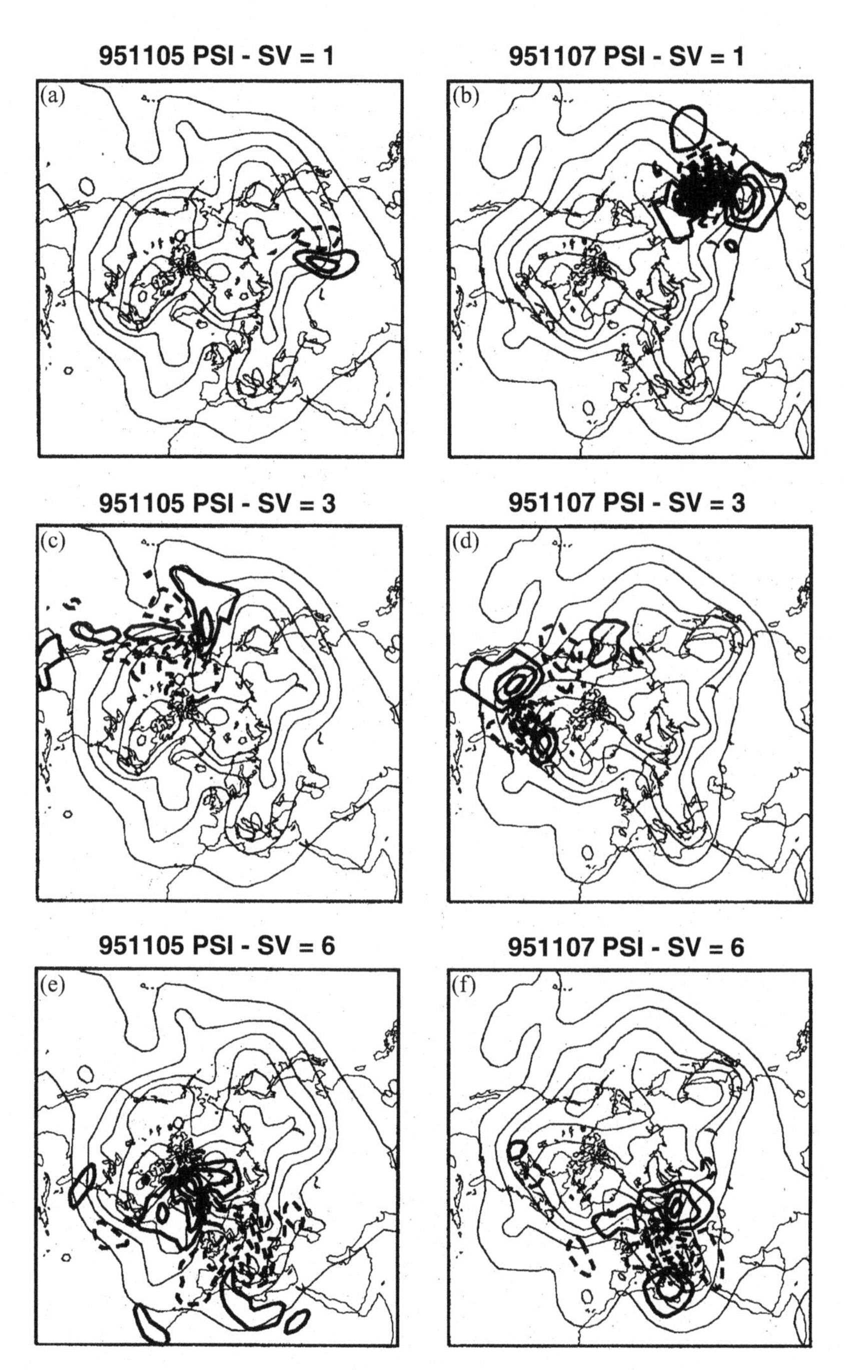

Figure 6.5.7: Singular vectors numbers 1 (top panels), 3 (middle panels), and 6 (bottom panels) at initial (left panels) and optimization time (right panels). Each panel shows the singular vector streamfunction at model level 11 (approximately 500 hPa), superimposed to the trajectory 500-hPa geopotential height field. Streamfunction contour interval is 0.5×10^{-8} m^2 s^{-1} for left panels and 20 times larger for the right panels; geopotential height contour interval is 80 m (from Buizza, 1997).

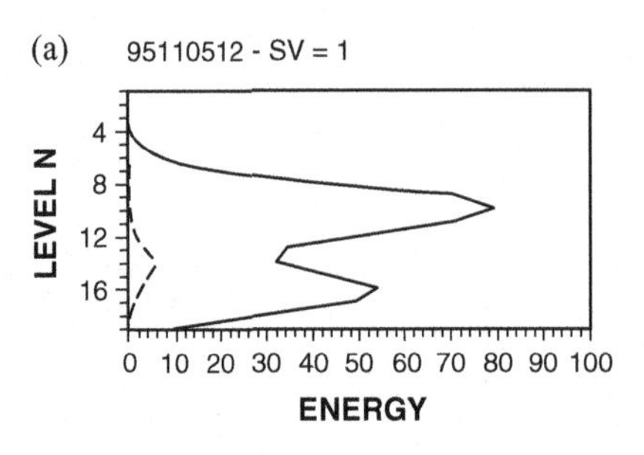

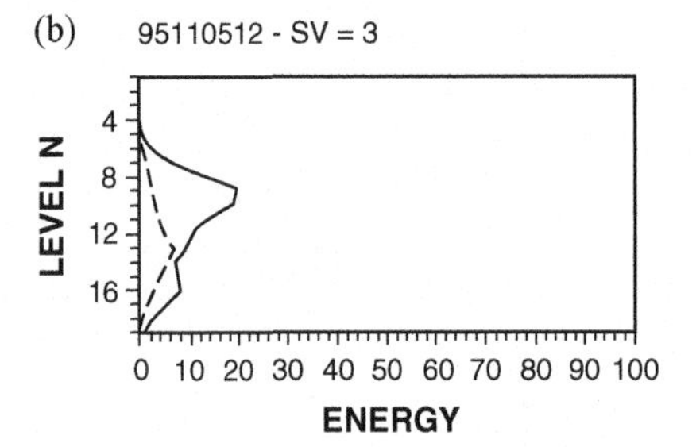

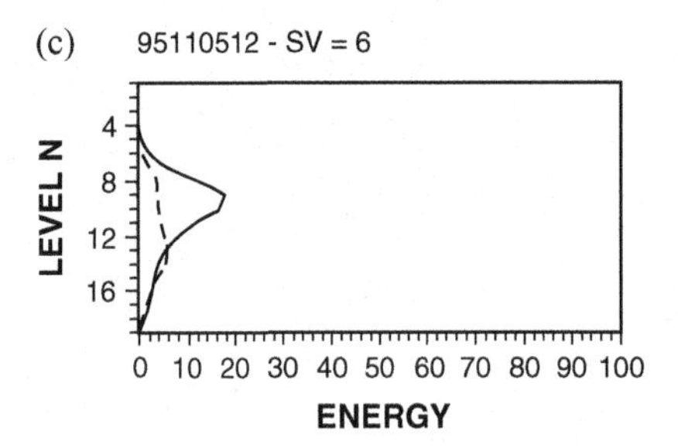

Figure 6.5.8: Total energy ($m^2\ s^{-2}$) vertical profile of the (a) first, (b) third, and (c) sixth singular vector of 5 November 1995, at the initial (dashed line, values multiplied by 100) and optimization (solid line) times. Note that singular vectors are normalized to have unit initial total energy norm. (From Buizza, 1997.)

with the total energy norm tend to have a maximum initial energy at low levels (about 700 hPa), and their final (evolved) energy at the tropopause level. In 1996, when they were using 36-h forecasts for the linear tangent model, ECMWF used 16 singular vectors selected from 38 leading singular vectors. This required a daily integration equivalent to about $3 \times 36 \times 2 \times 38$ hours of model integration with either $\mathbf{L}$ or $\mathbf{L}^T$ to create the perturbations. For this reason, the computation was done with a lower resolution (T42/19 level) than the operational model. A second set of perturbations was added for the Southern Hemisphere, which originally had no perturbations, requiring additional computations.

The selection of 16 singular vectors is such that the first four are always selected, and from the fifth on, each subsequent singular vector is selected if 50% of its energy is located outside the regions where the singular vectors already selected are localized. Once the 16 singular vectors are selected, an orthogonal rotation in phase space and a final rescaling are performed to construct the ensemble perturbations. The purpose of the phase-space rotation is to generate perturbations that have the same globally averaged energy as the singular vectors but smaller local maxima and more uniform spatial distribution. The rotated singular vectors are characterized by having similar growth rates (at least to the period of optimization). The rotation is defined to minimize the local ratio between the perturbation amplitude and the amplitude of the analysis error estimate of the ECMWF OI analysis. The rescaling allows local amplitudes up to $\sqrt{1.5}$ larger than the OI error.

The 16 rotated perturbations are three-dimensional fields of temperature, vorticity, divergence, and surface pressure (no moisture, since the propagator **L** is "dry", although there has been more recent work to include physical processes in the tangent linear model and adjoint). They are added and subtracted to the control initial conditions to create 33 initial conditions (32 + control), from which the ensemble forecast is run with the nonlinear model at T63 resolution.

In 1997 ECMWF changed the system to an ensemble of 50 members (plus control) run at a resolution of T156 (with a linear Gaussian grid, since their use of a semi-Lagrangian scheme allows the use of a more efficient linear rather than quadratic grid). This increase in resolution had a major positive effect on the quality of the ECMWF ensemble forecasting system. In March 1998 ECMWF added to the initial perturbations the *evolved* (or final) singular vectors from 48 h *before* the analysis time, which also resulted in improved results. The 2-day evolved singular vectors are much closer to the Lyapunov vectors (or bred vectors) (Barkmeijer *et al.*, 1998).

Initially both NCEP and ECMWF considered in their ensembles only the errors generated by uncertainties in the initial conditions, and neglected the additional errors due to the models themselves. This is a reasonable (but not perfect) assumption only for the extratropics (Reynolds *et al.*, 1994). In 1998 ECMWF tested an innovative way to account for the fact that the model has deficiencies (Buizza *et al.*, 1999). The time derivatives of the physical parameterizations are multiplied by Gaussian random numbers with a mean of 1.0 and a standard deviation of 0.2, which have a time lag correlation of several hours and horizontal correlation of a few hundred kilometers. This introduction of randomness in the "physics" had a very good impact on the ensemble. It increased the ensemble spread to levels similar to those of the control forecast error, which is a necessary condition if "nature" (the verifying analysis) is to be a plausible member of the ensemble (Toth and Kalnay, 1993).

6.5.3 Ensembles based on multiple data assimilation

Houtekamer *et al.* (1996) and Houtekamer and Mitchell (1998) have developed a very promising ensemble forecasting system based on running an ensemble of data assimilation systems to create the initial conditions. In their different data assimilation systems they add random errors to the observations (in addition to the original observational errors) and include different parameters in the physical parameterizations of the model in different ensembles. This is a promising approach, related to but more general than breeding. One novel approach introduced by Houtekamer *et al.* (1996) is the use of perturbations in the physical parameterizations in the models used in different analysis cycles. Through a careful combination of changes in major parameterizations, it is possible to use the ensemble forecasts to isolate the impact of particular parameterizations. As indicated by the results of Miller *et al.* (1994), the introduction of uncertainty in the model should improve the efficiency of the ensemble.

Hamill *et al.* (2000a) have shown in a quasi-geostrophic system that the multiple data assimilation ensemble system performs better than the singular vector or breeding approaches. The computational cost of creating the initial perturbations is comparable to that of the singular vector approach, whereas in the breeding method the perturbations are obtained as a by-product of the ensemble forecasts themselves and are therefore cost-free.

6.5.4 Multisystem ensemble approach

The ensemble forecasting approach should replicate in the initial perturbations the statistical uncertainty in the initial conditions: ideally, the initial perturbations should be the leading eigenvectors of the analysis error covariance (Ehrendorfer and Tribbia, 1997). Moreover, it should also reflect model imperfections and our uncertainty about model deficiencies. In the standard approaches discussed so far the uncertainty in the initial conditions is introduced through *perturbations added to the control analysis*, which is the best estimate of the initial conditions. As a result, the perturbed ensemble forecasts are, on the average, somewhat less skillful than the control forecast (see, e.g., Fig. 1.7.1). Similarly, when perturbations are introduced upon the control model parameterizations (Buizza *et al.*, 1999, Houtekamer *et al.*, 1996), the model is made slightly worse, since the control model has been tuned to best replicate the evolution of the atmosphere.

A different approach that has become more popular recently is that of a multisystem ensemble. It has long been known that an ensemble average of operational global forecasts from different operational centers is far more skilful than the best individual forecast (e.g., Kalnay and Ham, 1989, Fritsch *et al.*, 2000, and references therein). More recently, it has been shown that this is true also for shorter-range ensembles of regional models (Hou *et al.*, 2001), and that the use of multisystems can therefore extend the utility of ensemble forecasting to the short-range. Krishnamurti *et al.* (2000) have shown that if the multisystem ensemble includes correction of the systematic errors by regression, the quality of the ensemble system is further significantly improved. Krishnamurti *et al.* (2000) call this multiple system approach a "superensemble".

The advantages of a multisystem ensemble are not surprising. Instead of adding perturbations to the initial analysis, and introducing perturbations into the control model parameterizations, the multisystem approach takes the best (control) initial conditions and the best (control) model estimated at different operational centers that run competitive state-of-the-art operational analyses and model forecasts. Thus the multisystem probably samples the true uncertainty in both the initial conditions and the models better than any perturbation introduced *a posteriori* into a single operational system. The statistical correction of systematic errors introduced by Krishnamurti *et al.* (2000) is an added benefit of the method which can be considered an ideal "poor person's" approach to ensemble forecasting (Wobus and Kalnay, 1995).

6.6 Growth rate of errors and the limit of predictability in mid-latitudes and in the tropics

Lorenz (1963a, 1982) suggested that the limit of deterministic predictability was about 2 weeks. He obtained this empirical estimate from the doubling time of small errors derived from identical-twin model experiments, and from the rate of separation in time of atmospheric analogs (atmospheric states initially very similar to each other). Two weeks continues to be a good estimate of the limit of predictability despite the fact that different models provided different estimates (Charney *et al.*, 1966). The analog method cannot give a precise answer because it would take an exceedingly long time to find atmospheric analogs close enough to estimate the growth of small errors (Van den Dool, 1994).

As indicated in the introduction to this chapter, the doubling time for small errors in the mid-latitude synoptic (weather) scales, for which the dominant instability is baroclinic, was estimated in the 1960s to be 3 days or longer. Modern models have much more resolution and are less sluggish than the early primitive equation models. Identical-twin experiments with these models and measurements of actual numerical forecast error growth have lowered the estimate of the doubling time of small errors from 3 days to about 2 days (Lorenz, 1982, Dalcher and Kalnay, 1987, Simmons *et al.*, 1995, Toth and Kalnay, 1993).

Lorenz (1982) suggested a simple way to parameterize the evolution of small errors in a *perfect* model, in which the only source of errors is the unstable growth of small errors in the initial conditions, using the logistic equation:

$$\frac{d\varepsilon}{dt} = a\varepsilon(1-\varepsilon) \tag{6.6.1}$$

Here, ε represents the rms average forecast error scaled so that at long forecast leads $\varepsilon \to 1$, i.e., it is the rms forecast error divided by the square root of twice the variance of the atmosphere (cf. Section 6.4). Equation (6.6.1) indicates that very small errors grow exponentially with a growth rate a. When they reach finite amplitude, the error growth rate is lowered by the last factor on the right-hand side, which slows it down until it saturates at $\varepsilon \approx 1$. The solution of the logistic equation (6.6.1) is

$$\varepsilon(t) = \frac{\varepsilon_0 e^{at}}{1+\varepsilon_0(e^{at}-1)} \tag{6.6.2}$$

where ε_0 is the initial error.

Exercise 6.6.1: Derive eq (6.6.2) using separation of variables.

Figure 6.6.1 shows the solution for two values of the initial error, 10% and 1%, and an error growth rate $a = 0.35$/day, corresponding to a doubling time of about 2 days. The analysis error in the 500-hPa geopotential heights in current operational systems is of the order of 5–15 m, and the natural variability about 100 m, so that the current level of error in the initial conditions is ~10% or less. The upper limit for the best initial error achievable from data assimilation can be reasonably estimated to be no less than 1%. This is because, as pointed out by Lorenz, even if the observing system was essentially perfect at synoptic scales, errors in much smaller, unresolved scales would grow very fast and through non-linear interactions quickly introduce finite errors in the initial synoptic scales of the model. The solution of the logistic equation for initial errors of 10% and 1% (Fig. 6.6.1) suggests that 2 weeks is indeed a reasonable estimate of the time at which the forecast errors become so large that the ability to predict weather in mid-latitudes is lost. The range between the two curves can be taken as a simple upper estimate of how much forecasts could be improved by improving the initial conditions.

However, this is only an estimate of the *average* predictability in a perfect model. The actual predictability is quite variable and depends on the "atmospheric instabilities of the day". The 2-week "limit", which seemed huge during the 1960s when 2-day forecasts had little skill, is no longer large compared with what can be occasionally attained with current models. For example, during a very predictable period in December 1995, several numerical weather forecasts remained skillful for 15 days,

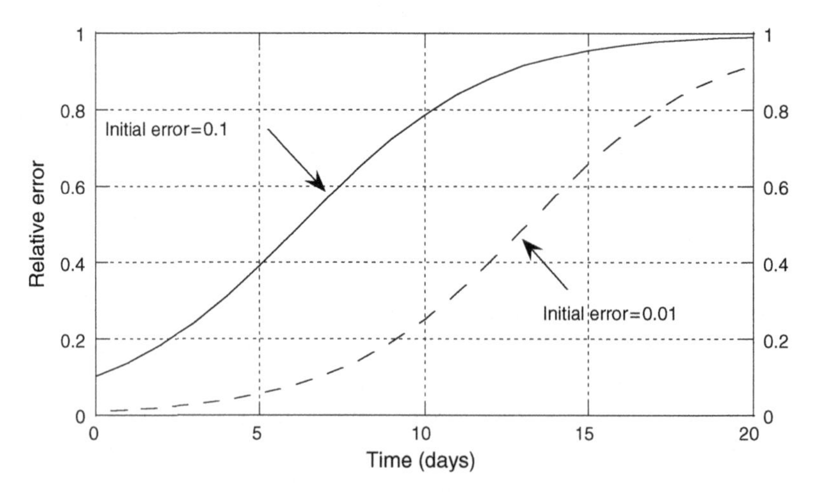

Figure 6.6.1: Time evolution of the rms forecast error divided by the square root of twice the climatological variance. It assumes that the forecast error growth satisfies the logistic equation (6.6.1), and that the growth rate of small errors is about 0.35/day, corresponding to a doubling of small errors in 2 days. Analysis errors in the initial conditions are estimated to be about 10% or less, but not smaller than 1%.

with a pattern correlation between the forecast and observed anomalies (known as *anomaly correlation*) of more than 70% for the Northern Hemisphere extratropics (Toth and Kalnay, 1996). There are also periods in which the atmospheric predictability is much lower than average, as indicated by the fact that different operational global forecasts tend to show large dips in the medium-range forecast skill on the same days. Because of the day-to-day variability of atmospheric predictability, it is important to use forecast ensembles. They provide a tool for estimating the day-to-day variations in predictability and allow human forecasters to extend the range of the forecasts provided to the public during periods of high predictability. Ensembles based on multiple models have proven to be especially useful to the forecasters.

From (6.6.2) it is clear that the limit of predictability depends on the rate of error growth, which is the inverse of the *e*-folding or exponential time scale. Generally the time scales of instabilities are related to their spatial scales, so that small-scale instabilities grow much faster than those with larger scales. For this reason short synoptic waves are typically less predictable than longer waves (e.g., Dalcher and Kalnay, 1987), and mesoscale phenomena, such as fronts, squall lines, mesoscale convective systems and tornadoes, are intrinsically predictable for shorter time scales, of the order of a day or less (e.g., Droegemeir, 1997). Convection is a typical example of a short time scale phenomenon: cumulus clouds grow with an exponential time scale of the order of 10 minutes or so. It is therefore impossible to predict the precipitation associated with an individual thunderstorm for more than about an hour. Nevertheless, *if convective activity is organized or forced by the larger scales*, then convective precipitation can remain predictable much longer than individual thunderstorms. For example, summer convective precipitation is notoriously difficult to predict. However, when summer mesoscale convection is *forced* by a synoptic scale system, convection can be predicted to occur when and where forced by the larger scales, and therefore becomes predictable well beyond its own short predictability time scale. Similarly, mesoscale phenomena forced by the interaction of synoptic scales with surface topography have a much longer predictability than when they are not subjected to this organizing influence from the larger scales.

Two types of surprisingly regular progression of smaller scale phenomena have been discovered. As indicated above, mesoscale summer convection in the USA, when "unforced" by upward motion associated with synoptic-scale waves, is extremely difficult to predict. However, an examination of the Doppler radar reflectivities has shown that the area of maximum convection has a tendency to propagate eastward with considerable regularity, and with its intensity modulated by the diurnal cycle. The individual maxima of this wave-like propagation can be traced on radar reflectivities for 1–3 days (Carbone *et al.*, 2000). This surprising discovery implies that such unforced convective activity, in principle, should be predictable for a day or two. Another example of regular propagation of convection is the Madden and Julian (1971, 1972) oscillation (MJO). The MJO has a zonal wavenumber 1 with maximum amplitude in the deep tropics, and it moves eastward around the Equator with a period of 30–60 days. The MJO is not always present, but there are periods of several months in which it is very prominent (Weickman *et al.*, 1985). Although current models are not yet able to reproduce well the intensity and speed of propagation of the MJO, its regularity indicates that, in principle, there should be predictability in the convective precipitation associated with the MJO that could be exploited by dynamical or statistical methods for time scales of a month or longer.

Both of these are examples of small-scale convection organized into a regular, longer lasting propagation. It may take many years before dynamical models are able to fully reproduce the observed quasi-regular motion. In the mean time, a combination of statistical and dynamical methods may be the best way to exploit this latent longer time scale predictability.

So far we have mostly discussed the predictability of weather in mid-latitudes. The dynamics of mid-latitudes is dominated by synoptic-scale baroclinic instabilities (Holton, 1992), and the limit of deterministic weather predictability is a reflection of their baroclinic instability rates of growth.

In the tropics, the situation is quite different. Baroclinic instability is generally negligible in the tropics, and barotropic and convective instabilities, and their interactions, are more dominant. Phenomena like easterly waves are a reflection of barotropic instability, and are less intense than baroclinic instabilities. Easterly waves are strongly modulated by convective precipitation, whereas in mid-latitudes, large-scale precipitation has a smaller effect on the evolution of the synoptic waves. Moreover, global atmospheric models are less accurate in the tropics, because their ability to parameterize realistically the subgrid scale processes such as convection, which are dominant in the tropics, is not as good as the numerical representation of the resolved baroclinic dynamics, which is dominant in the extratropics. In the tropics, the assumption of a perfect model is therefore much less justified than in the extratropics.

Equation (6.6.1) describes reasonably well error growth in a perfect model (an identical-twin experiment). The growth rates of random errors in an *imperfect* model have been parameterized by Dalcher and Kalnay (1987) and Reynolds *et al.* (1994) fitting operational forecast errors with an extension of the logistic equation (6.6.1) which includes growth of errors due not only to the presence of errors in the initial conditions but also to model deficiencies:

$$\frac{dv}{dt} = (bv + s)(1 - v) \tag{6.6.3}$$

Here v is the variance of the random error (systematic or time averaged errors having been separated beforehand), b is the growth rate for small error variances due to instabilities ("internal" source) and s is an "external" source of random error variance due to model deficiencies. The solution of (6.6.3) is given by

$$v(t) = 1 - \frac{1 + s}{1 + \mu} \tag{6.6.4}$$

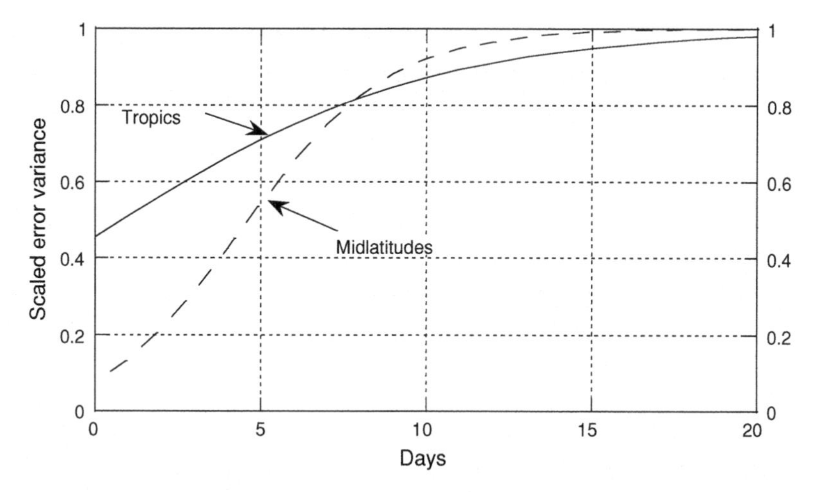

Figure 6.6.2: Parameterization of scaled forecast error variance in the presence of model deficiencies with values of the growth rate due to model deficiencies s and to instabilities b appropriate for mid-latitudes ($(b = 0.4, s = 0.05)$ and the tropics $(b = 0.1, s = 0.1)$. (From Reynolds *et al.*, 1994.)

where

$$\mu = \frac{v(0) + s/b}{1 - v(0)} e^{(b+s)t}$$

Reynolds *et al.* (1994) found that (6.6.4) fits observed errors fairly well, and estimated that the internal error growth rate is given by about 0.4/day in mid-latitudes and 0.1/day in the tropics. The external growth rate due to model deficiencies was found to be small, about 0.05/day in mid-latitudes, and considerably larger, about 0.1–0.2/day, in the tropics (Fig. 6.6.2). Although these estimates, which were obtained from fitting observed error growth, are only valid to the extent that the error variance growth for an imperfect model follows (6.6.3), they do reflect qualitatively the notion that in the tropics the synoptic scales are less unstable than in the extratropics, because of the absence of baroclinic instabilities. At the same time, the tropics are much harder to model, because of the difficulties associated with parameterizations of cumulus convection, which is much more influential in the tropics than in the extratropics. These results suggest that if convection did not play such a dominant role in the tropics, tropical weather forecasts would be skillful for longer periods than mid-latitudes predictions. In reality, the dominant role of tropical convection and the difficulties of its parameterization lead to the fact that currently tropical forecasts maintain useful skill only for about 3–5 days, whereas in the extratropics forecasts remain skillful on the average for 7 days or so.

6.7 The role of the oceans and land in monthly, seasonal, and interannual predictability

It is well known that there is considerable atmospheric variability not only in the day-to-day weather but also in longer time scales, such as weekly, monthly, seasonal, interannual, and even decadal averages. For example, the atmospheric circulation averaged for January 1987 (an El Niño year) was substantially different than that of January 1989 (a La Niña year). Similarly, July 1987, during El Niño, was quite different from July 1988, during La Niña. In this case, it is clear that the differences were strongly influenced by the ENSO, a much longer lasting phenomenon than individual weather events. However, seasonal and interannual atmospheric variability can also take place due to unpredictable weather "noise", not just to longer-lasting surface forcings such as ocean SST anomalies, soil moisture or snow cover anomalies. The monthly or seasonally averaged atmospheric anomalies due to weather noise (e.g., a monthly averaged cold and wet January in the eastern USA because of the passage of two or three very strong cyclones) are unpredictable beyond the first two weeks because weather itself is not predictable. On the other hand, the variability due to long-lasting surface anomalies, of which SST anomalies are the most important, is predictable, if we can predict the SST anomaly. *Potential predictability* beyond the limit of deterministic weather predictability can be defined as the difference between the total variance of the anomalies averaged over a month or a season, minus the variance that can be attributed to weather noise (Madden, 1989). The predictability associated with information within the initial conditions is sometimes referred to as "predictability of the first kind", whereas that associated with information contained in the slowly evolving boundary conditions is referred to as "predictability of the second kind".

In a paper based on global model simulations of the growth of perturbations in the presence of SST anomalies, Charney and Shukla (1981) pointed out that the tropics have a shorter limit of weather predictability than the extratropics due to the factors discussed in the previous section. At the same time they found that the tropics are much more responsive to the long-lasting ocean SST anomalies than the mid-latitudes. As a result, the potential predictability for the tropics at long time scales due to long-lasting ocean anomalies is much larger than that of the extratropics. The conclusions of this fundamental paper have been confirmed by many simulation experiments and actual dynamical and statistical forecasts.

This led to the search for methods to exploit the potential predictability associated with longer lasting lower boundary conditions, especially those associated with El Niño events. A major breakthrough occurred with the first successful prediction of El Niño by Cane *et al.* (1986), with a simplified coupled ocean–atmosphere model (Zebiak and Cane, 1987). ENSO is a complex interannual tropical oscillation due to the unstable coupling of the ocean and the atmosphere that has a profound effect in the global circulation even away from the tropics. As reviewed by Philander (1990),

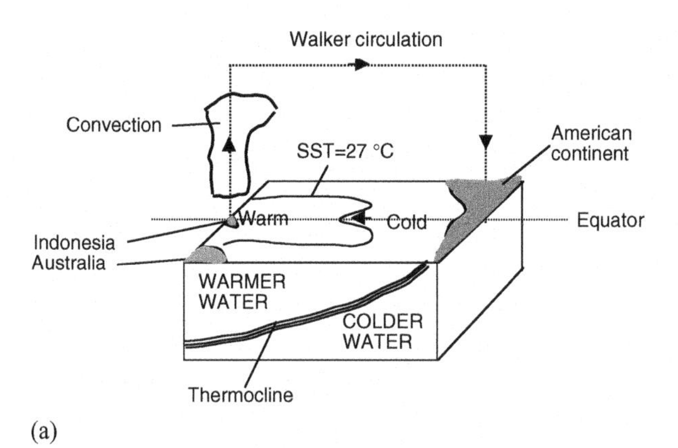

(a)

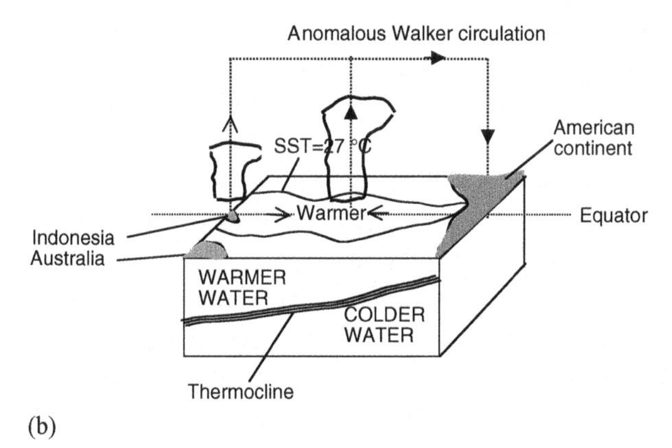

(b)

Figure 6.7.1: Schematic of the coupling of the ocean and atmosphere in the tropical Pacific: (a) normal conditions; (b) El Niño conditions.

there are several unstable coupled modes in the tropical Pacific, which can explain the different ways El Niño (warm central and eastern Pacific events) and La Niña (cold central and eastern Pacific events) occur.

Fig. 6.7.1(a) shows a schematic of the interaction of the tropical Pacific ocean and atmosphere during "normal" years. On the average, the trade winds (easterlies) in the tropics produce westward advection of warm temperatures, so that the SSTs are much warmer in the tropical western Pacific than in the eastern Pacific. For the same reason, the thermocline (an ocean layer of strong vertical gradients of temperature that separates the warm upper ocean from the colder lower layers) is also much deeper in the western Pacific. In the eastern Pacific the colder deeper layer below the thermocline may even surface close to the American continent. The surface easterlies also produce a cold equatorial tongue due to "Ekman pumping" at the Equator, where the Coriolis force acting on westward currents creates a poleward acceleration in both hemispheres, producing horizontal divergence and strong upwelling of cold water. The strongest atmospheric convection takes place in the "warm pool" of water close to the Indonesian region. This convection drives the east–west atmospheric circulation known as the Walker circulation, after Sir Gilbert Walker, who discovered a strong negative correlation between the sea level pressure in Darwin, Australia, and in Tahiti, in the mid-Pacific. He named this relationship the "Southern Oscillation" (Walker, 1928). Bjerknes (1969) discovered that El Niño and the Southern Oscillation were part of a single, coupled ocean–atmosphere phenomenon.

During an El Niño (or warm event, Fig. 6.7.1(b)), a weakening of the easterlies (westerly anomaly) will result in an eastward advection of warm SST by the eastward oceanic currents driven by the westerly atmospheric anomalies. The warm waters propagate eastward and in turn produce atmospheric low-level convergence, strengthening the warm anomaly depth and its eastward propagation. On other occasions the opposite effect occurs: La Niña (or cold event) takes place when the surface easterlies are stronger than normal, and because of this the central equatorial Pacific is colder than normal (not shown). The equatorial region's SST and surface winds, as measured with TOGA (Tropical Ocean, Global Atmosphere) buoys deployed in the equatorial Pacific are shown in Plate 3 for normal, El Niño and La Niña conditions. Both the actual fields and their anomalies are shown for each case, demonstrating the complexity of the coupling.

An explanation of why the ENSO episodes alternate between warm and cold events was offered by Schopf and Suarez (1988), Suarez and Schopf (1988), and by Battisti and Hirst (1989), who independently suggested the "delayed oscillator" mechanism for simple coupled models. In this mechanism, a westerly/warm anomaly in the equatorial central Pacific deepens because of the unstable coupling, and in the process of adjustment the anomaly generates Rossby waves moving westward (Gill, 1980). The Rossby waves elevate the thermocline in the western region. When they reach the western boundary of the equatorial Pacific, the Rossby waves are reflected as eastward moving Kelvin waves, which also elevate the thermocline. When the Kelvin waves reach the central Pacific, they counteract the effect of the thermocline deepening by the unstable coupling. When this delayed negative feedback becomes sufficiently strong, it reverses the sign of the anomaly, and a cold (La Niña) episode starts. The process then starts again with the opposite sign.[6]

However, the observed ENSO episodes are much more complex, and are not well represented by a simple model. There are different "flavors" of El Niño, with some

6 The delayed oscillator mechanism has been illustrated with the simple equation $\dot{T} = T - T^3 - rT(t - d)$. Here $T(t)$ would represent the SST anomaly in the central equatorial Pacific, and the left-hand side is its rate of change. The first term on the right-hand side represents the unstable coupling with the atmosphere. The second term represents damping effects due to dissipation. The last term on the right-hand side represents the negative feedback of the thermocline elevation, delayed by the time it takes the Rossby waves generated by the anomaly T to reach the western boundary and return as Kelvin waves. This mechanism clearly dominates the Zebiak and Cane (1987) model used for the first successful ENSO forecasts (Cane *et al.*, 1986). Cai *et al.* (2002) performed breeding experiments and showed that the perturbations of the forecasts grow fastest during the transitions between cold and warm episodes, and grow slowly or decay during the maxima of the warm and cold episodes, as would be expected from the linearized version of the delayed oscillator equation $\delta\dot{T} = (1 - 3T^2)\delta T - r\delta T(t - d)$. This suggests that the transitions between the ENSO episodes are the least predictable, at least for the Zebiak–Cane model.

propagating the SST warm anomaly eastward, and others propagating it westward (Philander, 1990). Much research is taking place to understand and model these differences in the evolution of the coupled system better. Despite these difficulties, the fact that there are coupled oscillations that have time scales of 3–7 years provides us with the hope that the interannual variability of the tropical climate, dominated by the interactions of the tropical ocean with the tropical atmosphere, could be predictable for seasons through years. Moreover, numerical experiments and analysis of past observations have indicated that the tropical anomalies, especially the anomalous location of major centers of precipitation, have a profound influence on the extra-tropical circulation (e.g., Horel and Wallace, 1981). A scientific program, TOGA was created to study this problem. Wallace *et al.* (1998) reviewed the results of the first decade of the TOGA research.

The hope that long-time scale coupling could form the basis for seasonal to interannual prediction has begun to become a reality. A number of "hindcast" experiments using *observed* (not *predicted*) SST forcing of the global atmosphere have been made. They are usually known as AMIP experiments, since this was the setup used in the Atmospheric Models Intercomparison Project (Gates *et al.*, 1999). In principle, they should give an upper limit to the predictability associated with SST anomalies, since the latter are "perfect". However, it is possible that the fact that uncoupled atmospheric runs using perfect SST do not include feedbacks from the atmosphere to the ocean may actually reduce the optimality of such an approach, since the atmosphere clearly has a profound effect on the oceans, especially in mid-latitudes (Peña *et al.*, 2002).

The AMIP experiments show that, indeed, in most models the tropical SST anomalies produce a reasonably realistic atmospheric response, especially during El Niño or La Niña years. In the extratropics, the situation is more complicated. There are regions, such as Europe, where there is generally little predictability due to ENSO, and others, like the winter northern extratropical Pacific and North America, where the response is stronger, indicating significant potential predictability (Shukla *et al.*, 2000). It should be noted that tropical ocean models, driven with observed surface wind stress, also give fairly realistic oceanic El Niño responses. The fact that both the tropical atmosphere and the tropical ocean respond realistically to one-way observed forcings (SST and wind stress, respectively), further corroborates that ENSO oscillations are the result of coupled ocean–atmosphere modes, either unstable, or marginally stable, forced by atmospheric stochastic noise due to the atmospheric weather or perhaps the MJO.

Two coordinated experiments with ensembles of global general circulation models were carried out in the late 1990s: PROVOST in Europe and Dynamical Seasonal Prediction (DSP) in the USA, both using "perfect" SSTs. Their results have been presented in a number of papers (Shukla *et al.*, 2000, Fischer and Navarra-Giotto, 2000, Brankovic and Palmer, 2000). Kobayashi *et al.* (2000) present the results obtained with similar experiments in Japan. Volume 126, No 567 of *The*

Quarterly Journal of the Royal Meteorological Society was dedicated to this topic and also contains papers on statistical predictions. These papers, and the special issue of the *Journal of Geophysical Research* (Volume 107, Issue C7, 1998) dedicated to the TOGA program contain a wealth of descriptions of the (late 1990s) state-of-the-art understanding and ability to predict atmospheric anomalies beyond 2 weeks.

Several operational and research centers have started issuing seasonal to interannual forecasts based on these ideas (Barnston *et al.*, 1994, Ji *et al.*, 1996, Latif *et al.*, 1994, 1998). At NCEP operational seasonal to interannual forecasts are based on coupled model integrations to predict SST anomalies, followed by ensembles of atmospheric forecasts forced with the predicted SST anomalies in the tropical Pacific, and with statistical predictions of SST in other oceans. NCEP also uses statistical prediction schemes (e.g., Van Den Dool, 1994). The final "official forecast' is subjectively determined from both the dynamical and the statistical predictions. The ECMWF predictions are computed with coupled global ocean–atmosphere models, and run every week for 6 months. Several of these forecasts are available on the web (http://www.cpc.ncep.noaa.gov/, http://www.ecmwf.int/services/seasonal/, http://iri.ldeo.columbia.edu/climate/forecasts/, http://grads.iges.org/nino/, and others).

In addition to the coupling of the atmosphere with the ocean, it is possible to have extended regional predictability from the coupling of the land and the atmosphere. The positive feedback within this coupling can be quite large (e.g., low precipitation results in low soil moisture, and this anomaly, in turn, reduces evaporation and precipitation during the spring and summer months). In subtropical regions associated with strong gradients in precipitation this mechanism can lead to long-lasting anomalies as large or larger than those due to SST anomalies (Koster *et al.*, 2000). Because of this, it should be in principle possible to predict the long-lasting nature of these anomalies for several months (e.g., Atlas *et al.*, 1993, Hong and Kalnay, 2000).

6.8 Decadal variability and climate change

We conclude this chapter by pointing out that in addition to interannual variability, there is also considerable climate variability in the decadal and longer time scales. Climate variability may be due to either natural causes or to long-term changes that can be attributed to anthropogenic sources of pollution or changes in the land surface. The impact of mankind on our environment (usually referred to as "global warming") is quite complex. Among the clearest examples of human impacts are the observed decreases of total ozone in Antarctic regions, and more recently in the Arctic regions, and the increase of CO_2 and other greenhouse gases to levels much higher than those reached in the past. The changes and expected impacts on climate change in the next decades have been reviewed by the International Panel on Climate Change (IPCC), a body of experts from many countries, which has in 2001 issued its Third Assessment Report (Intergovernmental Panel on Climate Change, 2001).

Plate 4 shows the variations of the earth's surface temperature over the last 140 years and over the last millennium. It is clear that there is climate variability at many time scales, and that the variability observed before the Industrial Revolution that took place last century is of natural and not of anthropogenic origin. Among the natural oscillations that would take place even in the absence of human forcings on long time scales are the North Atlantic Oscillation (NAO), associated with the Arctic Oscillation (AO), the Pacific Decadal Oscillation (PDO), the Antarctic Oscillation, and the Atlantic Subtropical Dipole.

We have seen that the variability associated with El Niño (ENSO) has a limit of predictability of the order of a few years, because the oscillations are the result of the chaotic dynamics of the coupled ocean–atmosphere with time scales of several years. Because of the long oceanic time scales associated with slow transports, it is possible that long-term coupling with the oceans may dominate decadal variability in, for example, the NAO. If this is true, then the NAO may be somewhat predictable, and

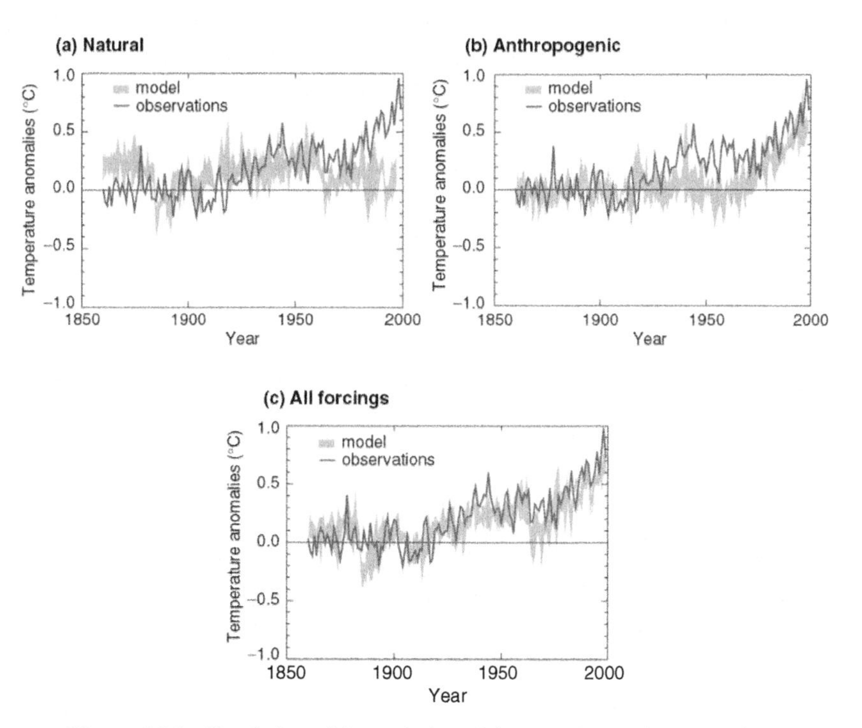

Figure 6.8.1: Simulation of the evolution of the annual global mean surface temperature and comparison with observations for different external forcings. The gray represents the range of four different runs with the same model. (a) Only natural forcings (solar variations and volcanic activity), (b) only anthropogenic forcings (greenhouse gases and an estimate of sulfate aerosols), (c) both. (From Intergovernmental Panel on Climate Change, 2001.)

its prediction will depend on the specification of the initial conditions of the ocean, and the ability of the models to reproduce the relevant physical mechanisms that dominate its coupled evolution. Rapid climate changes associated with transitions into ice ages may be completely unpredictable since they may be the result of unpredictable small changes, such as volcanic activity, resulting in major climate shifts.

Climate change of human origin is also predictable, but in a different sense than ENSO variability, which depends on the initial conditions of the coupled ocean–atmosphere. Anthropogenic climate change depends on an "external" forcing, such as the increase in greenhouse gases, rather than on the internal chaotic dynamics and the initial conditions. When the external forcing is known, the forced response of the climate change can be "predicted" fairly well with present-day models. Figure 6.8.1 shows the response of the global surface temperature to natural forcings (solar variations and volcanic activity), anthropogenic forcings (greenhouse gases and sulfate aerosols) and to both natural and anthropogenic forcings. It shows that some climate models are able to reproduce quite well the large-scale response to volcanic eruptions, and that the addition of greenhouse gas forcings results in fairly good agreement with observations. The fact that different climate models reproduce the global scale of the impact of increased greenhouse gases in a similar way suggests that their effects are to some extent predictable if we can predict the human forcing and its feedbacks on the environmental system. The impacts of climate change on regional scales are much harder to reproduce, because of the local influence of chaotic weather and climate dynamics. This is an area of research beyond the range of the subjects covered by this chapter.

Appendix A

The early history of NWP

Notes written from memory by Anders Persson (ECMWF) on 16 September 1999. The reader is encouraged to read P. D. Thompson's paper "Charney and the Revival of Numerical Weather Prediction", reproduced, together with Charney's letters to Thompson in Lindzen *et al.*, (1990).

History of NWP

In late 1945 Vladimir Zworykin, the "Father of Television", who worked at RCA, joined with John von Neumann, the "Father of the Computer", to suggest the use of the computer in meteorology. Zworykin's interest was in weather modification, and von Neumann's was in fluid dynamics. They also had the dream of connecting the TV and the computer into something we today know as a PC or Workstation. Their dream came partially true in Sweden in around 1955 when for the first time a forecast map that was made directly and automatically without any human intervention was produced on a screen (oscilloscope) (see Bergthorsson and Döös, (1955), Bergthorsson *et al.*, (1955), also the Rossby Memorial Volume).

In early 1946 von Neumann contacted Rossby's group. They told von Neumann why a zonally averaged dynamical model would not work, and instead suggested a barotropic model which had been manually tested by Victor Starr in his 1941 book on weather forecasting for a 72-h forecast at 700 hPa. Von Neumann was not satisfied with the simple barotropic approach and in speeches in the spring of 1946 presented more ambitious plans. Von Neumann and Zworykin also appeared at the annual meeting of the AMS (see *Bulletin of AMS* (1946)).

In the summer of 1946 the Princeton meeting took place. Few if any had any idea of what should be done. Not even the normally optimistic Rossby could see a solution to the problem. A working group was set up with Albert Cahn and Phil Thompson, with Hans Panofsky and Bernhard Haurwitz acting as advisors. By the autumn of 1946, there was still no clear idea of what to do. Cahn left meteorology to become a successful real estate agent in California, leaving Phil Thompson in despair. It was at this crucial state that Jule Charney moved to Chicago (on his way to Norway). Charney had attended the Princeton meeting, where he had offered some obscure ideas about having the whole atmosphere represented by a few singular levels.

In early 1947, Charney, now in Oslo, wrote to Phil Thompson that he indeed saw light at the end of the tunnel taking a completely new approach. It is important to realize that the *practical* (political/psychological) impact of L. F. Richardson's 1922 book was essentially to convince the meteorological community that NWP was impossible. This was further supported by the experience Phil Thompson and others had while trying to make use of Jack Bjerknes "tendency equation" (which was as much in vogue then as potential vorticity is today!).

For the 1948–50 events, I refer to the well-known literature.

Why Sweden?

The first real-time, operational NWP was run in Sweden in September 1954 (to 72 h at 500 hPa), half a year before the USA.

Two reasons:

(1) For a short period in 1954 the Swedes were in possession of the world's most powerful computer, BESK. In 1950 they had already constructed a more basic one, BARK. One must again realize the thinking at that time: even among the most radical, it was felt that having just one computer in Sweden for the coming 20–30 years was sufficient. Even in the USA they thought that four or five computers would be more than enough for the foreseeable future. The "explosion" only came in 1955 when IBM launched their first machine.

(2) Rossby moved to Sweden and wanted to repeat the ENIAC success of 1950 in his homeland. In this endeavor he was supported by: (a) the Swedish Airforce and other national institutions (but not the Meteorological Service!); (b) young enthusiastic scientists who worked at or visited his institutions, both Swedish and foreign; (c) the US Air Force and Woods Hole.

(See articles by Wiin-Nielsen in *Tellus*, 1991, and Bolin in *Tellus*, 1999).

The Swedish project was hampered or complicated by an internal political conflict. In 1954 a new Director of SMHI (the Swedish Meteorological Office) was to be elected by the government. Rossby would have been the obvious choice, but he was seen as a troublemaker. The "official" candidate was Alf Nyberg, who had taken a very skeptical attitude towards Rossby's project. Against him, Rossby lobbied Herrlin, head of the Military Meteorological Service. Unfortunately the run-up to the selection of new Director coincided with the launch of the first real-time operational NWP, 29 Sep–2 Oct 1954. Those who supported Nyberg took a negative attitude; those who supported Rossby took a positive one. In the end, the government chose Nyberg. SMHI began slowly to support NWP 5 day/week barotropic forecasts to 72 h at 500 hPa started in early December 1954. The US operational NWP started in May 1955, but it was not until 1958 that they reached the same quality standard as the Swedish. Japan started in 1959 along the same lines as Sweden.

More provocative ideas

Between 1950 (the first ENIAC run) and 1955 (the start of operational NWP) there was a long lapse of 5 years. Why? To what extent was the delay due to computer resources? To what degree to skepticism about NWP? Charney's presentation in his 1954 National Academy of Sciences paper is very political. My feeling from this paper and other sources is that he and the meteorological community were under strong pressure to present results, in particular with respect to the Thanksgiving Storm of 1950. The set up of the committee for NWP with George Cressman as Head seems to have been done in great haste. The Swedes were known to be progressing towards operational NWP.

During the "Dark Years" 1956–57 some influential persons relying on Norbert Wiener suggested that computers should be used for statistical forecasting of weather patterns. Ed Lorenz at MIT was given the task of finding out if nonlinear dynamic evolutions could be reproduced or simulated by statistical means. His report, of which I have a copy, was guardedly optimistic! If Phillips and Cressman in 1957–58 had not managed to develop *a functioning* NWP system, things might really have developed along other lines.... It was during this or related work that Lorenz discovered the Butterfly Effect.

As mentioned by Phillips in his 1990 monograph about energy dispersion (Phillips 1990a), if Charney *et al.* had run the ENIAC forecasts on a small area, the whole experiment would have had a severe setback, similar to Richardson's 1922 work. It is not commonly known that the UKMO lost 15 years (1950–65) by trying to run a (good) baroclinic model on an area that was too small.

Appendix B

Coding and checking the tangent linear and the adjoint models

We have seen in Chapters 5 and 6 that given a nonlinear model $\mathbf{x}(t) = M[\mathbf{x}(t_0)]$ integrated between t_0 and t, if we introduce a perturbation in the initial conditions $\delta\mathbf{x}(t_0)$, neglecting terms of order $O[\delta\mathbf{x}(t_0)]^2$, then

$$\mathbf{x}(t) + \delta\mathbf{x}(t) = M[\mathbf{x}(t_0) + \delta\mathbf{x}(t_0)] \approx M(\mathbf{x}(t_0)) + \frac{\partial M}{\partial \mathbf{x}}\delta\mathbf{x}(t_0) \tag{B.1.1}$$

so that the initial perturbation evolves like

$$\delta\mathbf{x}(t) \approx \mathbf{L}\delta\mathbf{x}(t_0) \tag{B.1.2}$$

The Jacobian $\mathbf{L} = \partial M/\partial\mathbf{x}$ is the tangent linear model that propagates the perturbation from t_0 to t. It is a matrix that for nonlinear models depends on the basic solution $\mathbf{x}(t)$. If there are n time steps between t_0 and $t_n = t$, this matrix is equal to the product of matrices corresponding to each time step:

$$\mathbf{L}(t_0, t_n) = \mathbf{L}(t_{n-1}, t_n) \cdots \mathbf{L}(t_1, t_2)\mathbf{L}(t_0, t_1) \tag{B.1.3}$$

The adjoint model $\mathbf{L}^T$ is frequently introduced in the context of 4D-Var (Chapter 5), with a cost function measuring the misfit of the model solution to observations:

$$J(\mathbf{x}(t_0)) = \tfrac{1}{2}\sum_{i=0}^{N}\left\{H[\mathbf{x}(t_i)] - y_i^o\right\}^T \mathrm{R}_i^{-1}\left\{H[\mathbf{x}(t_i)] - \mathrm{y}_i^o\right\} \tag{B.1.4}$$

Here the observation error covariance $\mathbf{R}_i$ at a given time t_i is assumed to be symmetric. The *control* variables (which we vary in order to find the minimum of the cost function) are the initial conditions $\mathbf{x}(t_0) = \mathbf{x}_0$. If we take an increment in the initial conditions $\delta\mathbf{x}(t_0)$, then (cf. Remark 5.4.1(d))

$$\delta J = (\nabla_{\mathbf{x}} J(\mathbf{x}_0), \delta\mathbf{x}(t_0)) = \sum_{i=0}^{N} \left(\mathbf{R}_i^{-1}\{H[\mathbf{x}(t_i)] - \mathbf{y}_i^o\}\right)^T \mathbf{H}[\delta\mathbf{x}(t_i)] \qquad \text{(B.1.5)}$$

Now, at a time t_i

$$\delta\mathbf{x}_i = \mathbf{L}(t_0, t_i)\delta\mathbf{x}(t_0) \qquad \text{(B.1.6)}$$

Therefore

$$\delta J = [\nabla_{\mathbf{x}} J(\mathbf{x}_0)]^T \delta\mathbf{x}(t_0) = \sum_{i=0}^{N} \left(\mathbf{L}^T \mathbf{H}^T \mathbf{R}_i^{-1}\{H[\mathbf{x}(t_i)] - \mathbf{y}_i^o\}\right)^T \delta\mathbf{x}(t_0) \qquad \text{(B.1.7)}$$

and the gradient of the cost function with respect to the initial conditions is

$$\nabla_{\mathbf{x}} J(\mathbf{x}_0) = \sum_{i=0}^{N} \mathbf{L}^T(t_i, t_0)\left(\mathbf{H}^T \mathbf{R}_i^{-1}\{H[\mathbf{x}(t_i)] - \mathbf{y}_i^o\}\right) \qquad \text{(B.1.8)}$$

As we saw in Chapters 5 and 6, the transpose $\mathbf{L}^T(t_i, t_0)$ of the matrix of the tangent linear model is the adjoint model. If the tangent linear model matrix is complex, then the adjoint is the complex conjugate of the transpose of the tangent linear model. The observational increments are "forcings" of the adjoint model in computing the gradient of the cost function for 4D-Var.

In addition to 4D-Var (Chapter 5), the adjoint model has several other important applications such as computing the vectors that grow fastest in a period of optimization. These are the leading singular vectors, i.e., the leading eigenvectors of $\mathbf{L}^T\mathbf{L}$ (Chapter 6). The adjoint of a model can also be used to find optimal parameters in a model, e.g., the diffusion coefficient that produces forecasts closest to a verification field (e.g., Caccuci, 1981, Zou *et al.*, 1992). All these applications require the definition of a norm, with respect to which the gradient is computed, and the choice of the appropriate control variables. In the case of singular vectors and 4D-Var, the control variables are the vectors of initial conditions; in the problem of finding an optimal parameter, the control variable is the parameter itself.

We now discuss briefly the rules for generating adjoint codes. More detailed discussions are available in Talagrand (1991), Talagrand and Courtier (1987), Navon *et al.* (1992), Yang and Navon (1995), and Giering and Kaminski (1998). There are presently two compilers available for the automatic generation of tangent linear and adjoint codes, given the FORTRAN code of the forward nonlinear model. (Odyssee, Rostaing *et al.*, 1996, and the Tangent and Adjoint Model Compiler (TAMC), Giering, 1994, Giering and Kaminski, 1998). TAMC is available on the web (http://puddle.mit.edu/~ralf/tamc/tamc.html) and has been widely used in recent years.

Taking the transpose of (B.1.3), we obtain that the adjoint model is

$$\mathbf{L}^T(t_n, t_0) = \mathbf{L}^T(t_1, t_0) \cdots \mathbf{L}^T(t_{n-1}, t_{n-2})\mathbf{L}^T(t_n, t_{n-1}) \qquad \text{(B.1.9)}$$

Within a time step, the TLM code is also composed by a number of subcodes or steps applied in succession:

$$\mathbf{L}(t_{i-1}, t_i) = \mathbf{S}_1 \cdots \mathbf{S}_{m-1}\mathbf{S}_m \tag{B.1.10}$$

For example S_1 could be initialization of the new time step, S_2–S_7 could be the computation of the time tendencies coming from horizontal advection, vertical advection, convection, large-scale precipitation, S_8 radiation, etc., and S_9 the update of the model variables for the next time step. Each of these steps may contain several DO loops.

In constructing the adjoint, because of the transposition, these steps (and the DO loops within the step) are also reversed:

$$\mathbf{L}^T(t_i, t_{i-1}) = \mathbf{S}_m \cdots \mathbf{S}_2\mathbf{S}_1 \tag{B.1.11}$$

We give now a simple example to illustrate the rules for constructing the adjoint from a nonlinear forward model. In the construction of the adjoint it is important to determine which are the "active" variables updated in the model.

Consider the simple forward model of the diffusion equation (in this case a linear model):

$$\frac{\partial u}{\partial t} = \sigma \frac{\partial^2 u}{\partial x^2} \tag{B.1.12}$$

It would be possible to derive first the analytic adjoint of this equation and then discretize it, but it is preferable to first discretize the original equation, code it, and then create the tangent linear model and adjoint of the code directly. This is because the adjoint of a discretized code is not necessarily identical to the discretization of the adjoint operator. We discretize (B.1.12) using finite differences and a scheme forward in time, centered in space, as discussed in Chapter 3:

$$u_i^{j+1} = u_i^j + \alpha\left(u_{i+1}^j - 2u_i^j + u_{i-1}^j\right) \tag{B.1.13}$$

where $\alpha = \sigma\Delta t/(\Delta x)^2$, $x_i = i\Delta x$, $i = 1, \ldots, I$, and $t_j = j\Delta t$. If we assume $u = u_b + \delta u$, where $u_b(t)$ is the basic solution, then the tangent linear model is

$$\begin{aligned} \delta u_i^{j+1} &= \delta u_i^j + \alpha\left(\delta u_{i+1}^j - 2\delta u_i^j + \delta u_{i-1}^j\right) \\ &= (1-2\alpha)\delta u_i^j + \alpha\delta u_{i+1}^j + \alpha\delta u_{i-1}^j \end{aligned} \tag{B.1.14}$$

or in matrix form

$$\delta u_i^{j+1} = \begin{pmatrix} 1-2\alpha & \alpha & \alpha \end{pmatrix} \begin{pmatrix} \delta u_i^j \\ \delta u_{i-1}^j \\ \delta u_{i+1}^j \end{pmatrix} \qquad i = 1, \ldots, I \tag{B.1.15}$$

In this computation, there are four "active" variables $\delta u_i^{j+1}, \delta u_i^j, \delta u_{i+1}^j, \delta u_{i-1}^j$ but only one of them has been modified. In preparation for the computation of the adjoint it is necessary to indicate explicitly that the other three active variables are *not* modified. The tangent linear model step is then

$$\begin{pmatrix} \delta u_{i-1}^j \\ \delta u_i^j \\ \delta u_{i+1}^j \\ \delta u_i^{j+1} \end{pmatrix} = \begin{pmatrix} 1 & 0 & 0 & 0 \\ 0 & 1 & 0 & 0 \\ 0 & 0 & 1 & 0 \\ \alpha & 1-2\alpha & \alpha & 0 \end{pmatrix} \begin{pmatrix} \delta u_{i-1}^j \\ \delta u_i^j \\ \delta u_{i+1}^j \\ \delta u_i^{j+1} \end{pmatrix} \qquad i = 1, \ldots, I \tag{B.1.16}$$

Equation (B.1.16) is the same as (B.1.15) but includes the additional identities (no modification) needed for the adjoint model. The adjoint model is the transpose of the tangent linear model matrix acting on the adjoint variables, so that this step becomes

$$\begin{pmatrix} \delta^* u_{i-1}^j \\ \delta^* u_i^j \\ \delta^* u_{i+1}^j \\ \delta^* u_i^{j+1} \end{pmatrix} = \begin{pmatrix} 1 & 0 & 0 & \alpha \\ 0 & 1 & 0 & 1-2\alpha \\ 0 & 0 & 1 & \alpha \\ 0 & 0 & 0 & 0 \end{pmatrix} \begin{pmatrix} \delta^* u_{i-1}^j \\ \delta^* u_i^j \\ \delta^* u_{i+1}^j \\ \delta^* u_i^{j+1} \end{pmatrix} \qquad i = I, \ldots, 1 \tag{B.1.17}$$

where the stars represent adjoint variables.

This can be written line by line as

$$\left. \begin{aligned} \delta^* u_{i-1}^j &= \delta^* u_{i-1}^j + \alpha \delta^* u_i^{j+1} \\ \delta^* u_i^j &= \delta^* u_i^j + (1-2\alpha)\delta^* u_i^{j+1} \\ \delta^* u_{i+1}^j &= \delta^* u_{i+1}^j + \alpha \delta^* u_i^{j+1} \\ \delta^* u_i^{j+1} &= 0 \end{aligned} \quad i = I, \ldots, 1 \right\} \tag{B.1.18}$$

Note that the equation for the adjoint variable on the left-hand side of (B.1.15) is executed last.

A second example is adapted form Giering and Kaminski (1998). Consider the nonlinear nth step of the following algorithm:

$$z^n = x^{n-1} \sin[(y^{n-1})^2] \tag{B.1.19}$$

where x, y, z are active variables. Using the chain rule, the tangent linear algorithm for this step is

$$\delta z^n = \sin[(y^{n-1})^2]\delta x^{n-1} + x^{n-1} \cos[(y^{n-1})^2] 2y^{n-1} \delta y^{n-1} \tag{B.1.20}$$

or in matrix form

$$\begin{pmatrix} \delta x \\ \delta y \\ \delta z \end{pmatrix}^n = \begin{pmatrix} 1 & 0 & 0 \\ 0 & 1 & 0 \\ \sin[(y^{n-1})^2] & x^{n-1}\cos[(y^{n-1})^2]2y^{n-1} & 0 \end{pmatrix} \begin{pmatrix} \delta x \\ \delta y \\ \delta z \end{pmatrix}^{n-1} \tag{B.1.21}$$

The adjoint operator is the transposed matrix of (B.1.21) acting on the adjoint variables:

$$\begin{pmatrix} \delta^* x \\ \delta^* y \\ \delta^* z \end{pmatrix}^{n-1} = \begin{pmatrix} 1 & 0 & \sin[(y^{n-1})^2] \\ 0 & 1 & x^{n-1}\cos[(y^{n-1})^2]2y^{n-1} \\ 0 & 0 & 0 \end{pmatrix} \begin{pmatrix} \delta^* x \\ \delta^* y \\ \delta^* z \end{pmatrix}^n \tag{B.1.22}$$

The FORTRAN statement for the nonlinear forward step is:

```
Z=X* SIN(Y**2)
```

The TLM FORTRAN statement is

```
DZ=(SIN(Y**2))*DX+(X*COS(Y**2)*2*Y)*DY,
```

and the adjoint FORTRAN statements for this step are

```
ADX=ADX+ (SIN(Y**2))*ADZ
ADY=ADY+ (X*COS(Y**2)*2*Y)*ADZ
ADZ=0.0
```

B.1 Verification

Finally, we discuss how to verify the correctness of the tangent linear and adjoint codes (Navon *et al.*, 1992). The verification of the tangent linear model is straightforward: for small increments in the initial conditions, the tangent linear model should reproduce the difference between two nonlinear integrations with quadratic errors:

$$\delta x(t) = \mathbf{L}\delta x(0) = M[x_0 + \delta x(0)] - M(x_0) + O(\|\delta x\|^2) \tag{B.1.23}$$

Therefore, Navon *et al.* (1992) suggested computing the pattern correlation of the left- and right-hand sides of (B.1.23), as well as the relative error, using an appropriate norm. For sufficiently small perturbation amplitudes, the relative error should be proportional to the amplitude of the initial perturbation.

To verify the correctness of the adjoint code, Navon *et al.* used the identity

$$(\mathbf{L}\delta x_0)^T(\mathbf{L}\delta x_0) = (\delta x_0)^T \mathbf{L}^T(\mathbf{L}\delta x_0) \tag{B.1.24}$$

This check can be applied to every single subroutine or DO loop:

$$(AQ)^T(AQ) = Q^T[A^T(AQ)] \tag{B.1.25}$$

Here Q represents the input of the original code, A represents either a single DO loop or a subroutine. The left-hand side involves only the TLM, whereas the right-hand side also involves the adjoint code. In practice, if the adjoint code is correct with respect to the tangent linear model, the identity (B.1.25) holds true up to machine accuracy.

B.2 Example of FORTRAN code

The codes for a complete model for the nonlinear Burgers equation, its tangent linear model and adjoint model, kindly provided by Seon Ki Park, are presented at the end of the appendix. The forward model uses the leapfrog scheme for the advection term and the DuFort–Frankel scheme for the diffusion. The continuous Burgers equation is

$$\frac{\partial u}{\partial t} = -u\frac{\partial u}{\partial x} + \frac{1}{R}\frac{\partial^2 u}{\partial x^2} \tag{B.2.26}$$

and the finite differences used in the code start with a single forward step followed by the leapfrog/DuFort–Frankel scheme:

$$u_i^{j+1} = u_i^{j-1} - \frac{\Delta t}{\Delta x}\left[u_i^j\left(u_{i+1}^j - u_{i-1}^j\right)\right] + \frac{2\Delta t}{\Delta x^2}\left[u_{i+1}^j - \left(u_i^{j+1} + u_i^{j-1}\right) + u_i^j\right] \quad \text{(B.2.27)}$$

Here the index j represents the time step and i represents space (x). Note that in the adjoint code, the order of the substeps and all the DO loops is reversed.

```
cccccccccccccccccccccccccccccccccccccccccccccccccccc
c                                                  c
c  Viscous BURGERS' Equation solved with Leap      c
c  Frog/DuFort-Frankel scheme (see Anderson        c
c  et. al. 1984).                                  c
c                                                  c
c  Leap Frog scheme has truncation erorr of        c
c  O[(dt)**2, (dx)**2]. Since the truncation error c
c  contains only odd derivative terms (lead by third c
c  order derivative term), the solution will exhibit c
c  dispersion errors but no dissipation errors. The c
c  stability condition is that the Courant number  c
c  (nu = c*dt/dx) be less than or equal to 1.      c
c  When nu = 1, there is no dispersion error.      c
c                                                  c
c  DuFort-Frankel scheme for diffusion process has c
c  truncation erorr of O[(dt)**2, (dx)**2,         c
c  (dt/dx)**2]. The truncation error has even terms c
c  and thus the solution will have dissipation     c
c  errors. But this error depends on r=            c
c  (1/R)*(dt/dx**2) and beta=k*dx. When we set r   c
c  and beta small, the dissipation error is not    c
c  significant.                                    c
c                                                  c
c  Author: Seon Ki Park (03/24/98)                 c
c                                                  c
c  Reference:                                      c
c                                                  c
c    Anderson, D.A., J.C., Tannehill, R.H. Pletcher, c
c      1984: Computational Fluid Mechanics and Heat c
c      Transfer. McGraw-Hill Book Company, 599 pp. c
```

```
c                                                   c
cccccccccccccccccccccccccccccccccccccccccccccccccccc
c
      SUBROUTINE BURGER(nx,n,ui,ub,uob,u,cost)
c
c     This subroutine integrates the model from a given
c     initial condition xinit.
c
      integer nx        ! Number of grid points
      integer n         ! Number of time steps
      real cost         ! Cost function
      real uob(nx,n)    ! Observations
      real ub(nx,n)     ! Basic states (u)
      real u(nx,n)      ! Model solutions (u)
      real ui(nx)       ! Initial conditions
      real dx           ! Space increment
      real dt           ! Time increment
      real R            ! Reynolds number (reciprocal of
                          diffusion coefficient)
c
      common /com_param/ dtdx, c1, c0
      real dtdx, c0, c1
c
      R = 1000.
c
      dx = 1.0
      dt = 0.1
      dtdx = dt/dx
      dtdxsq = dt/(dx**2)
c
      c1 = (2./R)*dtdxsq
      c0 = 1./(1.+c1)
c
c     Initialize the cost function:
c
      cost=0.
c
c     Set the initial conditions:
c
      do i=1, nx
         u(i,1)=ui(i)
      enddo
```

```
c
c     Set the boundary conditions:
c
      do j=1, n
         u(1,j) = 0.
         u(nx,j)= 0.
      enddo
c
c     Integrate the model numerically:
c
c.... FTCS for first time step integration
c
      do i=2,nx-1
        u(i,2)=u(i,1)-0.5*dtdx*u(i,1)*(u(i+1,1)-u(i-1,1))
      %          +0.5*c1*(u(i+1,1)-2.*u(i,1)+u(i-1,1))
      enddo
c
c.... Leap Frog/DuFort-Frankel afterwards
c
      do j=3, n
        do i=2,nx-1
          u(i,j) = c0*(u(i,j-2) + c1*(u(i+1,j-1)-u(i,j-2)
        %       +u(i-1,j-1)) - dtdx*u(i,j-1)*(u(i+1,j-1)
        %       -u(i-1,j-1)))
        enddo
      enddo
c
c     Cost function:
c
      do j = 1,n
        do i = 1,nx
          cost = cost + 0.5*(u(i,j)-uob(i,j))**2
        enddo
      enddo
c
c     Save nonlinear solutions to the basic fields:
c
      do j=1, n
        do i=1,nx
          ub(i,j)=u(i,j)
        enddo
      enddo
```

```
c
      return
      end
c
      SUBROUTINE BURGER_TLM(nx,n,ui,ubasic,u)
c
c     The tangent linear model of the burger's equations.
c
      real ubasic(nx,n)   ! Basic states
      real u(nx,n)        ! TLM solutions
      real ui(nx)         ! Initial conditions
      common /com_param/ dtdx, c1, c0
      real dtdx, c0, c1
c
c     Set the initial conditions:
c
      do i=1,nx
        u(i,1)=ui(i)
      enddo
c
c     Set the bundary conditions:
c
      do j=1, n
        u(1,j) = 0.
        u(nx,j)= 0.
      enddo
c
c     Integrate the model numerically:
c
c.... FTCS for first time step integration
c
      do i=2,nx-1
        u(i,2) = u(i,1)
      %         - 0.5*dtdx*(u(i,1)*(ubasic(i+1,1)
      %           -ubasic(i-1,1)) + ubasic(i,1)*(u(i+1,1)
      %           -u(i-1,1)))
      %         + 0.5*c1*(u(i+1,1)-2.*u(i,1) +u(i-1,1))
      enddo
c
c.... Leap Frog/DuFort-Frankel afterwards
c
      do j=3, n
```

```
        do i=2,nx-1
          u(i,j) = c0*(u(i,j-2) + c1*(u(i+1,j-1)-u(i,j-2)
     %        +u(i-1,j-1)) - dtdx*(u(i,j-1)
     %        *(ubasic(i+1,j-1)-ubasic(i-1,j-1))
     %        + ubasic(i,j-1)*(u(i+1,j-1)-u(i-1,j-1))))
        enddo
      enddo
c
c     Set the final value of u in ui:
c
      do i=1,nx
        ui(i)=u(i,n)
      enddo
c
      return
      end
c
      SUBROUTINE BURGER_ADJ(iforcing,nx,n,ui,ubasic,uob,u)
c
c     The adjoint model of the burger's equations.
c
      real ubasic(nx,n) ! Basic states
      real uob(nx,n)    ! Observations
      real u(nx,n)      ! Model solutions
      real ui(nx)       ! Initial conditions
      real dt           ! Time increment
      real dx           ! Space increment.
      integer iforcing  ! Index for forcing in the adjoint
      common /com_param/ dtdx, c1, c0
      real dtdx, c0, c1
c
c     Initialize adjoint variables:
c
      do j = 1, n
        do i = 1, nx
          u(i,j) = 0.
        enddo
      enddo
c
c     Set the final conditions:
c
      if (iforcing.eq.0) then
```

```
         do i=1, nx
           u(i,n)=ui(i)
           ui(i) = 0.
         enddo
       else      ! Add Cost Function Part as a Forcing
         do i=1, nx
           u(i,n)=ubasic(i,n)-uob(i,n)
           ui(i) = 0.
         enddo
       endif
c
c      Adjoint of Leap Frog/Dufort-Frankel
c
       do j = n, 3, -1
         do i=nx-1, 2, -1
           u(i-1,j-1) = u(i-1,j-1) + c0*(c1+dtdx*ubasic(i,
     %                  j-1))*u(i,j)
           u(i+1,j-1) = u(i+1,j-1) + c0*(c1-dtdx*ubasic(i,
     %                  j-1))*u(i,j)
           u(i,j-1) = u(i,j-1) - c0*dtdx*(ubasic(i+1,j-1)
     %                -ubasic(i-1, j-1))*u(i,j)
           u(i,j-2) = u(i,j-2) + c0*(1.-c1)*u(i,j)
c          u(i,j) = 0.
         enddo
         if (iforcing.eq.1) then
           do i=1,nx
             u(i,j-1) = u(i,j-1)+ubasic(i,j-1)-uob(i,j-1)
           enddo
         endif
       enddo
c
c      Adjoint of FTCS
c
       do i = nx-1, 2, -1
         u(i-1,1)=u(i-1,1)+0.5*(c1+dtdx*ubasic(i,1))*u(i,2)
         u(i+1,1)=u(i+1,1)+0.5*(c1-dtdx*ubasic(i,1))*u(i,2)
         u(i,1)=u(i,1)+u(i,2)*(1.-c1
     %         -0.5*dtdx*(ubasic(i+1,1)-ubasic(i-1,1)))
c        u(i,2)=0.
       enddo
c
       if (iforcing.eq.1) then
```

```
          do i=1,nx
            u(i,1) = u(i,1) + ubasic(i,1) - uob(i,1)
          enddo
        endif
c
c     Set the boundary conditions:
c
        do i=1, n
          u(1,i) = 0.
          u(nx,i)= 0.
        enddo
c
c     Set the final value of u in ui:
c
        do i=1,nx
          ui(i) =ui(i) + u(i,1)
        enddo
c
        return
        end
```

Appendix C

Post-processing of numerical model output to obtain station weather forecasts

If the numerical model forecasts are skillful, the forecast variables should be strongly related to the weather parameters of interest to the "person in the street" and for other important applications. These include precipitation (amount and type), surface wind, and surface temperature, visibility, cloud amount and type, etc. However, the model output variables are not optimal direct estimates of local weather forecasts. This is because models have biases, the bottom surface of the models is not a good representation of the actual orography, and models may not represent well the effect of local forcings important for local weather forecasts. In addition, models do not forecast some required parameters, such as visibility and probability of thunderstorms.

In order to optimize the use of numerical weather forecasts as guidance for human forecasters, it has been customary to use statistical methods to "post-process" the model forecasts and adapt them to produce local forecasts. In this appendix we discuss three of the methods that have been used for this purpose.

C.1 Model Output Statistics[1] (MOS)

This method, when applied under ideal circumstances, is the gold standard of NWP model output post-processing (Glahn and Lowry, 1972, Carter *et al.*, 1989). MOS is essentially multiple linear regression, where the predictors h_{nj} are model forecast variables (e.g., temperature, humidity, or wind at any grid point, either near the surface

1 I am grateful to J. Paul Dallavalle of the National Weather Service for information about MOS and Perfect Prog. The NWS homepage for statistical guidance is in http://www.nws.noaa.gov/tdl/synop/index.html

or in the upper levels), and may also include other astronomical or geographical parameters (such as latitude, longitude and time of the year) valid at time t_n. The predictors could also include past observations. The predictand y_n is a station weather observation (e.g., maximum temperature or wind speed) valid at the same time as the forecast. Here, as in any statistical regression, the quality of the results improves with the quality and length of the training data set used to determine the regression coefficients b_j.

The *dependent* data set used for determining the regression coefficients is

$$\left.\begin{aligned} y_n &= y(t_n) \qquad n = 1, \ldots, N \\ h_{nj} &= h_j(t_n) \qquad n = 1, \ldots, N; j = 1, \ldots, J \end{aligned}\right\} \tag{C.1.1}$$

where we consider one predictand y_n as a function of time t_n and J predictors h_{nj}.

The linear regression (forecast) equation is

$$\hat{y}_n = b_0 + \sum_{j=1}^{J} b_j h_{nj} = \sum_{j=0}^{J} b_j h_{nj} \tag{C.1.2}$$

where for convenience the predictors associated with the constant term b_0 are defined as $h_{n0} \equiv 1$. In linear regression the coefficients b_j are determined by minimizing the sum of squares of the forecast errors over the training period (e.g., Wilks, 1995). The sum of squared errors is given by:

$$SSE = \sum_{n=1}^{N} (y_n - \hat{y}_n)^2 = \sum_{n=1}^{N} e_n^2 \tag{C.1.3}$$

Taking the derivatives with respect to the coefficients b_j and setting them to zero we obtain:

$$\frac{\partial SSE}{\partial b_j} = 0 = \sum_{n=1}^{N} (y_n - \sum_{l=0}^{J} b_l h_{nl}) h_{nj} \qquad j = 0, 1, \ldots, J \tag{C.1.4}$$

or

$$\sum_{n=1}^{N} \left[h_{jn}^T y_n - h_{jn}^T \sum_{l=0}^{J} h_{nl} b_l \right] = 0 \qquad j = 0, \ldots, J \tag{C.1.5}$$

where $h_{jn}^T = h_{nj}$. Equations (C.1.5) are the "normal" equations for multiple linear regression that determine the linear regression coefficients b_j, $j = 0, \ldots, J$. In matrix form, they can be written as

$$\mathbf{H}^T\mathbf{H}\mathbf{b} = \mathbf{H}^T\mathbf{y} \qquad \text{or} \qquad \mathbf{b} = \left(\mathbf{H}^T\mathbf{H}\right)^{-1}\mathbf{H}^T\mathbf{y} \tag{C.1.6}$$

where

$$\mathbf{H} = \begin{bmatrix} 1 & h_{11} & \dots & h_{1J} \\ 1 & h_{21} & \dots & h_{2J} \\ \vdots & \vdots & h_{nj} & \vdots \\ 1 & h_{N1} & \dots & h_{NJ} \end{bmatrix} \quad \mathbf{b} = \begin{bmatrix} b_0 \\ b_1 \\ \vdots \\ b_J \end{bmatrix} \quad \mathbf{y} = \begin{bmatrix} y_1 \\ y_2 \\ \vdots \\ y_N \end{bmatrix} \tag{C.1.7}$$

are, respectively, the dependent sample predictor matrix (model output variables, geographical and astronomical parameters, etc.), the vector of regression coefficients, and the vector of predictands in the dependent sample. $\widehat{\mathbf{y}} = \mathbf{Hb}$, $\mathbf{e} = \mathbf{y} - \mathbf{Hb}$ are the linear predictions and the prediction error, respectively, in the dependent sample. The *dependent* estimate of the error variance of the prediction is $s_e^2 = SSE/(N - J - 1)$ since the number of degrees of freedom is $N - J - 1$. This indicates that one should avoid overfitting the dependent sample by ensuring that $N >> J$. For independent data, the expected error can be considerably larger than the dependent estimate s_e^2 because of the uncertainties in estimating the coefficients b_j. The best way to estimate the skill of MOS (or any statistical prediction) that can be expected when applied to *independent* data is to perform *cross-validation*. This can be done by reserving a portion (such as 10%) of the dependent data, deriving the regression coefficients from the other 90%, and then applying it to the unused 10%. The process can be repeated 10 times with different subsets of the dependent data to increase the confidence of the cross-validation, but this also increases the computational cost.

It is clear that for a MOS system to perform optimally, several conditions must be fulfilled:

(a) The training period should be as long as possible (at least several years).
(b) The model-based forecasting system should be kept unchanged to the maximum extent possible during the training period.
(c) After training, the MOS system should be applied to future model forecasts that also use the same unchanged model system.

These conditions, while favorable for the MOS performance, are not favorable for the continued improvement of the NWP model, since they require "frozen" models. The main advantage of MOS is that if the conditions stated above are satisfied, it achieves the best possible linear prediction. Another advantage is that it naturally takes into account the fact that forecast skill decreases with the forecast length, since the training sample will include, for instance, the information that a 1-day model prediction is on the average considerably more skillful than a 3-day prediction. The main disadvantage of MOS is that it is not easily adapted to an operational situation in which the model and data assimilation systems are frequently upgraded.

Typically, MOS equations have 10–20 predictors chosen by forward screening (Wilks, 1995). In the US NWS, the same MOS equations are computed for a few (4–10) relatively homogeneous regions in order to increase the size of the developmental database. In order to stratify the data into few but relatively homogeneous

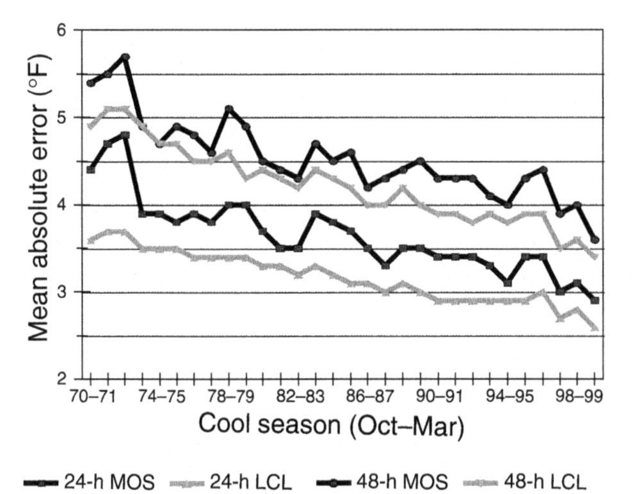

Figure C.1.1: Evolution of the mean absolute error of the MOS guidance and of the local official NWS forecasts (LCL) averaged over the USA. (Courtesy of J. Paul Dallavalle and Valery Dagostaro from the US NWS.)

time periods, separate MOS equations are developed for the cool season (October–March) and the warm season (April–September). As shown in Table C.3.1, MOS can reduce very substantially the errors in the NWP model forecasts, especially at short lead times. At long lead times, the forecast skill is lost, so that the MOS forecast becomes a climatological forecast and the MOS forecast error variance asymptotes to the climatology error. The error variance of an individual NWP forecast, on the other hand, asymptotes to twice the climatological error variance, plus the square of the model bias (see Section 6.5).

Figure C.1.1 shows the evolution of the error in predicting the maximum temperature by the statistical guidance (MOS) and by the local human forecasters (LCL). The human forecasters skill in the 2-day forecast is now as good as the one-day forecast was in the 1970s. The human forecasters bring added value i.e., make better forecasts than the MOS statistical guidance, which in turn is considerably better than the direct NWP model output. Nevertheless, the long-term improvements are driven mostly by the improvements in the NWP model and data assimilation systems, as discussed in Chapter 1.

In summary, the forecast statistical guidance (and in particular MOS) adds value to the direct NWP model output by objectively interpreting model output to remove systematic biases and quantifying uncertainty, predicting parameters that the model does not predict, and producing site-specific forecasts. It assists forecasters providing a first guess for the expected local conditions, and allows convenient access to information on local model and climatology conditions.

C.2 Perfect Prog

Perfect Prog is an approach similar to MOS, except that the regression equations are derived using as predictors, observations or analyses (rather than forecasts) valid at the prediction time, as if the forecasts were perfect. If station observations are used as predictors in the dependent sample, it is not possible to use the same variable for a predictor as for the predictand (e.g., Boston's observed maximum surface temperature could not be used as predictor for the maximum temperature in Boston). However, if model analyses are used as "perfect" forecasts, one can use like variables as predictors. For obvious reasons, Perfect Prog has not been much used except for very short forecasts. Perhaps it would be possible now to use the long homogeneous reanalyses that have been completed (Kistler *et al.,* 2001, Kalnay *et al.,* 1996, Gibson *et al.,* 1997) to derive very long and robust Perfect Prog statistics between model output and station data. After the regression between the reanalysis and station data is completed, the prediction of surface parameters could be done in two steps. In the first step, multiple regression would be used to predict the reanalysis field from model forecasts, which should be easier to achieve than predicting the station data directly, since a few parameters would be enough to represent the model bias and the decay of skill with time. In the second step, the Perfect Prog equations would be used to translate the predicted analysis into station weather parameters. In this approach, the disadvantage of Perfect Prog of not including the effective loss of skill associated with longer forecast lengths would be handled in the first step discussed above. This approach has yet to be thoroughly explored.

C.3 Adaptive regression based on a simple Kalman filter approach[2]

Adaptive regression based on Kalman filtering has also been widely used as a post-processor. In MOS and in other statistical prediction methods such as nonlinear regression or neural networks, the regression coefficients are computed from the dependent sample, and are not changed as new observations are collected until a new set of MOS equations are derived every 5 or 10 years. Because the regression coefficients are constant, the order of the observations is irrelevant in MOS, so that older data have as much influence as the newest observations used to derive the coefficients.

In adaptive regression, the Kalman filter equations (Section 5.6) are applied in a simple, sequential formulation to the multiple regression coefficients $\mathbf{b}_k = \mathbf{b}(t_k)$, whose values are *updated* every time step, rather than keeping them constant as in (C.1.2):

$$\hat{y}_k = \sum_{j=0}^{J} b_j(t_k) h_{kj} = \begin{bmatrix} 1 & h_{k1} & \dots & h_{kJ} \end{bmatrix} \begin{bmatrix} b_0 \\ b_1 \\ \dots \\ b_J \end{bmatrix}_k = \mathbf{h}_k^T \mathbf{b}_k \tag{C.3.1}$$

2 I am very grateful to Joaquim Ballabrera for insightful comments and suggestions that improved this section.

If we compare this equation with those in Chapter 5, we see that it has the form of a forecast of observations, $y_k^f = \mathbf{H}_k \mathbf{b}_k$, so that we can use the Kalman filter formulation with an "observation operator" $\mathbf{H} = \mathbf{h}_k^T$, a row vector in (C.1.7) corresponding to the time t_k. Recall that Kalman filtering consists of two steps (see (5.6.8)–(5.6.11)). In the first step, starting from the analysis at time t_{k-1}, we forecast the values of the model variables (in this case the coefficients $\mathbf{b}_k$) and their error covariance at time t_k. In the second step, the Kalman weight matrix is derived, and, after obtaining the observations at time t_k, the model variables and error covariance are updated, giving the analysis at time t_k. In adaptive regression, the "forecast" or first guess of the regression coefficients at t_k is simply that they are the same as the (analysis) coefficients at t_{k-1}, and their error covariance is the same as that estimated in the previous time step, plus an additional error introduced by this "regression forecast model":

$$\left.\begin{aligned} \mathbf{b}_k^f &= \mathbf{b}_{k-1}^a \\ \mathbf{P}_k^f &= \mathbf{P}_{k-1}^a + \mathbf{Q}_{k-1} \end{aligned}\right\} \tag{C.3.2}$$

Here $\mathbf{Q}_k = \mathbf{q}_k \mathbf{q}_k^T$ is the "regression model" error covariance (a matrix of tunable coefficients that is diagonal if we assume that the errors of the different coefficients are not correlated).

The Kalman gain or weight vector for adaptive regression is given by

$$\mathbf{k}_k = \mathbf{P}_k^f \mathbf{h}_k \left(\mathbf{h}_k^T \mathbf{P}_k^f \mathbf{h}_k + r_k\right)^{-1} \tag{C.3.3}$$

Note that for a single predictand, the forecast error covariance $\mathbf{h}_k^T \mathbf{P}_k^f \mathbf{h}_k$ and the observational error covariance $\mathbf{R}_k = r_k$ are both scalars, and computing the Kalman gain matrix does not require a matrix inversion.

At time t_k the observed forecast error or innovation $e_k = y_k^o - \mathbf{h}_k^T \mathbf{b}_k^f$ is used to *update* the regression coefficients:

$$\left.\begin{aligned} \mathbf{b}_k^a &= \mathbf{b}_k^f + \mathbf{k}_k \left(y_k^o - \mathbf{h}_k^T \mathbf{b}_k^f\right) \\ \mathbf{P}_k^a &= \left(\mathbf{I} - \mathbf{k}_k \mathbf{h}_k^T\right) \mathbf{P}_k^f \end{aligned}\right\} \tag{C.3.4}$$

In summary, the adaptive regression algorithm based on Kalman filtering can be written as:

$$\left.\begin{aligned} y_k^f &= \mathbf{h}_k^T \mathbf{b}_{k-1}^a \\ \mathbf{P}_k^f &= \mathbf{P}_{k-1}^a + \mathbf{Q}_{k-1} \\ e_k &= y_k^o - y_k^f \\ w_k &= \mathbf{h}_k^T \mathbf{P}_k^f \mathbf{h}_k + r_k \\ \mathbf{k}_k &= \mathbf{P}_k^f \mathbf{h}_k w_k^{-1} \\ \mathbf{b}_k^a &= \mathbf{b}_{k-1}^a + \mathbf{k}_k e_k \\ \mathbf{P}_k^a &= \mathbf{P}_k^f - \mathbf{k}_k w_k \mathbf{k}_k^T \end{aligned}\right\} \tag{C.3.5}$$

Table C.3.1. *rms error in the forecast of the surface temperature at 00Z averaged for eight US stations. In dependent regression and Kalman filtering, the only predictor used was the direct model prediction of the temperature interpolated to the station. The MOS prediction has more than ten predictors and several years of training.*

NWP (Aviation model)	Dependent Regression	Adaptive Regression	MOS
5.36 K	2.67 K	3.07 K	2.29 K

where w_k is a temporary scalar defined for convenience. The two tuning parameters in the algorithm are r_k, the observational error covariance (a scalar), and $\mathbf{Q}_k$, the "regression model" error covariance (a diagonal matrix with one coefficient for the variance of each predictor if the errors are uncorrelated). Unlike regression, MOS, or neural networks, adaptive regression is *sequential*, and gives more weight to recent data than to older observations. The larger $\mathbf{Q}_k$, the faster older data will be forgotten. It also allows for observational errors r_k. This method can be generalized to several predictands, in which case the observation error covariance matrix may also include observational error correlations.

Table C.3.1 compares a simple Kalman filtering applied to the 24-h surface temperature forecasts for July and August 1997 at 00Z, averaged for eight different US stations, using as a single predictor the global model output for surface temperature interpolated to each individual station. It was found that after only a few days of spin-up, starting with a climatological first guess, and with minimal tuning, the adaptive regression algorithm was able to reach a fairly steady error level substantially better than the numerical model error, and not much higher than regression on the dependent sample. Not surprisingly, MOS, using many more predictors and several years of training, provides an even better forecast than this simple AR.

In summary, Kalman filtering provides a simple algorithm for adaptive regression. It requires little training so that it is able to adapt rather quickly to changes in the model, and to long-lasting weather regimes. It is particularly good in correcting model biases. However, in general it is not as good as regression based on long dependent samples.

References

Aber, J. D., 1992: Terrestrial ecosystems. Chapter 6 in *Climate system modeling*, K. Trenberth, editor. Cambridge University Press, Cambridge.

Alligood, K. T., T. D. Sauer, and J. A. Yorke, 1997: *CHAOS: An introduction to dynamical systems,* Springer-Verlag, New York.

Anderson, D. A., J. C. Tannehill, and R. H. Pletcher, 1984: *Computational fluid mechanics and heat transfer.* McGraw-Hill Book Company. New York.

Anderson, J. L., 1996: A method for producing and evaluating probabilistic forecasts from ensemble model integrations. *J. Climate* **9**, 1518-1530.

Anderson, J. L., 2001: An ensemble adjustment Kalman Filter for data assimilation. *Mon. Wea. Rev.* **129**, 2884-2903.

Andersson, E., J. Haseler, P. Undén, P. Courtier, G. Kelly, D. Vasiljevic, C. Brankovic, C. Cardinali, C. Gaffard, A. Hollingsworth, C. Jakob, P. Janssen, E. Klinker, A. Lanzinger, M. Miller, F. Rabier, A. Simmons, B. Strauss, J.-N. Thepaut and P. Viterbo, 1998: The ECMWF implementation of three-dimensional variational assimilation (3D-Var). III: Experimental results. *Quart. J. Roy. Meteor. Soc.* **124**, 1831-1860.

Andersson, E., and H. Jarvinen, 1999: Variational quality control. *Quart. J. Roy. Meteor. Soc.* **125**, 697-722.

Anthes, R. A., 1970: Numerical experimenting with a two-dimensional horizontal variable grid. *Mon. Wea. Rev.* **98**, 810-822.

Anthes, R. A., 1983: Regional models of the atmosphere in middle latitudes. *Mon. Wea. Rev.* **111**, 1306-1335.

Anthes, R. A., E.-Y. Hsie, and Y.-H. Kuo, 1987: *Description of the Penn State/NCAR Mesoscale Model Version 4 (MM4).* NCAR Tech Note NCAR/TN-282+STR. (Available from the National Center for Atmospheric Research, PO Box 3000, CO 80307.)

Arakawa, A., 1966: Computational design for long-term numerical integrations of the equations of atmospheric motion. *J. Comp. Phys.* **1**, 119-143.

Arakawa, A., 1970: Numerical simulation of large-scale atmospheric motions, numerical solution of field problems in continuum physics, *SIAM-AMS Proc. Am. Math. Soc.* **2**, 24-40.

Arakawa, A., 1997: Adjustment mechanisms in atmospheric motions. *J. Meteor. Soc. Japan*, Special issue of collected papers. **75**, 155-179.

Arakawa, A., and C. S. Konor, 1996: Vertical differencing of the primitive equations based on the Charney–Phillips grid in hybrid σ–p coordinates. *Mon. Wea. Rev.* **124**, 511-528.

Arakawa, A., and V. Lamb, 1977: Computational design of the basic dynamical processes in the UCLA general circulation model. In *General circulation models of the atmosphere*, Methods in computational physics, Vol. 17, J. Chang, editor, Academic press, New York, pp. 174-264.

Arakawa, A., and S. Moorthi, 1988: Baroclinic instability in vertically discrete systems. *J. Atmos. Sci.* **45**, 1688-1707.

Arakawa, A., and W. H. Schubert, 1974: Interaction of a cumulus cloud ensemble with the large scale environment. Part I. *J. Atmos. Sci.* **31**, 674-701.

Arakawa, A., and M. J. Suarez, 1983: Vertical differencing of the primitive equations in sigma coordinates. *Mon. Wea. Rev.* **111**, 34-45.

Asselin, R. A., 1972: Frequency filter for time integrations. *Mon. Wea. Rev.* **100**, 487-490.

Atger, F., 1999: The skill of ensemble prediction systems. *Mon. Wea. Rev.* **127**, 1941-1953.

Atlas, R., N. Wolfson, and J. Terry., 1993: The effect of SST and soil moisture anomalies on GLA model simulations of the 1988 US Summer Drought. *J. Climate* **6**, 2034-2048.

Baer, F., 1961: The extended numerical integration of a simple barotropic model. *J. Meteor.* **18**, 319-339.

Baer, F., 1964: Integration with the spectral vorticity equation. *J. Atmos. Sci.* **17**, 635-644.

Baer, F., and J. Tribbia, 1977: On complete filtering of gravity modes through non-linear initialization, *Mon. Wea. Rev.* **105**, 1536-9.

Baker, W. E., R. Atlas, M. Halem, and J. Susskind, 1984: A case study of forecast sensitivity to data and data analysis techniques. *Mon. Wea. Rev.* **112**, 1544-1561.

Baker, W., S. Bloom, J. Woollen, M. Nestler, E. Brin, T. Schlatter, and G. Branstator, 1987: Experiments with a three-dimensional statistical objective analysis using FGGE data, *Mon. Wea. Rev.* **115**, 272-96.

Balgovind, R., A. Dalcher, M. Ghil, and E. Kalnay, 1983: A scholastic dynamic model for the spatial structure of forecast error statistics, *Mon. Wea. Rev.* **111**, 701-722.

Ballabrera-Poy, J., P. Brasseur, and J. Verron, 2001a: Dynamical evolution of the error statistics with the SEEK filter to assimilate data in eddy-resolving ocean models. *Quart. J. Roy. Meteor. Soc.* **127**, 233-253.

Ballabrera-Poy, J., A Busalacchi, and R Murtugudde, 2001b: Application of a reduced order Kalman filter to initialize coupled atmosphere–ocean model: impact on the prediction of El Niño. *J. Climate* **14**, 1720-1737.

Ballish, B., 1981: A simple test of initialization of gravity modes, *Mon. Wea. Rev.* **109**, 1318-1321.

Ballish, B., X. Cao, E. Kalnay, and M. Kanamitsu, 1992: Incremental nonlinear normal mode initialization. *Mon. Wea. Rev.* **120**, 1723-1734.

Barker, T. W., 1991: The relationship between spread and forecast error in extended-range forecasts. *J. Climate* **4**, 733-742.

Barkmeijer, J., R. Buizza, and T. N. Palmer, 1999: 3D-Var Hessian singular vectors and their potential use in the ECMWF Ensemble Prediction System. *Quart. J. Roy. Meteor. Soc.* **125**, 2333-2351.

Barkmeijer, J., M. Van Gijzen, and F. Bouttier, 1998: Singular vectors and the estimates of analysis-error covariance matrix. *Quart. J. Roy. Meteor. Soc.* **124**, 1695-1713.

Barnes, S., 1964: A techniques for maximizing details in numerical map analysis, *J. Appl. Meteor.* **3**, 395-409.

Barnes, S., 1978: Oklahoma thunderstorms on 29-30 April 1970, Part I: Morphology of a tornadic storm, *Mon. Wea. Rev.* **106**, 673-684.

Barnston, A. G., H. M. Van den Dool, D. R. Rodenhuis, C. R. Ropelewski, V. E. Kousky, E. A. O'Lenic, R. E. Livezey, S. E. Zebiak, M. A. Cane, T. P. Barnett, N. E. Graham, Ming Ji, A. Leetma, 1994: Long-lead seasonal forecasts – where do we stand? *Bull. Amer. Meteor. Soc.* **75**, 2097-2114.

Barrett, R. M. W. Berry, T. F. Chan, J. Demmel, J. Donato, J. Dongarra, V. Eijkhout, R. Pozo, C. Romine, and H. van der Vorst, 1995: *Templates for the solution of linear systems: Building blocks for iterative methods.* ISBN 0898713285.

Bates, J. R., and A. McDonald, 1982: Multiply upstream, semi-Lagrangian advective schemes: analysis and application to a multi-level primitive equation model. *Mon. Wea. Rev.* **110**, 1831-1842.

Bates, J. R., Y. Li, A. Brandt, S. F. McCormick, and J. Ruge, 1995: A global shallow water numerical model based on the semi-Lagrangian advection of potential vorticity, *Quart. J. Roy. Meteor. Soc.* **121**, 1981-2005.

Battisti, D. S., and A. C. Hirst, 1989: Interannual variability in a tropical atmosphere–ocean model: influence of the basic state, ocean geometry and nonlinearity. *J. Atmos. Sci.* **46**, 1687-1712.

Baumhefner, D. P., and D. J. Perkey, 1982: Evaluation of lateral boundary errors in a limited area model. *Tellus* **34**, 409-428.

Bell, R. S., and H. Le Core, 2000: Observation usage in our global NWP system. *NWP Gazette*, June. The Meteorological Office, Bracknell, UK. http://www.met-office.gov.uk/research/nwp/publications/

Bell, R. S., T. D. Dalby, D. Li, and F. W. Saunders, 1999: *The autumn 1999 global data assimilation upgrade package.* Forecasting Research Tech Rep No 280. The Meteorological Office, Bracknell, UK.

Bengtsson, L., 1975: *4-dimensional data assimsilation of meteorological observations*, World Meteorological Organization GARP Publication No. 15. World Meteorologival Organization, Geneva.

Bengtsson, L., 1991: Advances and prospects in numerical weather prediction. *Quart. J. Roy. Meteor. Soc.* **117**, 855–902.

Bengtsson, L., 1999: From short-range barotropic modelling to extended-range global weather prediction: a 40-year perspective. *Tellus* **51 A-B,** 13-32.

Bengtsson, L., and N. Gustavsson, 1971: An experiment in the assimilation of data in dynamic analysis, *Tellus* **23**, 328-336.

Bengtsson, L., M. Ghil, and E. Källén,1981: *Dynamic meteorology: data assimilation methods.* Springer-Verlag, New York.

Benjamin, S. G., G. A. Grell, J. M. Brown, R. Bleck, K. J. Brundage, T. L. Smith, and P. A. Miller, 1994: An operational isentropic/sigma hyrid forecast model and data assimilation system. In *Proceedings, The Life Cycles of Extratropical Cyclones*, Vol. III, Bergen, Norway, June 27–July 1, 1994. S. Gronas and M. A. Shapiro, editors, Geophysical Institute, University of Bergen, pp. 268-273.

Benjamin, S. G., G. A. Grell, K. J. Brundage, T. L. Smith, J. M. Brown, T. G. Smirnova, and Z. Yang, 1995a: The next version of the Rapid Update Cycle – RUC II. Preprints, 6th Conference on Aviation Weather Systems, 15–20 January, Dallas, AMS, 57-61.

Benjamin, S. G., D. Kim, and T. W. Schlatter, 1995b: The Rapid Update Cycle: A new mesoscale assimilation system in hybrid theta-sigma coordinates at the National Meteorological Center. Preprints, Second International Symposium on Assimilation of Observations in Meteorology and Oceanography, Tokyo, Japan, 13–17 March, pp. 337-342.

Benjamin, S. G., J. M. Brown, K. J. Brundage, D. Devenyi, B. Schwartz, T. G. Smirnova, T. L. Smith, and F.-J. Wang, 1996: The 40-km 40-level version of MAPS/RUC. Preprints, 11th Conference on Numerical Weather Prediction, AMS, Norfolk, pp. 161-163.

Benjamin, S. G., B. E. Schwartz, and R. E. Cole, 1999: Accuracy of ACARS wind and temperature observations determined by collocation. *Wea. Forecasting*, **14**, 1032-1038.

Bennett, A. F., 1992: *Inverse methods in physical oceanography*, Cambridge University Press, Cambridge.

Bennett, A. F., and B. S. Chua, 1999: Open boundary conditions for Lagrangian geophysical fluid dynamics. *J. Comput. Phys.*, **158**, 418-436.

Bennett, A. F., B. Chua, and L. Leslie, 1997. Generalized inversion of a global numerical weather prediction model (II). Analysis and implementation. *Meteorol. Atmos. Phys.* **61**, 129-140.

Benoit, R., J. Côté, and J. Mailhot, 1989: Inclusion of a TKE boundary layer parameterization in the Canadian regional finite-element model. *Mon. Wea. Rev*. **117**, 1726-1750.

Benoit, R., M. Desagne, P. Pellerin, S. Pellerin, Y. Chartier, and S. Desjardins, 1997: The Canadian MC2: A semi-lagrangian semi-implicit wide-band atmospheric model suited for finescale process studies and simulation. *Mon. Wea. Rev.* **125**, 2382-2415.

Bergman, K. H., 1979: Multivariate analysis of temperatures and winds using optimum interpolation. *Mon. Wea. Rev.*, **107**, 1423-1444.

Bergman, K., and W. Bonner, 1976: Analysis errors as a function of observation density for satellite temperature soundings with spatially correlated errors. *Mon. Wea. Rev.* **104**, 1308-1316.

Bergthorsson, P., and B. Döös, 1955: Numerical weather map analysis, *Tellus* **7**, 329-340.

Bergthorsson, P., B. Döös, S. Frykland, O. Hang, and R. Linquist, 1955: Routine forecasting with the barotropics model, *Tellus* **7**, 329-340.

Betts, A. K., and M. J. Miller, 1986: A new convective adjustment scheme. Part II: Single column tests using GATE wave,

BOMEX ATEX and arctic air mass data sets. *Quart. J. Roy. Meteor. Soc.* **112**, 693-709.

Bishop, C. H., and Z. Toth, 1999: Ensemble transformation and adaptive observations. *J. Atmos. Sci.* **56**, 1748-1765.

Bjerknes, J., 1969: Atmospheric teleconnections from the equatorial Pacific. *Mon. Wea. Rev.* **97**, 163-172.

Bjerknes, V., 1904: Das Problem der Wettervorhersage, betrachtet vom Stanpunkt der Mechanik und der Physik. *Meteor. Zeits.* **21**, 1-7.

Bjerknes, V., 1911: *Dynamic meteorology and hydrography,* Part II. *Kinematics*, Cargnegie Institute, Gibson Bros., New York.

Black, T. L., 1994: The new NMC mesoscale eta model: Description and forecast examples. *Wea. Forecasting* **9**, 265-278.

Black, T., D. Deaven, and G. DiMego, 1993: The step-mountain eta coordinate model: 80 km 'Early' version and objective verifications. NWS Technical Procedures Bull. 412, National Oceanic and Atmospheric Administration/National Weather Service. (Available from National Weather Service, Office of Meteorology, 1325 East-West Highway, Silver Spring, MD 20910.)

Bleck, R., 1978: On the use of hybrid vertical coordinates in numerical weather prediction. *Mon. Wea. Rev.* **106**, 1233-1244.

Bleck, R., and S. G. Benjamin, 1993: Regional weather prediction with a model combining terrain-following and isentropic coordinates. Part I: model description. *Mon. Wea. Rev.* **121**, 1770-1785.

Bloom, S. C., L. L. Takacs, A. M. da Silva, and D. Ledvina, 1996: Data assimilation using incremental analysis updates. *Mon. Wea. Rev.* **124**, 1256-1271.

Blumen, W., 1972: Geotrophic adjustment, *Rev. Geophys. Space Phys.* **10**, 485-528.

Boer, G., and T. Shepherd, 1983: Large-scale two-dimensional turbulence in the atmosphere, *J. Atmos. Sci.* **40**, 164-84.

Bolin, B., 1955: Numerical forecasting with the barotropic model, *Tellus* **7**, 27-49.

Bolin, B., 1956: An improved barotropic model and some aspects of using the balance equation for three-dimensional flow, *Tellus* **8**, 61-75.

Bolin, B., 1999: Carl-Gustaf Rossby – The Stockholm period 1947–1957, *Tellus* **51**, 4-12.

Bolin, B., and E. Eriksson, editors, 1959: *The atmosphere and the sea in motion: Scientific contributions to the Rossby Memorial Volume*, Rockefeller Institute Press, New York.

Borges, M. D., and D. L. Hartmann, 1992: Barotropic instability and optimal perturbations of observed non-zonal flows. *J. Atmos. Sci.* **49**, 335-354.

Bougeault, P., 1983: A non-reflective upper boundary condition for limited-height hydrostatic models. *Mon. Wea. Rev.* **111**, 420-429.

Bourke, W., 1972: An efficient, one-level, primitive-equation spectral model. *Mon. Wea. Rev.* **100**, 683-689.

Bourke, W., 1974: A multi-level spectral model. I. Formulation and hemispheric integrations. *Mon. Wea. Rev.* **102**, 687-701.

Bourke, W., and J. L. McGregor, 1983: A nonlinear vertical mode initialisation scheme for a limited area prediction model. *Mon. Wea. Rev.* **111**, 2285-2297.

Bourke, W., B. McAveney, K. Puri, and R. Thurling, 1977: Global modelling of atmospheric flow by spectral methods. In

General circulation models of the atmosphere, Methods in computation physics, Vol. 17, J. Chang, editor, Academic Press, New York, pp. 267-324.

Bouttier, F., and F. Rabier., 1997: The operational implementation of 4D-Var. *ECMWF Newsletter* No. 78, 2-5.

Bouttier, F., J. Derber, and M. Fisher, 1997: The 1997 revision of the Jb term in 3D/4D-Var. ECMWF Tech. Memo., 238, ECMWF, Shinfield Park, Reading, UK.

Boyer, J. F., 1991: *Review of recent advances in dynamical extended range forecasting for the extratropics. Proceedings NATO Workshop on Prediction and Interannual Climate Variations, July 1991, Trieste, Italy.* Springer-Verlag, New York.

Brandt, A., 1988: Multilevel computations: Review and recent developments. In *Multigrid methods: Theory, applications and supercomputing*, S. F. McCormick, editor, Marcel Dekker, Inc., New York, pp. 35-62.

Brankovic, C., and T. N. Palmer, 2000: Seasonal skill and predictability of ECMWF PROVOST ensembles. *Quart. J. Roy. Meteor. Soc.* **126**, 2035.

Bratseth, A., 1982: A simple and efficient approach to the initialization of weather prediction models, *Tellus* **34**, 352-357.

Bratseth, A., 1986: Statistical interpolation by means of successive corrections, *Tellus* **38A**, 439-447.

Bretherton, F., R. Davis, and C. Fandry, 1976: A technique for objective analysis and design of oceanographic experiments applied to MODE-73, *Deep Sea Research* **23**, 559-592.

Briggs, W. L., 1987: *A multigrid tutorial.* SIAM, Philadelphia.

Brooks, H. E., M. S. Tracton, D. J. Stensrud, G. DiMego, and Z. Toth, 1995: Short range ensemble forecasting: Report from a workshop, 25–27 July 1994, *Bull. Amer. Meteor. Soc.* **76**, 1617-1624.

Brooks, H. E., A. Witt, and M. D. Ellis, 1997: Verification of public weather forecasts available via the media. *Bull. Amer. Meteor. Soc.* **78**, 2167-2177.

Browning, G., and H. Kreiss, 1982: Initialization of the shallow water equations with open boundaries by the bounded derivative method, *Tellus* **34**, 334-351.

Bruaset, A. M., 1995: *Survey of preconditioned iterative methods*, Pitman Research Notes in Mathematics, No 328, Longman Science & Technology.

Bube, K., and M. Ghil, 1980: Assimilation of asynoptic data and the initialization problem. In *Dynamic meteorology data assimilation methods*, L. Bengtsson, M. Ghil, and E. Kallen, editors, Springer-Verlag, New York, pp. 111-138.

Buell, C., 1971: Two-point wind correlation on an isobaric surface in a non-homogeneous, non-isotropic atmosphere. *J. Appl. Meteor.* **10**, 1266-1274.

Buell, C., 1972a: Correlation function for wind and geopotential on isobaric surfaces, *J. Appl. Meteor.* **11**, 51-59.

Buell, C., 1972b: Variability of wind with distance and time on an isobaric surface, *J. Appl. Meteor.* **11**, 1085-1091.

Buizza, 1994: Sensitivity of optimal unstable structures. *Quart. J. Roy. Meteor. Soc.* **120**, 429-451.

Buizza, R., 1997: Potential forecast skill of ensemble prediction, and spread and skill distributions of the ECMWF Ensemble Prediction System. *Mon. Wea. Rev.* **125**, 99-119.

Buizza, R., 2000: Skill and economic value of the ECMWF Ensemble Prediction System. *Quart. J. Roy. Meteor. Soc.* **126**, 649-668.

Buizza, R. and A. Hollingsworth, 2001: Storm prediction over Europe using the ECMWF Ensemble Prediction System. *Mon. Wea. Rev.* in press.

Buizza R., and T. Palmer, 1995: The singular vector structure of the atmospheric general circulation. *J. Atmos Sci.* **52**, 1434-1456.

Buizza, R., J. Tribbia, F. Molteni, and T. Palmer, 1993: Computation of optimal unstable structures for a numerical weather prediction model. *Tellus* **45A**, 388-407.

Buizza R., R. Gelaro, F. Molteni, and T. N. Palmer, 1997: The impact of increased resolution on predictability studies with singular vectors. *Quart. J. Roy. Meteor. Soc.* **123**, 1007-1033.

Buizza, R., T. Petroliagis, T. N. Palmer, J. Barkmeijer, M. Hamrud, A. Hollingsworth, A. Simmons, and N. Wedi, 1998: Impact of model resolution and ensemble size on the performance of an ensemble prediction system. *Quart. J. Roy. Meteor. Soc.* **124**, 1935-1960.

Buizza, R., M. Miller, and T. N. Palmer, 1999: Stochastic representation of model uncertainties in the ECMWF Ensemble Prediction System. *Quart. J. Roy. Meteor. Soc.* **125**, 2887-2908.

Buizza, R., J. Barkmeijer, T. N. Palmer, and D. S. Richardson, 2000: Current status and future developments of the ECMWF Ensemble Prediction System. *Meteorol. Appl.* **7**, 163-175.

Bulovna, R., G. Hello, P. Bernard, and J. F. Geleyn, 1995: Integration of the fully elastic equations cast in the hydrostatic pressure terrain-following coordinate in the framework of the ARPEGE/Aladin NWP system. *Mon. Wea. Rev.* **123**, 515-535.

Burgers, G., P. J. Van Leewen, and G. Evensen., 1998: Analysis scheme in the ensemble Kalman filter. *Mon. Wea. Rev.* **126**, 1719-1724.

Burridge, D., 1975: A split semi-implicit reformulation of the Bushby–Timpson 10 level model. *Quart. J. Roy. Meteor. Soc.* **101**, 777-792.

Bushby, F., and V. Huckle, 1956: The use of a streamfunction in a two-parameter model of the atmosphere, *Quart. J. Roy. Meteor. Soc.* **82**, 409-418.

Caccuci, D. G. 1981: Sensitivity theory for non-linear systems. I: Nonlinear functional analysis approach. *J. Math. Phys.* **22**, 2794-2802.

Cahalan R. F., W. Ridgway, W. J. Wiscombe, T. L. Bell, and J. B. Snider, 1994: The albedo of fractal stratocumulus clouds. *J. Atmos. Sci.* **51**, 2434-2455.

Cai, M., E. Kalnay, and Z. Toth, 2002: Bred vectors of the Zebiak–Cane model and their application to ENSO predictions. *J. Climate* in press.

Caian, M. and J. F. Geleyn, 1997: Some limits to the variable mesh solution and comparison with the nested LAM solution. *Quart. J. Roy. Meteor. Soc.* **123**, 743-766.

Campana, K. A., 1994: Use of cloud analyses to validate and improve model-diagnostic cloud at NMC. *ECMWF/GWEX workshop. Modeling, validation, and assimilation of clouds*, Nov. 1994. ECMWF, Shinfield Park, Reading, UK.

Cane, M. A., 1992: Tropical Pacific ENSO models: ENSO as a mode of the coupled system. Chapter 18 in *Climate system modeling*, K. Trenberth, editor, Cambridge University Press, Cambridge.

Cane, M. A., S. E. Zebiak, and S. C. Dolan, 1986: Experimental forecasts of El Niño. *Nature* **321**, 827-832.

Caplan, P., and G. White, 1989: Performance of the National Meteorological Center's medium-range model. *Wea. Forecasting* **4**, 391-400.

Carbone, R. E., J. W. Wilson, T. D. Keenan, and J. M. Hacker, 2000: Tropical island convection in the absence of significant topography. Part I: Life cycle of diurnally forced convection. *Mon. Wea. Rev.* **128**, 3459-3480.

Carter, G. M., J. P. Dallavalle, and H. R. Glahn, 1989: Statistical forecasts based on the National Meteorological Center's numerical weather prediction system. *Wea. Forecasting* **4**, 401-412.

Charney. J., 1948: On the scale of atmospheric motions. *Geofys. Publikasjoner* **17**, 1-17.

Charney, J., 1949: On a physical basis for numerical prediction of large-scale motions in the atmosphere. *J. Meteor.* **6**, 371-385.

Charney, J. G., 1951: Dynamical forecasting by numerical process. *Compendium of meteorology.* American Meteorological Society, Boston, MA.

Charney, J. G., 1954: Numerical prediction of cyclogenesis. *Proc. Nat. Acad. Sci. US* **40**, 99-110.

Charney, J. G., 1955: The use of the primitive equations of motion in numerical prediction. *Tellus* **7**, 22-26.

Charney, J. G., 1962: Integration of the primitive and balance equations. In *Proc. of the international symposium on numerical weather predcition.* Tokyo, Nov. 1960, Meteorological Society of Japan.

Charney, J. G., and N. A. Phillips, 1953: Numerical integration of the quasi-geostrophic equations for barotropic and simple baroclinic flows. *J. Meteor.* **10**, 71-99.

Charney, J. G., and J. Shukla, 1981: Predictability of monsoons. In *Monsoon Dynamics. Proceedings of the Joint IUTAM/IUGG International Symposium on Monsoon Dynamics, New Delhi, India, 5-9 December 1977,* J. Lighthill and R. P. Pierce, editors, Cambridge University Press, Cambridge, pp. 99-110.

Charney, J. G., R. Fjørtoft, and J. von Neuman, 1950: Numerical integration of the barotropic vorticity equation. *Tellus* **2**, 237-254.

Charney, J. G., R. G. Fleagle, H. Riehl, V. E. Lally, and D. Q. Wark, 1966: The feasibility of a global observation and analysis experiment. *Bull. Amer. Phys. Soc.* **47**, 200-220.

Charney, J., M. Halem, and R. Jastrow, 1969: Use of incomplete historical data to infer the present state of the atmosphere, *J. Atmos. Sci.* **26**, 1160-1163.

Charnock, H., 1955: Wind stress on the water surface. *Quart. J. Roy. Meteor. Soc.* **81**, 639-640.

Chen, F., Z. Janjic, and K. Mitchell, 1997: Impact of atmospheric surface-layer parameterizations in the new land-surface scheme of the NCEP mesoscale Eta Model. *Bound.-Layer Meteor.* **85**, 391-421.

Chou, M. D., M. J. Suarez, C. H. Ho, M. M. H. Yan, and K. T. Lee, 1998: Parameterizations for cloud overlapping and shortwave single scattering properties for use in general circulation and cloud ensemble models. *J. Climate* **11**, 202-214.

Christensen, O. B., J. H. Christensen, B. Machenhauer, and M. Botzet, 1998: Very high-resolution regional climate simulations over Scandinavia – present climate. *J. Climate* **11**, 3204-3229.

Cohn, S. E., 1997: An introduction to estimation theory. *J. Meteor. Soc. Japan* **75** (1B), 257-288.

Cohn, S., and D. Dee, 1988: Observability of discretized partial differential equations, SIAM *J. Numer. Anal.* **25**, 586-617.

Cohn, S., and D. Dee, 1989: An analysis of the vertical structure function for arbitrary thermal profiles, *Quart. J. Roy. Meteor. Soc.* **115**, 143-71.

Cohn, S., and D. Parrish, 1991: The behavior of forecast error covariances for a Kalman Filter in two dimensions. *Mon. Wea. Rev.* **119**, 1757-1785.

Cohn, S. E., and R. Todling, 1996: Approximate data assimilation schemes for stable and unstable dynamics. *J. Meteor. Soc. Japan* **74**, 63-75.

Cohn, S. E., A. da Silva, J. Guo, M. Sienkiewicz, and D. Lamich, 1998: Assessing the effects of data selection with the DAO Physical-space Statistical Analysis System. *Mon. Wea. Rev.* **126**, 2913-2926.

Colella, P., and P. Woodward, 1984: The piecewise-parabolic method (PPM) for gas-dynamical simulations. *J. Comp. Phys.* **54**, 174-201.

Collins, W. G., 1998: Complex quality control of significant level radiosonde temperatures. *J. Atmos. Ocean. Tech.* **15**, 69-79.

Collins, W. G., 2001a: The operational complex quality control of radiosonde heights and temperatures at the national centers for environmental prediction. Part I: Description of the method. *J. Appl. Meteor.* **40**, 137-151.

Collins, W. G., 2001b: The operational complex quality control of radiosonde heights and temperatures at the national centers for environmental prediction. Part II: Examples of error diagnosis and correction from operational use. *J. Appl. Met.* **40**, 152-168.

Collins, W. G., and L. S. Gandin, 1990: Comprehensive hydrostatic quality control at the National Meteorological Center. *Mon. Wea. Rev.* **118**, 2752-2767.

Corazza, M., E. Kalnay, D. J. Patil, R. Morss, M. Cai, I. Szunyogh, B. R. Hunt, E. Ott, and J. A. Yorke, 2002: Use of the breeding technique to estimate the structure of the analysis "errors of the day". Submitted to *Nonlinear Processes in Geophysics.*

Corby, G. A., A. Gilchrist, and R. L. Newson, 1972: A general circulation model of the atmosphere suitable for long period integration. *Quart. J. Roy. Met. Soc.* **98**, 809-832.

Cote, J. and A. Staniforth, 1990: An accurate and efficient finite-element global model of the shallow-water equations. *Mon. Wea. Rev.* **118**, 2707-2717.

Cote, J., J. M. Roch, A. Staniforth, and L. Fillion, 1993: A variable resolution semilagrangian finite element global model of the shallow water equations. *Mon. Wea. Rev.* **121**, 231-243.

Cotton, W. R., G. Thompson, and P. W. Mielke Jr, 1994: Real-time mesoscale prediction on workstations. *Bull. Amer. Meteor. Soc.* **75**, 349-362.

Courant, R., and D. Hilbert, 1953: *Methods of mathematical physics,* Vol I, Interscience, New York.

Courant, R., and D. Hilbert, 1962: *Partial differential equations, Methods of mathematical physics,* Vol. II, Interscience, New York.

Courant, R., K. O. Friedrichs, and H. Lewy, 1928: Uber die Partielen Differenzengleichungen der Mathematischen Physik. *Math. Annalen* **100**, 32-74.

Courtier, P., 1997: Variational methods. *J. Meteor. Soc. Japan* **75**, 211-218.

Courtier, P., and J.-F. Geleyn, 1987: A global numerical weather prediction model with variable resolution: application to the shallow-water equations. *Quart. J. Roy. Meteor. Soc.* **114**, 1321-1346.

Courtier, P., and J.-F. Geleyn, 1988: A global numerical weather prediction model with variable resolution: Application to the shallow-water equations. *Quart. J. Roy. Meteor. Soc.* **114**, 1321-1346.

Courtier, P., and O. Talagrand, 1987: Variational assimilation of meteorological observations with the adjoint vorticity equations, Part II, Numerical results. *Quart. J. Roy. Meteor. Soc.* **113**, 1329-1347.

Courtier, P., and O. Talagrand, 1990: Variational assimilation of meteorological observations with the direct and adjoint shallow water equations. *Tellus* **42**A, 531-549.

Courtier, P., C. Freydier, J.-F. Geleyn, F. Rabier, and M. Rochas, 1991: The Arpege project at Meteo France. In *Numerical methods in atmospheric models, ECMWF Seminar Proceedings, 9–13 September 1991*, ECMWF, Shinfield Park, Reading, UK, pp. 193–231.

Courtier, P., E. Andersson, W. Heckley, J. Pailleux, D. Vasiljevic, M. Hamrud, A. Hollingsworth, C. Jakob, P. Janssen, F. Rabier, and M. Fisher, 1998: The ECMWF implementation of three-dimensional variational assimilation (3D-Var). I: Formulation. *Quart. J. Roy. Meteor. Soc.* **124**, 1783-1807.

Courtier, P., J. Derber, R. M. Errico, J. F. Louis and T. Vukicevic, 1993: Review of the use of adjoint, variational methods and Kalman filters in meteorology. *Tellus* **45**A, 343-357.

Courtier, P., J.-N. Thepaut, and A. Hollingsworth., 1994: A strategy for operational implementation of 4d-Var using an incremental approach. *Quart. J. Roy. Meteor. Soc.* **120**, 1367-1387.

Cressman, G. P., 1959: An operational objective analysis system. *Mon. Wea. Rev.* **87**, 367-374.

Cullen, M. J. P., 1979: The finite element method. In *Numerical methods used in atmospheric models.* Garp. Publ. Ser., Vol. 17, pp. 301-338.

Cullen, M. J. P., and C. Hall, 1979: Forecasting and general circulation results from finite element models. *Quart. J. Roy. Meteor. Soc.*, **105**, 571-592.

Da Silva, A., J. Pfaendtner, J. Guo, M. Sienkiewicz, and S. Cohn, 1995: *Assessing the effects of data selection with DAO's physical-space statistical analysis system. Proceedings of the second international symposium on the assimilation of observations in meteorology and oceanography, Tokyo, Japan,* World Meteorological Organization and Japan Meteorological Agency, Tokyo, Japan.

Dahlquist, G., and A. Björk, 1974: *Numerical methods*. Prentice-Hall, New Jersey.

Dalcher, A., and E. Kalnay, 1987: Error growth and predictability in operational ECMWF forecasts. *Tellus* **39A**, 474-491.

Dalcher, A., E. Kalnay, and R. N. Hoffmann, 1988: Medium range lagged average forecasts. *Mon. Wea. Rev.* **116**, 402-416.

Daley, R., 1978: Variational non-linear normal mode initialization. *Tellus* **30**, 201-218.

Daley, R., 1980: On the optimal specification of the initial state for deterministic forecasting, *Mon. Wea. Rev.* **108**, 1719-1735.

Daley, R., 1983: Linear non-divergent mass-wind laws on the sphere, *Tellus* **35A**, 17-27.

Daley, R., 1991: *Atmospheric data analysis.* Cambridge University Press, Cambridge.

Daley, R., and K. Puri, 1980: Four dimensional data assimilation and the slow manifold, *Mon. Wea. Rev.* **108**, 85-99.

Davies, H. C., 1976: A lateral boundary formulation for multi-level prediction models. *Quart. J. Roy. Meteor. Soc.* **102**, 405-418.

Davies, H. C., 1983: Limitations on some lateral boundary schemes used in regional NWP models. *Mon. Wea. Rev.* **111**, 1002-1012.

Deardorff, J. W., 1972: Parameterization of the planetary boundary layer for use in general circulation models. *Mon. Wea. Rev.* **100**, 93-106.

Dee, D. P., and A. Da Silva, 1986: Using Hough harmonics to validate and assess nonlinear shallow-water models. *Mon. Wea. Rev.* **114**, 2191-2196.

Dee, D., and A. Da Silva, 1998: Data assimilation in the presence of forecast bias. *Quart. J. Roy. Meteor. Soc.* **124**, 269-295.

Dee, D. P., and A. Da Silva, 1999: Maximum-likelihood estimation of forecast and observation error covariance parameters. Part I: Methodology. *Mon. Wea. Rev.* **127**, 1822-1834.

Deque, M., and J. P. Piedelievre, 1995: High resolution climate simulation over Europe. *Climate Dynamics* **11**, 321-339.

Derber, J., 1987: Variational four-dimensional analysis using quasi-geostrophic constraints, *Mon. Wea. Rev.* **115**, 998-1008.

Derber, J., 1989: A variational continuous assimilation technique, *Mon. Wea. Rev.* **117**, 2437-2446.

Derber, J., and F. Bouttier, 1999: A reformulation for the background error covariance in the ECMWF global data assimilation system. *Tellus* **51A**, 195-222.

Derber, J., and A. Rosati, 1989: A global oceanic data assimilation system. *J. Phys. Oceanogr.* **19**, 1333-1347.

Derber, J. C., D. F. Parrish, and S. J. Lord, 1991: The new global operational analysis system at the National Meteorological Center. *Wea. Forecasting* **6**, 538-547.

Derber, J. C., W.-S. Wu, 1998: The use of TOVS cloud-cleared radiances in the NCEP SSI analysis system. *Mon. Wea. Rev.* **126**, 2287-2302.

Dey, C., and L. Morone, 1985: Evolution of the National Meteorological Center global assimilation system: January 1982–December 1983. *Mon. Wea. Rev.* **113**, 304-18.

Dickinson, R. E., 1984: Modeling evapotranspiration for three-dimensional global climate models: climate processes and climate sensitivity. *Geophys, Monogr.* **29**, 58-72.

Dickinson, R. E., 1992: Land surface. Chapter 5 in *Climate system modeling*, K. Trenberth, editor. Cambridge University Press, Cambridge.

Dickinson, R. E., and D. Williamson, 1972: Free oscillations of a discrete stratified fluid with application to numerical weather prediction. *J. Atmos. Sci.* **29**, 623-640.

Dickinson, R. E., A. Henderson-Sellers, and P. J. Kennedy, 1993: *Biosphere-atmosphere transfer scheme (BATS) version 1E as coupled to the NCAR community climate model.* National Center for Atmospheric Research, Boulder, CO/ TN-387+STR.

Dietachmayer, G. S., 1992: Application of continuous dynamic grid adaptation techniques to meteorological modeling. Part II: Efficiency. *Mon. Wea. Rev.* **120**, 1707-1722.

Dietachmayer, G. S., and K. K. Droegemeier, 1992: Application of continuouis dynamic gred adaption techniques to mereorological modeling. Part I: Basic formulation and accuracy. *Mon. Wea. Rev.* **120**, 1675-1706.

DiMego, G. J., 1988: The National Meteorological Center regional analysis system. *Mon. Wea. Rev.* **116**, 977-1000.

DiMego, Geoffrey, P. A. Phoebus, and J. E. McDonell, 1985: *Data processing and quality control for optimum interpolation analyses at the National Meteorological Center.* Washington, DC, US Dept of Commerce, National Oceanic and Atmospheric Administration, National Weather Service. Office Note 306.

DiMego, G. J., K. E. Mitchell, R. A. Petersen, J. E. Hoke, J. P. Gerrity, J. J. Tuccillo, R. L. Wobus, and H. M. H. Juang, 1992: Changes to NMC's regional analysis and forecast system. *Wea. Forecasting* **7**, 185-198.

Donner, L., 1988: An initialization for cumulus convection in numerical weather prediction models. *Mon. Wea. Rev.* **116**, 377-385.

Döös, B., and M. Eaton, 1957: Upper air analysis over ocean areas, *Tellus* **9**, 184-194.

Dormand, J. R., 1996: *Numerical methods for differential equations: a computational approach.* CRC.

Droegemeier, K. K., 1997: The numerical prediction of thunderstorms: Challenges, potential benefits, and results from real-time operational tests. *WMO Bulletin* **46**, 324-336.

Droegemeier, K. K., S. M. Lazarus, R. Davies-Jones, 1993: The influence of helicity on numerically simulated convective storms. *Mon. Wea. Rev.* **121**, 2005-2029.

Duchon, C., 1979: Lanczos filtering in one and two dimensions. *J. Appl. Meteor.* **18**, 1016-1022.

Dudhia, J., 1993: A non-hydrostatic version of the Penn State/NCAR mesoscale model: Validation tests and simulation of an Atlantic cyclone and cold front. *Mon. Wea. Rev.* **121**, 1493-1513.

Durran, D. R., 1989: Improving the anelastic approximation. *J. Atmos. Sci.* **46**, 1453-1461.

Durran, D. R., 1991: The third-order Adams–Bashforth method: an attractive alternative to leapfrog time differencing. *Mon. Wea. Rev.* **119**, 702-720.

Durran, D. R., 1999: *Numerical methods for wave equations in geophysical fluid dynamics.* Springer, New York.

Durran, D. R., and J. B. Klemp, 1983: A compressible model for the simulation of moist mountain waves. *Mon. Wea. Rev.* **111**, 2341-2361.

Durran, D. R., M. J. Yang, D. N. Slinn, and R. G. Brown, 1993: Toward more accurate wave-permeable boundary conditions. *Mon. Wea. Rev.* **121**, 604-620.

Ebisuzaki, W., and E. Kalnay, 1991: Ensemble experiments with a new lagged average forecasting scheme. WMO, Research activities in atmospheric and oceanic modeling. Report #15, pp. 6.31-6.32. (Available from WMO, C.P. No 2300, CH1211, Geneva, Switzerland.)

ECMWF, 1999: *A strategy for ECMWF 1999–2008.* ECMWF, Shinfield Park, Reading, UK.

Edwards, A. W. F., 1984: *Likelihood.* Cambridge University Press, Cambridge.

Egbert, G. D., A. F. Bennett, and M. G. G. Foreman, 1994: TOPEX/POSEIDON tides estimated using a global inverse model, *J. Geophys. Res.* **99**, 24821-24852.

Ehrendorfer, M., 1997: Predicting the uncertainty of numerical weather forecasts: a review. *Meteorol. Zeitschrift N. F.* **6**, 147-183.

Ehrendorfer, M., and R. Errico, 1995: Mesoscale predictability and the spectrum of optimal perturbations. *J. Atmos. Sci.* **52**, 3475-3500.

Ehrendorfer, M., and J. J. Tribbia, 1997: Optimal prediction of forecast error covariances through singular vectors. *J. Atmos. Sci.* **54**, 286-313.

Eliasen, E., and B. Machenhauer, 1965: A study of fluctuations of the atmosheric flow patterns represented by spherical harmonics, *Tellus* **17**, 220-238.

Eliasen, E., B. Machenauer, and E. Rasmussen, 1970: *On a numerical method for the integration of the hydrodynamical equations with spectral representation of the horizontal fields*, Report #2, Institute for Theoretisk Meteorologi, University of Copenhagen.

Eliassen, A., 1949: The quasi-static equations of motion with pressure as independent variable. *Geofys. Publikasjoner* **17**, No. 3.

Eliassen, A., and E. Raustein, 1968: A numerical integration experiment with a model atmosphere based on isentropic coordinates. *Meteorologiske Annaler* **5**, 45-63.

Eliassen, A., J. S. Sawyer, and J. Smagorinsky, 1954: *Upper air network requirements for numerical weather prediction.* Technical Note No 29. World Meteorological Organization, Geneva.

Emanuel, K. A., and D. J. Raymond, editors, 1993: *The representation of cumulus convection in numerical modeling of the atmosphere.* Meteorological Monographs, Vol 24, American Meteorological Society, Boston.

Engquist, B., and A. Majda, 1979: Radiation boundary conditions for acoustic and elastic wave calculations. *Comm. Pure Appl. Math.* **32**, 313-357.

Epstein, E. S., 1969: Stochastic-dynamic prediction. *Tellus* **21**, 739-759.

Errico, R., 1982: National mode initialization and the generation of gravity waves by quasi-geostrophic forcing, *J. Atmos. Sci.* **39**, 573-86.

Errico, R. M., 1997: What is an adjoint model? *Bull. Amer. Meteor. Soc.* **78**, 2577-2591.

Errico, R., and D. Baumhefner, 1987: Predictability experiments using a high-resolution limited area model. *Mon. Wea. Rev.* **115**, 488-504.

Errico, R., and R. Langland, 1999a: Notes on the appropriateness of bred modes for generating initial perturbations used in ensemble forecasting. *Tellus* **51A**, 431-441.

Errico, R. M., and R. Langland, 1999b: Response to comments by Toth *et al. Tellus* **51A**, 450-451.

Errico, R. M., and T. Vukicevic, 1992: Sensitivity analysis using an adjoint of the PSU-NCAR mesoscale model. *Mon. Wea. Rev.* **120**, 1644-1660.

Ertel, H., 1942: Ein neuer hydrodynamischer Wirbelsatz. *Met. Zeitsch.* **59**, 271-281.

Evensen, G., 1994: Sequential data assimilation with a nonlinear quasigeostrophic model using Monte Carlo methods to forecast error statistics. *J. Geophys. Res.* **99** (C5), 10143-10162.

Evensen, G., and P. J. Van Leeuwen, 1996: Assimilation of GEOSAT altimeter data for the Aghulas current using the ensemble Kalman filter with a quasigeostrophic model. *Mon. Wea. Rev.* **124**, 85-96.

Falkovich, A., E. Kalnay, S. Lord, and M. B. Mathur, 2000a: A new method of observed rainfall assimilation in forecast models. *J. Appl. Meteor.* **39**, 1282-1298.

Falkovich A., S. Lord, and R. Treadon, 2000b: A new methodology of rainfall retrievals from indirect measurements. *Meteorol. Atmos. Phys.* **75**, 217-232.

Farrell, B. F., 1982: The initial growth of disturbances in a baroclinic flow. *J. Atmos. Sci.* **39**, 1663-1686.

Farrell, B. F., 1988: Optimal excitation of neutral Rossby waves. *J. Atmos. Sci.* **45**, 163-172.

Ferziger, J. H., and M. Peric, 2001: *Computational methods for fluid dynamics.* Springer-Verlag, Berlin.

Fillion, L., and C. Temperton, 1989: Variational implicit normal mode initialization, *Mon. Wea. Rev.* **117**, 2219-2229.

Fischer, M., 1998: *Development of a simplified Kalman Filter.* ECMWF Tech Memo 260, ECMWF, Shinfield Park, Reading, UK.

Fischer, M., and P. Courtier, 1995: *Estimating the covariance matrices of analysis and forecast error in variational data assimilation,* ECMWF Tech. Memo. 220, ECMWF, Shinfield Park, Reading, UK.

Fischer, M., and A. Navarra-Giotto, 2000: A coupled atmosphere-ocean general-circulation model: The tropical Pacific. *Quart. J. Roy. Meteor. Soc.* **126** No. 567 DSP/PROVOST Issue.

Fischer, C., A. Joly, and F. Lalaurette, 1998: Error growth and Kalman filtering within an idealized baroclinic flow. *Tellus* **50A**, 596-615.

Fjørtoft, R., 1952: On a numerical method of integrating the barotropic vorticity equation. *Tellus* **4**, 179-94.

Fjørtoft, R., 1953: On the changes in the spectral distribution of kinetic energy for two-dimensional nondivergent flow. *Tellus* **5**, 225-230.

Flattery, T. W., 1971: Spectral models for global analysis and forecasting. Tech. Report 243, Air Weather Service, US Air Force, 42-54. (Available from the National Centers for Environmental Prediction, 5200 Auth Rd., Rm. 100. Camp Springs, MD 20746.)

Fletcher, C. A., 1988: *Computational techniques for fluid dynamics.* Volume I: *Fundamental and general techniques.* Springer-Verlag, New York.

Fox-Rabinovitz, M. S., Stenchikov, G. L., Suarez, M. J., and Takacs, L. L., 1997: A finite-difference GCM dynamical core with a variation-resolution stretched grid. *Mon. Wea. Rev.* **125**, 2493-2968.

Frederiksen, J. S., 1997: Adjoint sensitivity and finite-time normal mode disturbances during blocking. *J. Atmos. Sci.* **54**, 1144-1165.

Fritsch, J. M., J. Hilliker, J. Ross, and R. L. Vislocky, 2000: Model consensus. *Wea. Forecasting* **15**, 571-582.

Gal-Chen, T., 1983: Initialization of mesoscale models: The possible impact on remotely sensed data. In: *Mesoscale meteorology: Theories, observations and models*, D. Lilly and T. Gel-Chen, editors, Reidel, Dordrecht.

Gal-Chen, T., and R. Somerville, 1975: On the use of a coordinate transformation for the solution of the Navier–Stokes equations. *J. Comp. Phys.* **17**(2), 209-228.

Galerkin, B., 1915: Rods and plates. Series occurring in various questions concerning the elastic equilibrium of rods and plates. *Vestnik Inzhenerov* **19**, 897-908.

Gandin, L. S., 1963: Objective analysis of meteorological fields, *Gidrometerologicheskoe Izdatelstvo,* Leningrad. English translation by Israeli Program for Scientific Translations, Jerusalem, 1965.

Gandin, L. S., 1988: Complex quality control of meteorological observations. *Mon. Wea. Rev.* **116**, 1137-1156.

Gandin, L. S., L. L. Morone, W. G. Collins, 1993: Two years of operational comprehensive hydrostatic quality control at the NMC. *Wea. Forecasting* **8** No 1, 57-72.

Garratt, J. R., 1994: *The atmospheric boundary layer.* Cambridge University Press, Cambridge.

Gates, W. L., 1992: AMIP: The Atmospheric Model Intercomparison Project. *Bull. Amer. Meteor. Soc.* **73**, 1962-1970.

Gates, W. L., J. S. Boyle, C. Covey, C. G. Dease, C. M. Doutriaux, R. S. Drach, M. Fiorino, P. J. Gleckler, J. J. Hnilo, S. M. Marlais, T. J. Phillips, G. L. Potter, B. D. Santer, K. R. Sperber, K. E. Taylor, D. N. Williams, 1999: An overview of the results of the Atmospheric Model Intercomparison Project (AMIP). *Bull. Amer. Meteor. Soc.* **80**, 29-56.

Gelaro, R., R. Buizza, T. N. Palmer, and E. Klinker, 1998: Sensitivity analysis of forecast errors and the construction of optimal perturbations using singular vectors. *J. Atmos. Sci.* **55**, 1012-1037.

Ghil, M., 1997: Advances in sequential estimation for atmospheric and oceanic flows. *J. Meteor. Soc. Japan* **75**, 289-304.

Ghil, M., and S. Childress, 1987: *Atmospheric dynamics, dynamo theory and climate dynamics*, Topics in geophysical fluid dynamic. Springer-Verlag, New York.

Ghil, M., and P. Malanotte-Rizzoli, 1991: Data assimilation in meteorology and oceanography. *Adv. Geophys.* **33**, 141-266.

Ghil, M., and A. W. Robertson, 2002: "Waves" vs "particles" in the atmosphere phase space: a pathway to long-range forecasting? Submitted to *Proceedings of the National Academy of Sciences.*

Ghil, M., and R. Todling, 1996: Tracking atmospheric instabilities with the Kalman filter. Part II: Two-layer results, *Mon. Wea. Rev.* **124**, 2340-2352.

Ghil, M., S. Cohn, J. Tavantzis, K. Bube, and E. Isaacson, 1981: Applications of estimation theory to numerical weather prediction. In *Dynamic Meteorology: Data Assimilation Methods*, L. Bengtsson, M. Ghil, and E. Kallen, editors. Springer-Verlag, New York.

Ghil, M., R. Benzi, and G. Parisi (editors), 1985: *Turbulence and predictability in geophysical fluid dynamics and climate dynamics*, North-Holland Publ. Co., Amsterdam.

Ghil, M., M. Kimoto, and J. D. Neelin, 1991: Nonlinear dynamics and predictability in the atmospheric sciences, *Rev. Geophys.* Supplement (US Nat. Rept to Int. Union of Geodesy. & Geophys. 1987–1990), **29**, 46-55.

Ghil, M., K. Ide, A. Bennett, P. Courtier, M. Kimoto, M. Nagata, M. Saiki, and N. Sato, editors, 1997: *Data assimilation in meteorology and oceanography.* Meteor. Soc. Japan, Tokyo, Japan.

Gibson, J. K., P. Kallberg, S. Uppala, A. Nomura, A. Hernandez, and E. Serrano, 1997: *ECMWF reanalysis description.* ECMWF Reanalysis Report Series #1, ECMWF, Shinfield Park, Reading, UK.

Giering, R., 1994: Adjoint model compiler. Users manual (Available online at http://klima47.dkrz.de/giering/amc.)

Giering R., and T. Kaminski, 1998: Recipes for adjoint code construction. *Trans. Math. Software* **4**, 437-474. (Available from R. Giering, Dept. of Earth, Atmospheric and Planetary Sciences, Massachusetts Institute of Technology, 77 Massachusetts Ave., Cambridge, MA 02139.)

Gilchrist, A., 1979: The Meteorological Office 5-Layer General Circulation Model. In *Report of the JOC Conference on climate models: Performance, intercomparison, and sensitivity studies*, Vol. 1, W. L. Gates, editor, WMO/ICSU, Washington, DC, pp. 254-295.

Gilchrist, B., and G. Cressman, 1954: An experiment in objective analysis. *Tellus* **6**, 309-18.

Giles, M. B., and W. T. Thompkins, 1985: Propagation and stability of wavelike solutions of finite-difference equations with variable coefficients. *J. Comput. Phys.* **58**, 349-360.

Gill, A. E., 1980: Some simple solutions for heat-induced tropical circulation. *Quart. J. Roy. Meteor. Soc.* **106**, 447-462.

Gill, A. E., 1982: *Atmosphere–ocean dynamics*, Academic Press, New York.

Gill, P., W. Murray, and M. Wright, 1981: *Practical optimization*, Academic Press, London.

Giorgi, F., 1990: Simulation of regional climate using a limited area model nested in a general circulation model. *J. Climate* **3**, 941-963.

Giorgi, F., and L. O. Mearns, 1999: Introduction to special section: regional climate modeling revisited. *J. Geophys. Res.* **104**, 6335-6352.

Giorgi, F., M. R. Marinucci, G. T. Bates, and G. DeCanio, 1993: Development of a second generation regional climate model. Part 2: Convective processes and assimilation of lateral boundary conditions. *Mon. Wea. Rev.* **121**, 2815-2832.

Glahn H. R., and D. A. Lowry, 1972: The use of model output statistics in objective weather forecasting. *J. Appl. Meteor.* **11**, 1203-1211.

Gleick, J., 1987: *Chaos: making a new science.* Viking Penguin, New York.

Golub, G., and C. Van Loan, 1996: *Matrix computations*, 3rd edition, The Johns Hopkins University Press Ltd, London.

Grant, W. F., 1975: Initialization Procedures for Primitive Equation Models, MS thesis, MIT, Cambridge, MA.

Gravel, S., and A. Staniforth, 1992: Variable resolution and robustness. *Mon. Wea. Rev.* **120**, 2633-2640.

Greenbaum, A., 1997: *Iterative methods for solving linear systems.* SIAM, Philadlephia.

Gregory, D., J.-J. Morcrette, C. Jakob, A. Beljaars, and T. Stockdale, 2000: Revision of convection, radiation and cloud schemes in the ECMWF Integrated Forecasting System. *Quart. J. Roy. Meteor. Soc.* **126**, 1685-1710.

Grell, G. A., 1993: Prognostic evaluation of assumptions used by cumulus parameterizations. *Mon. Wea. Rev.* **121**, 764-87.

Grell, G., J. Dudhia, and D. R. Stauffer, 1994: *A description of the fifth-generation Penn State/NCAR Mesoscale Model (MM5).* NCAR/TN-398+STR (Available from MMM Division, NCAR, PO Box 3000, Boulder CO 80307.)

Gustafsson, B., 1981: The convergence rate for difference approximations to general mixed initial-boundary value problems, *SIAM J. Numer. Anal.* **18**, 179-190.

Gustafsson B., and P. Olsson, 1995: Fourth-order difference methods for hyperbolic IBVPs. *J. Comp. Phys.* **117**, 300-317.

Gustafsson, B., H. O. Kreiss, and J. Oliger, 1972: Stability theory of difference approximations for initial boundary value problems. II. *Math. Com.* **26**, 649-686.

Gustafsson, N., 1990: Sensitivity of limited area model data assimilation to lateral boundary condition fields. *Tellus* **42**A, 109-115.

Hackbusch, W., 1985: *Multigrid methods and applications.* Springer, Berlin.

Hageman, L. A., and D. M. Young, 1981: *Applied iterative methods.* Wiley, New York.

Haidvogel, D. B., and F. O. Bryan, 1992: Ocean general circulation modeling. Chapter 11 in *Climate system modeling*, K. Trenberth, editor. Cambridge University Press, Cambridge.

Halem, M., E. Kalnay, W. E. Baker, and R. Atlas, 1982: An assessment of the FGGE satellite observing system during SOP-1. *Bull. Amer. Met. Soc.* **63**, 407-426.

Haltiner, G. J., and R. T. Williams, 1980: *Numerical prediction and dynamic meteorology.* John Wiley and Sons, New York.

Hamill, T. M., and S. J. Colucci, 1997: Verification of Eta-RSM short-range ensemble forecasts. *Mon. Wea. Rev.* **125**, 1322-1327.

Hamill, T. M., and S. J. Colucci, 1998: Evaluation of Eta-RSM ensemble probabilistic precipitation forecasts. *Mon. Wea. Rev.* **126**, 711-724.

Hamill, T. M., and C. Snyder, 2000: A hybrid ensemble Kalman filter-3D variational analysis scheme. *Mon. Wea. Rev.* **128**, 2905-2919.

Hamill, T. M., S. L. Mullen, C. Snyder, Z. Toth, and D. P. Baumhefner, 2000a: Ensemble forecasting in the short to medium range: Report from a workshop. *Bull. Amer. Meteor. Soc.* **81**, 2653-2664.

Hamill, T. M., C. Snyder, and R. E. Morss, 2000b: A comparison of probabilistic forecasts from bred, singular-vector, and perturbed observation ensembles. *Mon. Wea. Rev.* **128**, 1835-1851.

Hamill, T. M., J. S. Whitaker, and C. Snyder, 2001: Distance-dependent filtering of background error covariance estimates in an ensemble Kalman filter. *Mon. Wea. Rev.* **129**, 2776-2790.

Hane, C., R. Wilhelmson, and T. Gal-Chen, 1981: Retrieval of thermodynamic variables within deep convective clouds: Experiments in three dimensions. *Mon. Wea. Rev.* **109**, 564-576.

Hardiker, V., 1997: A global numerical weather prediction model with variable resolution. *Mon. Wea. Rev.* **125**, 59-73.

Harrison, E. J., and R. L. Elsberry, 1972: A method for incorporating nested finite grids in the solution of systems of geophysical equations. *J. Atmos. Sci.* **29**, 1235-1245.

Harshvardhan, D., A. Randall, T. G. Corsetti, and D. A. Dazlich, 1989: Earth radiation budget and cloudiness simulations with a global general circulation model. *J. Atmos. Sci.* **46**, 1922-1942.

Hartmann, D. L., R. Buizza., and T. N. Palmer, 1997: Singular vectors: the effect of spatial scale on linear growth of disturbances. *J. Atmos. Sci.* **52**, 3885-3894.

Heikes, R., and D. A. Randall, 1995: Numerical integration of the shallow-water equations on a twisted icosahedral grid. Part I: Basic design and results of tests. *Mon. Wea. Rev.* **123**, 1862-1880.

Held, I. M., and M. Suarez, 1994: A proposal for the intercomparison of the dynamical cores of atmospheric general circulation models. *Bull. Amer. Meteor. Soc.* **75**, 1825-1830.

Henderson-Sellers, A., Z. L. Yang, and R. E. Dickinson, 1993: The project for intercomparison of land-surface parameterization schemes. *Bull. Amer. Meteor. Soc.* **74**, 1335-1350.

Hibler, W. D., and G. M. Flato, 1992: Sea ice models. Chapter 12 in *Climate system modeling*, K. Trenberth, editor. Cambridge University Press, Cambridge.

Hinkelmann, K., 1951: Der mechanismus des meteorologischen larmes, *Tellus* **3**, 285-296.

Hinkelmann, K., 1959: Ein numerisches experiment mit den primitiven gleichungen. In: *The atmosphere and the sea in motion*, Rockfeller Institute Press, New York.

Hodur, R., 1997: The Naval Research Laboratory Coupled Ocean/Atmosphere Mesoscale Prediction System (COAMPS). *Mon. Wea. Rev.* **125**, 1414-1430.

Hoffman, R., 1984: SASS wind ambiguity removal by direct minimization, Part II: Use of smoothness and dynamical constraints. *Mon. Wea. Rev.* **112**, 1829-1852.

Hoffman, R., 1986: A four-dimensional analysis exactly satisfying equations of motion. *Mon. Wea. Rev.* **114**, 388-397.

Hoffman, R. N., and E. Kalnay, 1983: Lagged average forecasting, an alternative to Monte Carlo forecasting. *Tellus* **35A**, 100-118.

Hoffman, R. N., Z. Liu, J.-F. Louis, and C. Grassotti, 1995: Distortion representation of forecast errors. *Mon. Wea. Rev.* **123**, 2758-2770.

Hoke, J., and R. Anthes, 1976: The initialization of numerical models by a dynamic relaxation technique. *Mon. Wea. Rev.* **104**, 1551-1556.

Hollingsworth, A., 1980: *An experiment in Monte Carlo forecasting. Workshop on stochastic-dynamic forecasting.* ECMWF, Shinfield Park, Reading, UK.

Hollingsworth, A., and P. Lönnberg, 1986: The statistical structure of short-range forecast errors as determined from radiosonde data, Part I: The wind field. *Tellus* **38A**, 111-136.

Hollingsworth, A., D. Shaw, P. Lönnberg, L. Illari, K. Arpe, and A. Simmons, 1986: Monitoring of observation and analysis quality as a data assimilation system. *Mon. Wea. Rev.* **114**, 861-879.

Holloway, G., and B. West, 1984: *Predictability of fluid motions*, American Institute of Physics, New York.

Holton, J. R., 1992: *An introduction to dynamic meteorology*, 3rd edition, Academic Press, San Diego, CA.

Hong, S. Y., and E. Kalnay, 2000: Role of sea surface temperature and soil-moisture feedback in the 1998 Oklahoma–Texas drought. *Nature* **408** (6814), 842-844.

Hong, S.-Y., and A. Leetmaa, 1999: An evaluation of the NCEP RSM for regional climate modeling. *J. Climate* **12**, 592-609.

Hong, S. Y., and H. L. Pan, 1996: Nonlocal boundary layer vertical diffusion in a medium-range forecast model. *Mon. Wea. Rev.* **124**, 2322-2339.

Horel, J. D., and J. M. Wallace, 1981: Planetary-scale atmospheric phenomena associated with the Southern Oscillation. *Mon. Wea. Rev.* **109**, 813-829.

Hoskins, B. J., M. E. McIntyre, and A. W. Robertson, 1985: On the use and significance of isentropic potential vorticity maps. *Quart. J. Roy. Meteor. Soc.* **111**, 877-946.

Hou, D., E. Kalnay, and K. K. Droegemeier, 2001: Objective verification of the SAMEX '98 ensemble forecasts. *Mon. Wea. Rev*. **129**, 73-91.

Hough, S., 1898: On the application of harmonic analysis to the dynamical equations, *Phil. Trans. Roy. Soc. London, Ser. A* **191**, 139-185.

Houghton, J., F. Taylor, and C. Rodgers, 1985: *Remote sounding of atmospheres*, Cambridge University Press, Cambridge.

Houtekamer, P. L., and J. Derome, 1994: Prediction experiments with two-member ensembles. *Mon. Wea. Rev.* **122**, 2179-2191.

Houtekamer, P. L., and H. L. Mitchell, 1998: Data assimilation using an ensemble Kalman filter technique. *Mon. Wea. Rev.* **126**, 796-811.

Houtekamer, P. L., and H. L. Mitchell, 2001: A sequential ensemble Kalman filter for atmospheric data assimilation. *Mon. Wea. Rev.* **129**, 796-911 with 123-137.

Houtekamer, P. L., L. Lefaivre, J. Derome, H. Ritchie, and H. L. Mitchell, 1996: A system simulation approach to ensemble prediction. *Mon. Wea. Rev.* **124**, 1225-1242.

Howcroft, J. G., 1971: *Local forecast model: present status and preliminary verification.* NMC Office note 50. National Weather Service, NOAA, US Dept of Commerce. (Available from the National Centers for Environmental Prediction, 5200 Auth Rd, Rm 100, Camp Springs, MD 20746.)

Huang, X. Y., and P. Lynch, 1993: Diabatic digital-filtering initialization: Application to the HIRLAM model. *Mon. Wea. Rev.* **121**, 589-603.

Hughes, F. D., 1987: *Skill of medium-range forecast group.* Office Note #326. National Meteorological Center, NWS, NOAA, US Dept of Commerce. (Available from the National Centers for Environmental Prediction, 5200 Aurk Rd, Rm 100, Camp Springs, MD 20746.)

Ide, K., P Courtier, M. Ghil, and A. Lorenc, 1997: Unified notation for data assimilation: Operational, sequential and variational. *J. Meteor. Soc. Japan* **75**, 181-189.

Ingleby, N.B., 2000: The statistical structure of forecast errors and its representation in The Met. Office 3-dimensional variational data assimilation scheme. *Quart. J. Roy. Meteor. Soc.* in press.

Ingleby, N. B., and A. C. Lorenc, 1993: Bayesian quality control using multivariate normal distributions. *Quart. J. Roy. Meteor. Soc.* **119**, 1195-1225.

Intergovernmental Panel on Climate Change. (Available at *http://www.ipcc.ch/*, Geneva, IPCC, 2001: Switzerland.)

Isaacson, E., and H. B. Keller, 1966: *Analysis of numerical methods*, John Wiley and Sons, New York.

James, I. N., 1994: *Introduction to circulating atmospheres*, Cambridge University Press, Cambridge.

Janjic, Z. I., 1974: A stable centered difference scheme free of the two-grid-interval noise. *Mon. Wea. Rev.* **102**, 319-323.

Janjic, Z. I., 1984: Nonlinear advection schemes and energy cascade on semi-staggered grids. *Mon. Wea. Rev.* **112**, 1234-1245.

Janjic, Z. I., 1989: On the pressure gradient force error in coordinate spectral models. *Mon. Wea. Rev.* **117**, 2285-2292.

Janjic, Z. I., 1990: The step-mountain eta coordinate model: physical package. *Mon. Wea. Rev.* **118**, 1429-1443.

Janjic, Z. I., 1994: The step-mountain eta coordinate model: Further developments of the convection, viscous sublayer, and turbulence closure schemes. *Mon. Wea. Rev.* **122**, 927-945.

Janjic, Z. I., J. P. Gerrity Jr, and S. Nickovic, 2001: An alternative approach to nonhydrostatic modeling. *Mon. Wea. Rev.* in press.

Janssen, P., 1999: *Wave modeling and altimeter wave height data.* ECMWF Tech Memo No. 269, ECMWF, Shinfield Park, Reading, UK.

Janssen, P., 2000: Potential benefits of ensemble prediction of waves. *ECMWF Newsletter No 86 – Winter 1999–2000.*

Jazwinski, A. H., 1970: *Stochastic processes and filtering theory.* Academic Press, New York.

Ji, M., and A. Leetmaa, 1997: Impact of data assimilation on ocean initialization and El Niño prediction. *Mon. Wea. Rev.* **125**, 742-753.

Ji, Y., and A. C. Vernekar, 1997: Simulation of the Asian Summer Monsoons of 1987 and 1988 with a regional model nested in a Global GCM. *J. Climate* **10**, 1965-1979.

Ji, M., A. Kumar, and A. Leetmaa, 1994: An experimental coupled forecast system at the national meteorological center: some early results. *Tellus* **46A**, 398-418.

Ji, M., A. Leetmaa, and V. E. Kousky, 1996: Coupled model predictions of ENSO during the 1980s and the 1990s at the National Centers for Environmental Prediction. *J. Climate* **9**, 3105-3120.

Johnson, D. R., 1980: A generalized transport equation for use with meteorological coordinate systems. *Mon. Wea. Rev.* **108**, 733-745.

Johnson, D. R., T. M. Zapotocny, F. M. Reams, B. J. Wolf, and R. B. Pierce, 1993: A comparison of simulated precipitation by hybrid isentropic-sigma and sigma models. *Mon. Wea. Rev.* **121**, 2088-2114.

Joiner, J., and A. Da Silva, 1998: Efficient methods to assimilate satellite retrievals based on information content. *Quart. J. Roy. Meteor. Soc.* **124**, 1669-1694.

Jones, R. G., J. M. Murphy and N. Noguer, 1995: Simulation of climate change over Europe using a nested regional-climate model: I: Assessment of control climate, including sensitivity to location of lateral boundaries. *Quart. J. Roy. Meteor. Soc.* **526**, 1413-1449.

Jones, R. W., 1977: A nested grid for a three-dimensional model of a tropical cyclone. *J. Atmos. Sci.* **34**, 1528-1551.

Juang, H.-M. H., 1992: A spectral fully compressible non-hydrostatic mesoscale model in hydrostatic sigma coordinates: Formulation and preliminary results. *Meteor. Atmos. Phys.* **50**, 75-88.

Juang, H.-M. H., and M. Kanamitsu, 1994: The NMC nested regional spectral model. *Mon. Wea. Rev.* **122**, 3-26.

Juang, H.-M. H., and S.-Y. Hong, and M. Kanamitsu, 1997: The NCEP Regional Spectral Model: An Update. *Bull. Amer. Meteor. Soc.* **78**, 2125-2143.

Julian, P., 1984: Objective analysis in the tropics: A proposed scheme. *Mon. Wea. Rev.* **112**, 1752-1767.

Julian, P., and H. J. Thiebaux, 1975: On some properties of correlation functions used in optimum interpolation schemes. *Mon. Wea. Rev.* **103**, 605-616.

Juvanon du Vachat, R., 1986: A general formulation of normal modes for limited-area models: Applications to initialization, *Mon. Wea. Rev.* **114**, 2478-2487.

Kaas, E., A. Guldberg, W. May, and M. Decque., 1999: Using tendency errors to tune the parameterization of unresolved dynamical scale interactions in atmospheric general circulation models. *Tellus* **51A**, 612-629.

Kain, J. S., and J. M. Fritsch, 1990: A one-dimensional entraining/detraining plume model and its application in convective parameterization. *J. Atmos. Sci.* **47**, 2784-2802.

Kallberg, P., 1977: *Test of a boundary relaxation scheme in a barotropic model.* ECMWF Research Dept Internal Report number 3, ECMWF, Shinfield Park, Reading, UK.

Kalman, R., 1960: A new approach to linear filtering and prediction problems, *Trans. ASME, Ser. D, J. Basic Eng.* **82**, 35-45.

Kalman, R., and R. Bucy, 1961: New results in linear filtering and prediction theory, *Trans. ASME, Ser. D, J Basic Eng.* **83**, 95-108.

Kalnay, E., and A. Dalcher, 1987: Forecasting forecast skill. *Mon. Wea. Rev.* **115**, 349-356.

Kalnay, E., and M. Ham, 1989: Forecasting forecast skill in the Southern Hemisphere. Preprints of the 3rd International Conference on Southern Hemisphere Meteorology and Oceanography, Buenos Aires, 13–17 November 1989. Boston, MA: Amer. Meteor. Soc.

Kalnay, E., and M. Kanamitsu, 1988: Time schemes for strongly non-linear damping equations. *Mon. Wea. Rev.* **116**, 1945-1958.

Kalnay, E., and Z. Toth, 1994: Removing growing errors in the analysis cycle. Preprints of the Tenth Conference on Numerical Weather Prediction, Amer. Meteor. Soc., 1994, pp. 212-215, Boston, MA.

Kalnay, E., and Z. Toth, 1996: *The breeding method, Proceedings of the Seminar on Predictability*, held at ECMWF on 4–8 September 1995. (Available from ECMWF, Shinfield Park, Reading, Berkshire RG2 9AX, e.kooij@ecmwf.int)

Kalnay, E., M. Kanamitsu, R. Kistler, W. Collins, D. Deaven, L. Gandin, M. Iredell, S. Saha, G. White, J. Woollen, Y. Zhu, M. Chelliah, W. Ebisuzaki, W. Higgins, J. Janowiak, K. C. Mo, C. Ropelewski, J. Wang, A. Leetmaa, R. Reynolds, R. Jenne, and D. Joseph, 1996: The NCEP/NCAR 40-year Reanalysis Project. *Bull. Amer. Meteor. Soc.* **77**, 437-471.

Kalnay, E. D., L. T. Anderson, A. F. Bennett, A. J. Busalacchi, S. E. Cohn, P. Courtier, J. Derber, A. C. Lorenc, D. Parrish, J. Purser, N. Sato, and T. Schlatter, 1997: Data assimilation in the ocean and in the atmosphere: What should be next? *J. Meteor. Soc. Japan* **75**, 489-496.

Kalnay, E., S. Lord, and R. McPherson, 1998: Maturity of operational numerical weather prediction: the medium range. *Bull. Amer. Meteor. Soc.* **79**, 2753-2769.

Kalnay, E., S. K. Park, Z. X. Pu, and J. Gao, 1999: Applications of the quasi-inverse method to data assimilation. *Mon. Wea. Rev.* **128**, 864-875.

Kalnay, E., M. Corazza and M. Cai, 2002: are bred vectors the same as Lyapunov vectors? *AMS Sympos. on observations, data assimilation and probabilistic prediction* Am. Meteor. Soc., 2002, pp. 173–177, Boston, MA.

Kalnay-Rivas, E., and L.-O. Merkine, 1981: A simple mechanism for blocking. *J. Atmos. Sci.* **38**, 2077-2091.

Kalnay-Rivas, E., A. Bayliss, and J. Storch, 1977: The 4th order GISS model of the global atmosphere. *Beitr. Phys. Atm.* **50**, 299-311.

Kanamitsu, M., 1989: Description of the NMC global data assimilation and forecast system. *Wea. Forecasting* **4**, 335-342.

Kanamitsu, M., J. C. Alpert, K. A. Campana, P. M. Caplan, D. G. Deaven, M. Iredell, B. Katz, H.-L. Pan, J. Sela, and

G. H. White, 1991: Recent changes implemented into the global forecast system at NMC. *Wea. Forecasting* **6**, 425-436.

Kaplan, J. L., and J. A. Yorke, 1979: Chaotic behavior of multidimensional difference equations. In *Functional differential equations and approximation of fixed points,* H.-O. Peitgen and H. O. Walther, editors, pp. 204-227. Lecture notes in mathematics, 730. Springer-Verlag, Berlin.

Karin, J. S., and J. M. Fritsch, 1990: A one-dimensional entraining/detraining plume model and its application in convective parameterization. *J. Atmos. Sci.* **47**, 2784, 2802.

Kasahara, A., 1974: Various vertical coordinate systems used for numerical weather prediction. *Mon. Wea. Rev.* **102**, 509-522.

Kasahara, A., 1976: Normal modes of ultra-long waves in the atmosphere, *Mon. Wea. Rev.* **104**, 669-690.

Kasahara, A., 1979: *Numerical methods used in atmospheric models,* Vol. 2, GARP Publications Series No. 17, WMO and ICSU, Geneva.

Kasahara, A., 1982a: Non-linear normal mode initialization and the bounded derivative method, *Rev. Geophys. Space Phys.* **19**, 450-468.

Kasahara, A., 1982b: Significance of non-elliptic regions in balanced flows of the tropical atmosphere, *Mon. Wea. Rev.* **110**, 1956-1967.

Kasahara, A., R. Balgovind, and B. Katz, 1988: Use of satellite radiometric imagery data for improvement in the analysis of divergent winds in the tropics, *Mon. Wea. Rev.* **116**, 866-883.

Kelley, C. T., 1995: *Iterative methods for linear and nonlinear equations.* SIAM, Philadelphia.

Kiehl, J. T., 1992: Atmospheric general circulation modeling. Chapter 10 in *Climate system modeling*, K. Trenberth, editor. Cambridge University Press, Cambridge.

Kiehl, J. T., J. J. Hack, G. B. Bonan, B. A. Boville, D. L. Williamson, and P. J. Rasch, 1998: The national center for atmospheric research community climate model CCM3. *J. Climate* **11**, 1131-1149.

Kim, Y.-J., and A. Arakawa, 1995: Improvement of orographic gravity wave parameterization using a mesoscale gravity wave model. *J. Atmos. Sci.* **52**, 1875-1902.

Kimoto, M., I. Yoshikawa, and M. Ishii, 1997: An ocean data assimilation system for climate monitoring. *J. Meteor. Soc. Japan* **75**, 471-487.

Kirtman, B. P., J. Shukla, B. Huang, Z. Zhu, and E. K. Schneider, 1997: Multiseasonal predictions with a coupled tropical ocean global atmosphere system. *Mon. Wea. Rev.* **125**, 789-808.

Kistler, R. E., 1974: A study of data assimilation techniques in an autobarotropic primitive equation channel model. MS Thesis, Dept of Meteorology, Penn State University.

Kistler, R., E. Kalnay, W. Collins, S. Saha, G. White, J. Woollen, M. Chelliah, W. Ebisuzaki, M. Kanamitsu, V. Kousky, H. van den Dool, R. Jenne, and M. Fiorino, 2001: The NCEP/NCAR 50-Year reanalysis: monthly means CD-ROM and documentation. *Bull. Amer. Meteor. Soc.* **82**, 247-268.

Kitade, T., 1983: Non-linear normal mode initialization with physics, *Mon. Wea. Rev.* **111**, 2194-2213.

Klein, W. H., B. M. Lewis, and I. Enger, 1959: Objective prediction of 5-day mean temperature during winter. *J. Meteor.* **16**, 672-682.

Klemp, J. B., and D. R. Durran., 1983: An upper boundary condition permitting internal gravity wave radiation in numerical mesoscale models. *Mon. Wea. Rev.* **111**, 430-444.

Klemp, J. B., and D. K. Lilly, 1978: Numerical simulation of hydrostatic mountain waves. *J. Atmos. Sci.* **35**, No. 1, 78-107.

Klemp, J. B., and R. B. Wilhelmson, 1978: The simulation of three-dimensional convective storm dynamics. *J. Atmos. Sci.* **35**, 1070-1096.

Kobayashi, C., K. Takano, S. Kusunoki, M. Sugi, and A. Kitoh, 2000: Seasonal predictability in winter over eastern Asia using the JMA global model. *Quart. J. Roy. Meteor. Soc.* **126** No. 567 DSP/PROVOST Issue.

Koch, S., M. Desjardins, and P. Kocin, 1983: An interactive Barnes objective map analysis scheme for use with satellite and conventional data, *J. Climate – Appl. Meteor.* **22**, 1487-1503.

Kogan, Z. N., D. K. Lilly, Y. L. Kogan, and V. Filyushkin, 1995: Evaluation of radiative parametrizations using an explicit cloud microphysical model. *Atmos. Res.* **35**, 157-172.

Konor, C. S., and A. Arakawa, 1997: Design of an atmospheric model based on a generalized vertical coordinate. *Mon. Wea. Rev.* **125**, 1649-1673.

Koster, R. D., M. J. Suarez, and M. Heiser, 2000: Variance and predictability of precipitation at seasonal-to-interannual timescales. *J. Hydromet.* **1**, 26-46.

Kreiss, H., and J. Oliger, 1973: *Methods for the approximate solution of time dependent problems.* GARP Publications Series No. 10 WMO and ICSU, Geneva.

Krishnamurti, T. N., 1999: Improved weather and seasonal climate forecasts from multimodel superensemble. *Science* **285**, 1548-1550.

Krishnamurti, T. N., and L. Bounoa, 1996: *An introduction to numerical weather prediction techniques.* CRC Press, Boca Raton, FA.

Krishnamurti, T., H. Bedi, W. Heckley, and K. Ingles, 1988: On the reduction of spin up time for evaporation and precipation in a global spectral model. *Mon. Wea. Rev.* **116**, 907-920.

Krishnamurti, T. N., C. M. Kishtawal, Z. Zhang, T. LaRow, D. Bachiochi, E. Williford, S. Gadgil, and S. Surendran, 2000: Multimodel ensemble forecasts for weather and seasonal climate. *J. Climate* **13**, 4196-4216.

Kuo, H. L., 1965: On the formation and intensification of tropical cyclones through latent heat release by cumulus convection. *J. Atmos. Sci.* **22**, 40-63.

Kuo, H. L., 1974: Further studies of the parameterization of the influence of cumulus convection on the large-scale flow. *J. Atmos. Sci.* **31**, 1232-1240.

Kuo, Y.-H., and R. Anthes, 1984: Accuracy of diagnostic heat and moisture budgets using SESAME-79 field data as revealed by observing system simulation experiments. *Mon. Wea. Rev.* **112**, 1465-81.

Kuo, Y.-H., E. Donall, and M. Shapiro, 1987: Feasibility of short-range numerical weather prediction using observations from a network of profilers. *Mon. Wea. Rev.* **115**, 2402-2427.

Kurihara, Y., 1965: Numerical integration of the primitive equations on a spherical grid. *Mon. Wea. Rev.* **93**, 3998-4150.

Kurihara, Y., and M. Bender, 1980: Use of a movable nested-mesh model for tracking a small vortex. *Mon. Wea. Rev.* **108**, 1792-1809.

Kurihara, Y., G. J. Tripoli, and M. A. Bender, 1979: Design of a movable nested-mesh primitive equation model. *Mon. Wea. Rev.* **107**, 239-249.

Lacarra, R., and O. Talagrand, 1988: Short-range evolution of small perturbations in a barotropic model. *Tellus* **40A**, 81-95.

Langland, R. H., R. L. Elsberry, and R. M. Errico, 1995: Evaluation of physical processes in an idealized extratropical cyclone using adjoint sensitivity. *Quart. J. Roy. Meteor. Soc.* **121**, 1349-1386.

Langland, R., Z. Toth, R. Gelaro, I. Szunyogh, M. Shapiro, S. Majumdar, R. Morss, G. D. Rohaly, C. Velden, N. Bond, and C. Bishop, 1999: The North Pacific experiments (NORPEX-98): Targetted observations for improved North American weather forecasts. *Bull. Amer. Meteor. Soc.* **80**, 1363-1384.

Langlois, W. E., and H. C. W. Kwok, 1969: Description of the Mintz–Arakawa Numerical General Circulation Model, Dept of Meteorology, University of California at Los Angeles, 1969.

Laprise, R., D. Caya, G. Bergeron, and M. Giguere, 1997: The formulation of Andre Robert MC^2 (Mesoscale Compressible Community) Model. *Atmosphere-Ocean* Special Vol. XXXV No1, 195-220.

Latif, M., T. P. Barnett, M. A. Cane, M. Flugel, N. E. Graham, H. Von Storch, J. S. Xu, and S. E. Zebiak, 1994: A review of ENSO prediction studies. *Climate Dyn.* **9**, 167-179.

Latif, M., D. Anderson, T. Barnett, M. A. Cane, R. Kleeman, A. Leemaa, J. J. O'Brien, A. Rosati, and E. Schneider, 1998: A review of predictability and prediction of ENSO, *J. Geophys. Res.* **103**, 14375-14393.

LeDimet, F. X., and O. Talagrand, 1986: Variational algorithms for analysis and assimilation of meteorological observations: theoretical aspects. *Tellus* **38A**, 97-110.

Leetmaa, A., and J. Derber, 1995: An ocean analysis system for seasonal to interannual climate studies. *Mon. Wea. Rev.* **123**, 460-481.

Legras, B., and R. Vautard, 1996: A guide to Lyapunov vectors. *Proceedings of the ECMWF Seminar on Predictability.* September 4–8, 1995, Reading, England, Vol 1, ECMWF, Shinfield Park, Reading, UK, pp. 143-156.

Leith, C. E., 1965: Numerical simulation of the earth's atmosphere. In *Methods in computational physics*, B. Alder, S. Fernbach, and M. Rotenberg, editors, Academic Press, New York, pp. 1-28.

Leith, C. E., 1974: Theoretical skill of Monte Carlo forecasts. *Mon. Wea. Rev.* **102**, 409-418.

Leith, C. E., 1980: Non-linear normal mode initialization and quasi-geostrophic theory. *J. Atmos. Sci.* **37**, 958-968.

Leslie, L. M., and R. J. Purser, 1995: 3-Dimensional mass-conserving semilagrangian scheme employing forward trajectories. *Mon. Wea. Rev.* **123**, 2551-2566.

Lewis, J., 1972: An operational analysis using the variational method. *Tellus* **24**, 514-530.

Lewis, J., and S. Bloom, 1978: Incorporation of time continuity into sub-synoptic analysis by using dynamical constraints. *Tellus* **30**, 496-515.

Lewis, J., and J. Derber, 1985: The use of adjoint equations to solve a variational adjustment problem with advective constraint. *Tellus* **37A**, 309-322.

Li, Y., J. Ruge, J. R. Bates, and A. Brandt, 2000: A proposed adiabatic formulation of 3-d global atmospheric models based on potential vorticity. *Tellus* **52A**, 129-139.

Lilly, D. K., 1965: On the computational stability of numerical solutions of time dependent non-linear geophysical fluid dynamics problems. *Mon. Wea. Rev.* **93**, 11-26.

Lilly, D., 1983: Mesoscale variability of the atmosphere. In *Mesoscale meteorology: Theories, observations and models*, D. Lilly and T. Gal-Chen, editors, Reidel, Dordrecht, pp. 13-24.

Lilly, D. K., and P. J. Kennedy, 1973: Observations of a stationary mountain wave and its associated momentum flux and energy dissipation. *J. Atmos. Sci.* **30**, 1135-1152.

Lilly, D. K., and J. B. Klemp, 1979: The effects of terrain shape on nonlinear hydrostatic mountain waves. *J. Fluid Mech.* **95**, 241-261.

Lilly, D. K., and J. B. Klemp, 1980: Comments on the evolution and stability of finite-amplitude mountain waves. Part II: Surface wave drag and severe downslope windstorms. *J. Atmos. Sci.* **37**, 2119-2121.

Lilly, D. K., J. M. Nickolls, R. M. Chervin, P. J. Kennedy, and J. B. Klemp, 1982: Aircraft measurements of wave momentum flux over the Colorado Rocky Mountains. *Quart. J. Roy. Meteor. Soc.* **108**, 625-642.

Lin, S.-J., 1997: A finite-volume integration method for computing pressure gradient forces in general vertical coordinates. *Quart. J. Roy. Meteor. Soc.* **123**, 1749-1762 Part B.

Lin, S. J., and R. B. Rood, 1996: Multidimensional flux-form semi-Lagrangian transport schemes. *Mon. Wea. Rev.* **124**, 2046-2070.

Lin, S.-J., and R. B. Rood, 1997: An explicit flux-form semi-Lagrangian shallow-water model on the sphere. *Quart. J. Roy. Meteor. Soc.* **123**, 2477-2498.

Lindzen, R. S., 1988: Supersaturation of vertically propagating internal gravity waves. *J. Atmos. Sci.* **45**, 705-711.

Lindzen, R. S., E. S. Batten, and J. W. Kim, 1968: Oscillations in atmospheres with tops. *Mon. Wea. Rev.* **96**, 133-140.

Longuet-Higgens, M. S., 1968: The eigenfunctions of Laplaces tidal equations over the sphere. *Phil. Trans. Roy. Soc. London series A* **262**, 511-607.

Lönnberg, P., and A. Hollingsworth, 1986: The statistical structure of short-range forecast errors as determined from radiosonde data, Part II: The covariance of height and wind errors. *Tellus* **38A**, 137-161.

Lorenc, A., 1981: A global three-dimensional multivariate statistical interpolation scheme. *Mon. Wea. Rev.* **109**, 701-721.

Lorenc, A., 1986: Analysis methods for numerical weather prediction. *Quart. J. Roy. Meteor. Soc.* **112**, 1177-1194.

Lorenc, A., 1988a: A practical approximation to optimal four-dimensional data assimilation. *Mon. Wea. Rev.* **116**, 730-745.

Lorenc, A., 1988b: Optimal nonlinear objective analysis. *Quart. J. Roy. Meteor. Soc.* **114**, 205-240.

Lorenc, A. C., 1996: Development of an operational variational assimilation scheme. *J. Met. Soc. Japan* **75**, 339-346.

Lorenc, A. C., 1997: Development of an Operational variational assimilation scheme. *J. Met. Soc. Japan* **75**, 339-346.

Lorenc, A., and O. Hammon, 1988: Objective quality control of observations using Bayesian methods, Theory and a practical implementation. *Quart. J. Roy. Meteor. Soc.* **114**, 515-543.

Lorenc, A. C., R. S. Bell, and B. Macpherson, 1991: The Meteorological Office analysis correction data assimilation scheme. *Quart. J. Roy. Meteor. Soc.* **117**, 59-89.

Lorenc, A. C., S. P. Ballard, R. S. Bell, N. B. Ingleby, P. L. F. Andrews, D. M. Barker, J. R. Bray, A. M. Clayton, T. Dalby, D. Li, T. J. Payne, and F. W. Saunders, 2000: The Met. Office global 3-dimensional variational data assimilation scheme. Submitted to *Quart. J. Roy. Meteor. Soc.*

Lorenz, E. N., 1955: Available potential energy and the maintenance of the general circulation. *Tellus* **7**, 157-167.

Lorenz, E. N., 1960: Energy and numerical weather prediction. *Tellus* **12**, 364-373.

Lorenz, E. N., 1962: The statistical prediction of solutions of dynamic equations. Proc. Intern. Symp. Numer. Weather Pred. Tokyo: *J. Meteor. Soc.* **647**, 629-635.

Lorenz, E. N., 1963a: Deterministic non-periodic flow. *J. Atmos Sci.* **20**, 130-141.

Lorenz, E. N., 1963b: The predictability of hydrodynamic flow. *Trans. NY Acad. Sci., Series II* **25**, 409-432.

Lorenz, E. N., 1965: A study of the predictability of a 28-variable atmospheric model. *Tellus* **17**, 321-333.

Lorenz, E. N., 1968: The predictability of a flow which possesses many scales of motion. *Tellus* **21**, 289-307.

Lorenz, E. N., 1971: An *N*-cycle time differencing scheme for step-wise numerical integration. *Mon. Wea. Rev.* **99**, 644-648.

Lorenz, E. N., 1977: An experiment in nonlinear statistical forecasting. *Mon. Wea. Rev.* **105**, 590-602.

Lorenz, E., 1980: Attractor sets and quasi-geostrophic equilibrium. *J. Atmos. Sci.* **37**, 1685-99.

Lorenz, E. N., 1982: Atmospheric predictability experiments with a large numerical model. *Tellus* **34**, 505-513.

Lorenz, E. N., 1990: Charney – a remarkable colleague. In *The atmosphere – a challenge*, R. Lindzen, E. Lorenz, and G. Platzman, editors. American Meteorological Society, Boston, pp. 89-92.

Lorenz, E. N., 1993: *The essence of chaos.* University of Washington Press, Seattle.

Lorenz, E. N., 1996: *Predictability – A problem partly solved.* Proceedings of the ECMWF Seminar on Predictability. September 4–8, 1995, Reading, England, Vol. 1, ECMWF, Shinfield Park, Reading, England, pp. 1-18.

Lorenz, E., and V. Krishnamurthy, 1987: On the existence of a slow manifold. *J. Atmos. Sci.* **44**, 2940-2950.

Lott, F., and M. J. Miller, 1997: A new sub-grid scale orography drag parameterization: its formulation and testing. *Quart. J. Roy. Meteor. Soc.* **123**, 101-127.

Louis, J. F., 1979: A parametric model of vertical eddy fluxes in the atmosphere. *Boundary-layer Meteor.* **17**, 187-202.

Lynch, P., 1997: The Dolph–Chebyshev window: a simple optimal filter. *Mon. Wea. Rev.* **125**, 1976-1982.

Lynch, P., and X.-Y. Huang, 1992: Initialization of the HIRLAM model using a digital filter. *Mon. Wea. Rev.* **120**, 1019-1034.

Lynch, P., and X.-Y. Huang, 1994: Diabatic initialization using recursive filters. *Tellus* **46A**, 583-597.

Lynch, P., R. McGrath, and A. McDonald, 1999: Digital filter initialization for HIRLAM. HIRLAM 4 Project, c/o Met Eireann, Glasnevin Hill, Dublin 9, Ireland.

Machenhauer, B., 1977: On the dynamics of gravity oscillations in a shallow water model with applications to normal mode initialization. *Contrib. Atmos. Phys.* **50**, 253-271.

Machenhauer, B., 1979: The spectral method. In *Numerical Methods Used in Atmospheric Models*, Volume II, GARP Publication Series No. 17, World Meteorological Organization, Geneva, pp. 121-275.

Machenhauer, B., 1991: Spectral methods. In *Numerical Methods in Atmospheric Models* Vol. 1, ECMWF, Shinfield Park, Reading, UK, pp. 3-86.

Madden, R. A., and P. R. Julian, 1971: Detection of a 40–50 day oscillation in the zonal wind in the tropical Pacific. *J. Atmos. Sci.* **28**, 702-708.

Madden, R. A., and P. R. Julian, 1972: Description of global-scale circulation cells in the tropics with a 40–50 day period. *J. Atmos. Sci.* **29**, 1109-1123.

Madden, R. A., 1986: Seasonal variations of the 40–50 day oscillation in the tropics. *J. Atmos. Sci.* **43**, 3138-3158.

Madden, R. A., 1989: On predicting probability distributions of time-averaged meteorological data. *J. Climate* **2**, 922-928.

Mahfouf, J.-F., 1991: Analysis of soil moisture from near surface parameters: a feasibility study. *J. Appl. Meteor.* **30**, 1534-1547.

Malanotte-Rizzoli, P., and W. Holland, 1986: Data constraints applied to models of the ocean general circulation, Part I: The steady case. *J. Phys. Oceanogr.* **16**, 1665-1682.

Manabe, S., 1969: Climate and ocean circulation, 1. The atmospheric circulation and hydrology of the earth's surface. *Mon. Wea. Rev.* **97**, 739-774.

Manabe, S., and R. Wetherald, 1967: Thermal equilibrium of the atmosphere with a given distribution of relative humidity, *J. Atmos. Sci.* **24**, 241-259.

Manabe, S., J. Smagorinsky, and R. F. Strickler, 1965: Simulated climatology of a general circulation model with a hydrological cycle. *Mon. Wea. Rev.* **93**, 769-798.

Marchuk, G., 1974: *Numerical solution of the problems of the dynamics of the atmosphere and ocean*, in Russian, Gidrometeoizadt, Leningrad.

Mass, C. F., and Y.-H. Kuo, 1998: Regional real-time numerical weather prediction: current status and future potential. *Bull. Amer. Meteor. Soc.* **71**, 792-805.

Masuda, Y., and H. Ohnishi, 1986: An integration scheme of the primitive equation model with an icosahedral–hexagonal grid system and its application to the shallow water equations. Short- and medium-range numerical weather prediction. *J. Meteor. Soc. Japan*, Special volume, 317-326.

Matsuno, T., 1966a: False reflection of waves at the boundary due to the use of finite differences. *J. Met. Soc. Series II* **44**, 145-157.

Matsuno, T., 1966b:. Numerical integrations of the primitive equations by a simulated backward difference method. *J. Meteor. Soc. Japan Series* II **44**, 76-84.

McCormick, S. F., 1992: *Multilevel projection methods in partial differential equations.* CBMS-NSF Series, SIAM, Philadelphia.

McDonald, A., 1984: Accuracy of multiply-upstream and semi-lagrangian advective schemes. *Mon. Wea. Rev.* **112**, 1267-1275.

McDonald, A., 1997: Lateral boundary conditions for operational regional forecast models: a review. HIRLAM Technical Report No. 32.

McDonald, A., and J. Bates, 1989: Semi-Lagrangian integration of a gridpoint shallow-water model on the sphere. *Mon. Wea. Rev.* **117**, 130-137.

McDonald, A., and J. Haugen, 1992: A two-time-level, three-dimensional semi-Lagrangian, semi-implicit, limited-area gridpoint model of the primitive equations. *Mon. Wea. Rev.* **120**, 2603-2621.

McFarlane, N. A., 1987: The effect of orographically excited gravity-wave drag on the general circulation of the lower stratosphere and troposphere. *J. Atmos. Sci.* **44**, 1775-1800.

McGregor, J. L., 1997: Regional climate modeling. *Meteorol. Atmos. Phys.* **63**, 105-117.

McNally, A., J. Derber, W. Wu, and B. Katz, 1999: The use of raw TOVSA level 1b radiances in the NCEP SSI analysis system. *Quart. J. Roy. Met. Soc.* **126**, 689-724.

McPherson, R. D., K. H. Bergman, R. E. Kistler, G. E. Rasch, and D. S. Gordon, 1979: The NMC operational global data assimilation system. *Mon. Wea. Rev.* **107**, 1445-1461.

Meehl, G. A., 1992: Global coupled models, atmosphere, ocean, sea ice. Chapter 17 in *Climate system modeling*, K. Trenberth, editor, Cambridge University Press, Cambridge.

Mellor, G. L., and T. Yamada, 1974: A hierarchy of turbulent closure models for planetary boundary layer. *J. Atmos. Sci.* **31**, 1791-1806.

Mellor, G. L., and T. Yamada, 1982: Development of a turbulent closure model for geophysical fluid problems. *Rev. Geophys. Space Phys.* **20**, 851-875.

Menard, R., and R. Daley, 1996: The application of Kalman smoother theory to the estimation of 4DVAR error statistics. *Tellus* **48A**, 221-237.

Merilees, P., 1968: On the linear balance equation in terms of spherical harmonics. *Tellus* **20**, 200-202.

Merilees, P. E., and S. A. Orszag, 1979: The pseudospectral method. In *Numerical methods used in atmospheric models*, Volume II, GARP Publication Series No. 17, World Meteorological Organization, Geneva, pp. 276-299.

Mesinger, F., 1977: Forward–backward scheme and its use in a limited area model. *Beitr. Phys. Atmos.* **40**, 200-210.

Mesinger, F., 1984: A blocking technique for representation of mountains in atmospheric models. *Riv. Meteor. Aeronautica* **44**, 195-202.

Mesinger, F., 1996: Improvements in quantitative precipitation forecasts with the Eta regional model at the National Centers for Environmental Prediction: the 48-km upgrade. *Amer. Meteor. Soc.* **77**, 2637–2649. *Corrigendum*: **78**, 506.

Mesinger, F., 1997: Dynamics of limited-area models: Formulation and numerical methods, *Meteorol. Atmos. Phys.* **63** (1-2) 3-14.

Mesinger, F., and A. Arakawa., (editors) 1976: *Numerical methods used in atmospheric models*. GARP Publication series No. 17, 1, World Meteorological Organization, Geneva.

Mesinger, F., and Z. I. Janjic, 1985: Problems and numerical methods of the incorporation of mountains in atmospheric models. *Lectures in Appl. Math.* **22**, 81-120.

Mesinger, F., Z. I. Janjic, S. Nickovic, D. Gavrilov, and D. G. Deaven, 1988: The step-mountain coordinate: model description and performance for cases of Alpine lee cyclogenesis and for a case of an Appalachian redevelopment. *Mon. Wea. Rev.* **116**, 1493-1518.

Meurant, G., 1999: Computer solution of large linear systems. Elsevier, North Holland.

Miller, M., and M. Moncrieff, 1983: The dynamics and simulation of organized deep convection In: *Mesoscale meteorology: Theories, observations and models*, D. Lilly and T. Gal-Chen, editors, Reidel, Dordrecht, pp. 451-495.

Miller, M. J., and A. J. Thorpe, 1981: Radiation conditions for the lateral boundary conditions of limited area models. *Quart. J. Roy. Meteor. Soc.* **107**, 615-628.

Miller, M. J., and A. A. White, 1984: On the non-hydrostatic equations in pressure and sigma coordinates. *Quart. J. Roy. Meteor. Soc.* **110**, 515-533.

Miller, R. N., M. Ghil, and F. Gauthiez, 1994: Advanced data assimilation in strongly nonlinear dynamical systems, *J. Atmos. Sci.* **51**, 1037-1056.

Mitchell, H. L., C. Charrette, C. Chouinard, and B. Brasnett, 1990: Revised interpolation statistics for the Canadian data assimilation procedure: Their derivation and application. *Mon. Wea. Rev.* **118**, 1591-1614.

Miyakoda, K., and R. Moyer, 1968: A method of initialization for dynamical weather forecasting, *Tellus* **20**, 115-128.

Miyakoda, K., and A. Rosati, 1977: One-way models: the interface conditions and the numerical accuracy. *Mon. Wea. Rev.* **105**, 1092-1107.

Miyakoda, K., and J. Sirutis, 1977: Comparative integrations of global models with various parameterized processes of subgrid-scale vertical transports. *Beitr. Phys. Atmos.* **50**, 445-487.

Miyakoda, K., L. Umscheid, D. Lee, J. Sirutis, R. Lusen, and F. Pratte, 1976: The near real time, global, four dimensional analysis experiment during the GATE period, Part I. *J. Atmos. Sci.* **33**, 561-591.

Moeng, C.-H., and J. C. Wyngaard,1989: Evaluation of turbulent transport and dissipation closures in second-order modeling. *J. Atmos. Sci.* **46**, 2311-2330.

Molteni, F., and T. N. Palmer, 1993: Predictability and finite time instability of the northern winter circulation. *Quart. J. Roy. Meteor. Soc.* **119**, 269-298.

Molteni, F., R. Buizza, T. N. Palmer, and T. Petroliagis, 1996: The new ECMWF Ensemble Prediction System: Methodology and validation. *Quart. J. Roy. Meteor. Soc.* **122**, 73-119.

Monin, A. S., and A. M. Obukhov, 1954: Basic laws of turbulent mixing in the ground layer of the atmosphere. *Akad. Nauk SSR Geoffiz. Inst. Tr.* **151**, 63-187.

Monin, A., and A. Yaglom, 1975: *Statistical fluid mechanics*, Vol. 2, translation of *Statisticheskaia gidromekhanika*, MIT Press, Cambridge, MA.

Moorthi, S., and M. Suarez, 1992: Relaxed Arakawa–Schubert: a parameterization of moist convection for general circulation models. *Mon. Wea. Rev.* **120**, 978-1002.

Morss, R. E., 1999: Adaptive observations: Idealized sampling strategies for improving numerical weather prediction. PhD thesis,

Massachussetts Institute of Technology, Cambridge, MA.

Morss, R. E., K. A. Emanuel, and C. Snyder, 2001: Idealized adaptive observation strategies for improving numerical weather prediction. *J. Atmos. Sci.* **58**, 210-234.

Mullen, S. L., and D. P. Baumhefner, 1994: Monte Carlo simulation of explosive cyclogenesis. *Mon. Wea. Rev.* **122**, 1548-1567.

Mureau, R., F. Molteni, and T. N. Palmer, 1993: Ensemble prediction using dynamically-conditioned perturbations. *Quart. J. Roy. Meteor. Soc.*, 119, 299-323.

Murphy, J. M., and T. N. Palmer, 1986: Experimental monthly long-range forecasts for the UK. Part II: A real time long-range forecast by an ensemble of numerical integrations. *Meteorol. Mag.* **115**, 337-349.

Navon, I., and D. Legler, 1987: Conjugate-gradient methods for large-scale minimization in meteorology, *Mon. Wea. Rev.* **115**, 1479-1502.

Navon, I. M., X. Zou, J. Derber, and J. Sela, 1992: Variational data assimilation with an adiabatic version of the NMC spectral model. *Mon. Wea. Rev.* **120**, 1433-1446.

Niiler, P. P., 1992: The ocean circulation. Chapter 4 in *Climate system modeling*, K. Trenberth, editor, Cambridge University Press, Cambridge.

Nitta, T., and J. Hovermale, 1969: A technique of objective analysis and initialization for the primitive forecast equations. *Mon. Wea. Rev.* **97**, 652-658.

O'Brien, J., 1970: Alternative solutions to the classical vertical velocity problem, *J. Appl. Meteor.* **9**, 197-203.

Ogura, Y., and N. A. Phillips, 1962: A scale analysis of deep and shallow convection in the atmosphere. *J. Atmos. Sci.* **19**, 173-179.

Oliger, J., and A. Sundstrom, 1978: Theoretical and practical aspects of some initial boundary problems in fluid dynamics. *SIAM J. Appl. Math.* **35**, 418-446.

Oortwijn, J., and J. Barkmeijer, 1995: Perturbations that optimally trigger weather regimes. *J. Atmos. Sci.* **52**, 3952-3944.

Ooyama, K., 1971: A theory on parameterization of cumulus convection. *J. Meteor. Soc. Japan* **49**, Special Issue, 744-756.

Orlanski, I., 1975: A rational subdivision of scales for atmospheric processes. *Bull. Amer. Meteor. Soc.* **56**, 527-530.

Orlanski, I., 1976: A simple boundary condition for unbounded hyperbolic flows. *J. Comp. Phys.* **21**, 251-269.

Orszag, S. A., 1970: Transform method for calculation of vector coupled sums: Application to the spectral form of the vorticity equation. *J. Atmos. Sci.* **27**, 890-895.

Orszag, S., 1971: On the elimination of aliasing in finite-difference schemes by filtering high wavenumber components. *J. Atmos. Sci.* **28**, 1974.

Ott E., T. Sauer, and J. A. Yorke., 1994: *Coping with chaos: analysis of chaotic data and the exploitation of chaotic systems.* John Wiley and Sons, New York.

Ott, E., M. Corazza, B. Hunt, E. Kalnay, D. J. Patil, I. Szunyogh, and J. Yorke, 2002: Exploiting local low dimensionality of the atmospheric dynamics for efficient ensemble Kalman filtering. http://arxiv.org/abs/physics/0203058

Paegle, J., 1989: A variable resolution global model based upon Fourier and finite element representation. *Mon. Wea. Rev.* **117**, 583-606.

Paegle, J., and J. N. Paegle, 1976: On geopotential data and ellipticity of the balance equation: a data study. *Mon. Wea. Rev.* **1104**, 1279-1288.

Paegle, J., and E. B. Wayman, 1983: The influence of the tropics on the prediction of ultralong waves, Part II, Latent heating. *Mon. Wea. Rev.* **111**, 1356-1371.

Paegle, J., C.-D. Zhang, and D. P. Baumhefner, 1987: Atmospheric response to tropical thermal forcing in real data integrations. *Mon. Wea. Rev.* **115**, 2975-2995.

Palmer, T. N., 1997: *Predictability of the atmosphere and oceans: from days to decades. Proceedings of the ECMMFW Seminar on Predictability*, Vol 1, ECMFW, Shinfield Park, Reading, UK.

Palmer, T. N., and S. Tibaldi, 1988: On the prediction of forecast skill. *Mon. Wea. Rev.* **116**, 2453-2480.

Palmer, T. N., G. J. Schutts, and R. Swinbank, 1986: Alleviation of a systematic westerly bias in circulation and numerical weather prediction models through an orographic gravity-wave drag parameterization. *Quart. J. Roy. Meteor. Soc.* **112**, 1001-1039.

Palmer, T. N., F. Molteni, R. Mureau, R. Buizza, P. Chapelet, and J. Tribbia, 1993: *Ensemble prediction. ECMWF Seminar Proceedings on Validation of models over Europe.* Vol 1. ECMWF, Shinfield Park, Reading, UK.

Palmer, T. N., R. Gelaro, J. Barkmeijer, and R. Buizza, 1998: Singular vectors, metrics and adaptive observations. *J. Atmos. Sci.* **55**, 633-653.

Pan, D.-M., and D. A. Randall, 1998: A cumulus parameterization with a prognostic closure. *Quart. J. Roy. Meteor. Soc.* **124**, 949-981.

Pan, H.-L., 1990: A simple parameterization scheme of evapotranspiration over land for the NMC Medium-Range Forecast Model. *Mon. Wea. Rev.* **118**, 2500-2512.

Pan, H.-L., 1999: Parameterization of subgrid-scale processes. In *Global energy and water cycles*, K. A. Browning and R. J. Gurney, editors, Cambridge University Press, Cambridge, pp. 44-47.

Pan, H.-L., and L. Mahrt, 1987: Interaction between soil hydrology and boundary-layer development. *Boundary-Layer Meteor.* **38**, 185-202.

Pan, H.-L., and W.-S. Wu., 1995: Implementing a mass flux convection parameterization package for the NMC Medium-Range Forecast Model. Office note 409, National Meteorological Center.

Panofsky, H., 1949: Objective weather-map analysis. *J. Appl. Meteor.* **6**, 386-392.

Parrish, D. F., and J. D. Derber, 1992: The National Meteorological Center spectral statistical interpolation analysis system. *Mon. Wea. Rev.* **120**, 1747-1763.

Parrish, D., J. Purser, E. Rogers, and Y. Lin, 1996: *The regional 3D-variational analysis for the Eta Model.* Preprints, 11th Conf. Numerical Weather Prediction, Norfolk, VA, Amer. Meteor. Soc. pp. 454-455.

Patil, D. J. S., B. R. Hunt, E. Kalnay, J. A. Yorke, and E. Ott, 2001: Local low dimensionality of atmospheric dynamics. *Phys. Rev. Lett.* **86**, 5878.

Pedlosky, J., 1979: *Geophysical fluid dynamics*, Springer-Verlag, New York.

Peña, M., E. Kalnay, and M. Cai, 2002: Statistics of locally coupled ocean and atmosphere intraseasonal anomalies in Reanalysis and AMIP data. Submitted to *Nonlinear Processes in Geophysics.*

Perkey, D. J., and C. Kreitzberg, 1976: A time dependent lateral boundary scheme for limited area primitive equations models. *Mon. Wea. Rev.* **104**, 744-755.

Pham, D. T., J. Verron, and M.-C. Roubaud, 1998: Singular evolutive extended Kalman filter with EOF initialization for data assimilation in oceanography. *J. Mar. Syst.* **16**, 323-340.

Philander, S. G., 1990: *El Niño–La Niña and the Southern Oscillation.* Academic Press, New York.

Phillips, N. A., 1956: The general circulation of the atmosphere, a numerical experiment. *Quart. J. Roy. Met. Soc.* **82**, 123-164.

Phillips, N. A., 1957: A coordinate system having some special advantages for numerical forecasting. *J. Met. Soc.* **14**, 184-185.

Phillips, N. A., 1959: An example of nonlinear computational instability. In *The atmosphere and the sea in motion. Rossby Memorial Volume*, Rockefeller Institute Press, New York, pp. 501-504.

Phillips, N., 1960a: On the problem of the initial data for the primitive equations, *Tellus* **12**, 121-126.

Phillips, N. A., 1960b: Numerical weather prediction. *Adv. Computers* **1**, 43-91.

Phillips, N., 1963: Geostrophic motion. *Rev. Geophys.* **1**, 123-176.

Phillips, N. A., 1966: The equations of motion for a shallow rotating atmosphere and the traditional approximation. *J. Atmos. Sci.* **23**, 626-630.

Phillips, N., 1973: Principles of large scale numerical weather prediction. In *Dynamic meteorology*, P. Morel, editor, Riedel, Dordrecht, pp. 1-96.

Phillips, N., 1979: The nested grid model. NOAA. Tech. Report NWS 22, US Dept of Commerce, Washington DC, 80 pp. Available from the National Centers for Environmental Prediction, 5200 Auth Rd, Rm. 100. Camp Springs, MD 20746.

Phillips, N., 1981: Variational analysis and the slow manifold. *Mon. Wea. Rev.* **109**, 2415-2426.

Phillips, N., 1982: On the completeness of multi-variate optimum interpolation for large-scale meteorological analysis. *Mon. Wea. Rev.* **110**, 1329-1334.

Phillips, N., 1986: The spatial statistics of random geostrophic modes and first-guess errors. *Tellus* **38**A, 314-322.

Phillips, N., 1990a: *Dispersion processes in large-scale weather.* World Meteorological Organization, No 700, World Meteorological Organization, Geneva.

Phillips, N., 1990b: The emergence of quasi-geostrophic theory. In *The atmosphere – a challenge*, R. Lindzen, E. Lorenz and G. Platzman, editors. American Meteorological Society, Boston, MA, pp. 177-206.

Phillips, N., 1998: Carl-Gustav Rossby: His times, personality and actions. *Bull. Amer. Meteor. Soc.* **79**, 1079-1112.

Phillips, N. and J. Shukla, 1973: On the strategy of combining coarse and fine grid meshes in numerical weather prediction. *J. Appl. Meteor.* **12**, 763-770.

Pielke, R. A., W. R. Cotton, R. L. Walko, C. J. Tremback, W. A. Lyons, L. D. Grasso, M. E. Nicholls, M. D. Moran, D. A. Wesley, T. J. Lee, and J. H. Copeland, 1992: A comprehensive meteorological modeling system – RAMS. *Meteor. Atmos. Phys.* **49**, 69-91.

Pires, C., R. Vautard, and O. Talagrand, 1996: On extending the limits of variational assimilation in chaotic systems. *Tellus* **48**A, 96-121.

Platzman, G. W., 1960: The spectral form of the vorticity equation. *J. Meteor.* **17**, 635-644.

Platzman, G., 1967: A retrospective view of Richardson's book on weather prediction, *Bull. Amer. Meteor. Soc.* **48**, 514-550.

Platzman, G., 1979: The ENIAC computations of 1950: Gateway to numerical weather prediction, *Bull. Amer. Meteor. Soc.* **60**, 302-312.

Poincaré, H., 1897: *Les Méthodes Nouvelles de la Méchanique Céleste*, Vols. I and II. Gauthier-Villars, Paris.

Pu, Z.-X., and E. Kalnay, 1999: Targeting observations with the quasi-inverse linear and adjoint NCEP global models: Performance during FASTEX. *Quart. J. Roy. Meteor. Soc.* **125**, 3329-3338.

Pu, Z.-X., E. Kalnay, D. Parrish, W. Wu and Z. Toth, 1997a: The use of the bred vectors in the NCEP operational 3-dimensional variational system. *Wea. Forecasting* **12**, 689-695.

Pu, Z.-X., E. Kalnay, J. Sela, and I. Szunyogh., 1997b: Sensitivity of forecast errors to initial conditions with a quasi-inverse linear model. *Mon. Wea. Rev.* **125**, 2479-2503.

Pu, Z.-X., S. J. Lord, and E. Kalnay, 1998: Forecast sensitivity with dropwindsonde data and targeted observations. *Tellus* **50**A, 391-410.

Puri, K., 1981: Local geostrophic wind correction in the assimilation of height data and its relationship to the slow manifold. *Mon. Wea. Rev.* **109**, 52-55.

Puri, K., 1987: Some experiments on the use of tropical heating information for initial state specification, *Mon. Wea. Rev.* **115**, 1394-1406.

Puri, K., and W. Bourke, 1982: A scheme to retain the Hadley circulation during normal mode initialization. *Mon. Wea. Rev.* **110**, 327-335.

Purser, R. J., 1984: *A new approach to the optimal assimilation of meteorological data by iterative Bayesian analysis*. Preprint of the 10th AMS Conference on Weather Forecasting and Analysis, Clearwater Beach, FL, American Meteorological Society, pp. 102-105.

Purser, R. J., and L. M. Leslie, 1991: An efficient interpolation procedure for high-order 3-dimensional semilagrangian models. *Mon. Wea. Rev.* **119**, 2492-2498.

Purser, R. J., and L. M. Leslie, 1996: Generalized Adams–Bashforth time integration schemes for a semilagrangian model employing the second derivative form of the horizontal momentum equations. *Quart. J. Roy. Meteor. Soc.* **122**, 737-763.

Purser, J., and L. M. Leslie, 1997: High-order generalized Lorenz *N*-cyle schemes for semilagrangian models employing second derivatives in time. *Mon. Wea. Rev.* **125**, 1276.

Rabier, F., E. Klinker, P. Courtier, and A. Hollingsworth, 1996: Sensitivity of forecast errors to initial conditions. *Quart. J. Roy. Meteor. Soc.* **122**, 121-150.

Rabier, F., J.-F. Mahfouf, M. Fisher, H. Järvinen, A. Simmons, E. Andersson, F. Bouttier, P. Courtier, M. Hamrud, H. Haseler, A. Hollingsworth, L. Isaksen, E. Klinker, S. Saarinen, C. Temperton, J.-N. Thepaut, P. Undén, and D. Vasiljevic, 1997: The ECMWF operational implementation of four-dimensional variational assimilation. ECMWF Research Department Tech. Memo. No. 240.

Rabier F, A. McNally, E. Andersson, P. Courtier, P. Unden, J. Eyre, A. Hollingsworth, and F. Bouttier, 1998: The ECMWF implementation of three-dimensional variational assimilation (3D-Var). II: Structure function. *Quart. J. Roy. Meteor. Soc.* **124**, 1809-1830.

Rancic, M., R. J. Purser, and F. Mesinger, 1996: A global shallow-water model using an expanded spherical cube: gnomonic versus conformal coordinates. *Quart. J. Roy. Meteor. Soc.* **122**, 959-982.

Randall, D. A., editor, 2000: *General circulation model development: Past, present and future*, International Geophysics Series, Vol 70, Academic Press, New York.

Randall, D. A., K.-M. Xu, R. J. C. Somerville, and S. Iacobellis, 1996: Single-column models and cloud ensemble models as links between observations and climate models. *J. Climate* **9**, 1683-1697.

Rasch, P., 1986: Towards atmospheres without tops: absorbing upper boundary conditions for numerical models. *Quart. J. Roy. Meteor. Soc.* **112**, 1195-1218.

Rasch, P. J., and D. L. Williamson, 1990: Computational aspects of moisture transport in global models of the atmosphere. *Quart. J. Roy. Meteor. Soc.* **116**, 1071-1090.

Ray, P., C. Ziegler, W. Bumgarner, and R. Seraphin, 1980: Single and multiple-Doppler radar observations of tornadic storms, *Mon. Wea. Rev.* **108**, 1607-1625.

Reynolds, C. A., and T. N. Palmer, 1998: Decaying singular vectors and their impact on analysis and forecast correction. *J. Atmos. Sci.* **55**, 3005-3023.

Reynolds, C. A., P. J. Webster, and E. Kalnay, 1994: Random error growth in NMC's global forecasts. *Mon. Wea. Rev.* **122**, 1281-1305.

Reynolds, O., 1895: On the dynamical theory of incompressible viscous fluids and the determination of the criterion. *Phil. Trans. Roy. Soc. London* **186**, 124-64.

Richardson, L. F., 1922: *Weather prediction by numerical process.* Cambridge University Press, Cambridge. Reprinted by Dover (1965, New York) with a new introduction by Sydney Chapman.

Richtmeyer, R., and K. Morton, 1967: Difference methods for initial value problems, Interscience, New York.

Ringler, T. D., R. P. Heikes, and D. A. Randall, 1999: *Modeling the atmospheric general circulation using a spherical geodesic grid: A new class of dynamical cores.* Department of Meteorology, Colorado State University, Fort Collins, CO.

Ritchie, H., 1986: Eliminating the interpolation associated with the semi-Lagrangian scheme. *Mon. Wea. Rev.* **114**, 135-146.

Ritchie, H., 1987: Semi-Lagrangian advection on a Gaussian grid. *Mon. Wea. Rev.* **115**, 608-619.

Ritchie, H., C. Temperton, A. Simmons, M. Hortal, T. Davies, D. Dent, and M. Hamrud, 1995: Implementation of the semi-Lagrangian method in a high-resolution version of the ECMFW forecast model. *Mon. Wea. Rev.* **123**, 489-514.

Rivest, C., A. Staniforth, and A. Robert, 1994: Spurious resonant response of semi Lagrangian discretizations to orographic forcing: diagnosis and solution. *Mon. Wea. Rev.* **122**, 366-376.

Robert, A., 1966: The integration of a low order spectral form of the primitive meteorological equations. *J. Met. Soc. Japan* **44**, 237-244.

Robert, A., 1969: The integration of a spectral model of the atmosphere by the implicit method. *Proc. WMO/IUGG Symposium on NWP,* Japan Meteorological Society, Tokyo, Japan, pp. 19-24.

Robert, A., 1979: The semi-implicit method. In *Numerical Methods Used in Atmospheric Models*, Vol. II, World Meteorological Organization, Geneva, pp. 419-437.

Robert, A. J., 1981: A stable numerical integration scheme for the primitive meteorological equations. *Atmosphere–Ocean* **19**, 35-46.

Robert, A., 1982: A semi-implicit and semi-Lagrangian numerical integration scheme for the primitive meteorological equations. *J. Meteor. Soc. Japan* **60**, 319-325.

Robert, A., and E. Yakimiw, 1986: Identification and elimination of an inflow boundary computational solution in limited area model integrations. *Atmosphere Ocean* **24**, 369-385.

Robert, A., T. L. Yee, and H. Ritchie, 1985: A semi-Lagrangian and semi-implicit numerical integration scheme for multilevel atmospheric models. *Mon. Wea. Rev.* **113**, 388-394.

Rogers, E., D. G. Deaven, and G. J. DiMego, 1995: The regional analysis system for the operational early Eta Model. *Wea. Forecasting* **10**, 810-825.

Rogers, E., T. L. Black, D. G. Deaven, G. J. DiMego, Q. Zhao, M. Baldwin, N. W. Junker, and Y. Lin, 1996: Changes to the operational early Eta analysis/forecast system at the National Centers for Environmental Prediction. *Wea. Forecasting* **11**, 319-413.

Rogers, E., D. Parrish, and G. J. DiMego, 1998: Data assimilation experiments with the regional 3-D variational analysis at the National Centers for Environmental Prediction. Preprints, 12th Conf. on Numerical Weather Prediction, Phoenix, AZ, Amer. Meteor. Soc., pp. 32-33.

Rohaly, G. D., R. H. Langland, and R. Gelaro, 1998: Identifying regions where the forecast of tropical cyclone tracks is most sensitive to initial condition uncertainty using adjoint methods. Preprints, 12th Conf. on Numerical Weather Prediction, 11–16 January 1998, Phoenix, AZ, Amer. Meteor. Soc., pp. 337-340.

Rood, R. B., 1987: Numerical advection algorithms and their role in atmospheric transport and chemistry models. *Rev. Geophys.* **25**, 71-100.

Ropelewski, C., and M. Halpert, 1987: Global and regional scale precipitation patterns associated with the El Niño/Southern Oscillation. *Mon. Wea. Rev.* **114**, 2352-2362.

Rosati, A., R. Budgel, and K. Miyakoda, 1995: Decadal analysis produced from an ocean data assismilation system. *Mon. Wea. Rev.* **123**, 2206-2228.

Rossby, C. G., 1936: Dynamics of steady ocean currents in light of experimental fluid dynamics. *Papers in physical oceanography and meteorology.* Vol. 1, No.1, Massachusetts Inst. Of Techn. and Woods Hole Oceanographic Institution.

Rossby, C., 1938: On the mutual adjustment of pressure and velocity distribution in certain simple current systems, *J. Mar. Res.* **2**, 239-63.

Rossby, C.-G., 1940: Planetary flow patterns in the atmosphere. *Quart. J. Roy. Meteor. Soc.* **66** (Supp.), 68-87.

Rossby, C.-G., 1945: On the propagation of frequencies and energies in certain types of oceanic and atmospheric waves. *J. Meteor.* **2**, 187-204.

Rossby, C.-G., *et al.*, 1939: Relation between variations in the intensity of the zonal circulation of the atmosphere and the displacements of the semi-permanent centers of action. *J. Mar. Res.* **2**, 38-55.

Rostaing, N., S. Dalmas, and A. Galligo, 1996: Automatic differentiation in Odyssee. In *Computational differentiation – techniques, applications, and tools*, M. Berz, C. Bischof, G. Corliss, and A. Griewank, editors, SIAM, Philadelphia, PA, pp. 558-568.

Ruelle, D., 1989: *Chaotic evolution and strange attractors.* Cambridge University Press, Cambridge.

Ruelle, D., 1993: *Chance and chaos*, Princeton University Press., Princeton, NJ.

Rutherford, I., 1972: Data assimilation by statistical interpolation of forecast error fields, *J. Atmos. Sci.* **29**, 809-815.

Rutherford, I., and R. Asselin, 1972: Adjustment of the wind field to geopotential data in a primitive equations model, *J. Atmos. Sci.* **29**, 1059-1063.

Sadourny, R., A. Arakawa, and Y. Mintz, 1968: Integration of the non-divergent barotropic vorticity equation with an icosahedral-hexagonal grid for the sphere. *Mon. Wea. Rev.* **96**, 351-356.

Saltzman, B., 1962: Finite amplitude free convection as an initial value problem. *J. Atmos. Sci.* **19**, 329-341.

Sameh, A., and V. Sarin, 1999: Hybrid parallel linear solvers, *Int. J. Comput. Fluid Dyn.* **12**, 213-223.

Sasaki, Y., 1958: An objective analysis based on the variational method. *J. Meteor. Soc. Japan* **36**, 77-88.

Sasaki, Y., 1969: Proposed inclusion of time variation terms, observational and theoretical, in numerical variational objective analysis. *J. Meteor. Soc. Japan* **47**, 115-124.

Sasaki, Y., 1970: Some basic formalisms in numerical variational analysis. *Mon. Wea. Rev.* **98**, 875-883.

Sauer, T., J. Yorke, and M. Casagli, 1991: Embedology. *J. Stat. Phys*, **65**, 579-616.

Savijarvi, H., 1995: Error growth in a large numerical forecast system. *Mon. Wea. Rev.* **123**, 212-221.

Schlatter, T., 1975: Some experiments with a multivariate statistical objective analysis scheme, *Mon. Wea. Rev.* **103**, 246-257.

Schlatter, T., and G. Branstator, 1979: Estimation of errors in NIMBUS 6 temperature profiles and their spatial correlation, *Mon. Wea. Rev.* **107**, 1402-1413.

Schmidt, F., 1977: Variable fine mesh in a spectral global model. *Beitr. Phys. Atmos.* **50**, 211-217.

Schopf, P. S., and M. J. Suarez., 1988: Vacillations in a coupled ocean-atmosphere model. *J. Atmos. Sci.* **45**, 549-568.

Schubert, S., R. Rood, and J. Pfaendtner, 1993: An assimilated data set for earth science applications. *Bull. Amer. Meteor. Soc.* **74**, 2331-2342.

Schwartz, B. E., S. G. Benjamin, S. M. Green, and M. R. Jardin, 2000: Accuracy of RUC-1 and RUC-2 wind and aircraft trajectory forecasts by comparison with ACARS observations. *Wea. Forecasting* **15**, 313-326.

Seaman, R., and F. Gauntlett, 1980: Directional dependence of zonal and meridional wind correlation coefficients. *Aust. Met. Mag.* **28**, 217-221.

Segami, A., K. Kurihara, H. Nakamura, M. Ueno, I. Takano, and Y. Tatsumi, 1989: Operational mesoscale weather prediction with Japan spectral model. *J. Met. Soc. Japan* **67**, 907-923.

Sela, J. G., 1980: Spectral modeling at the National Meteorological Center. *Mon. Wea. Rev.* **108**, 1279-1292.

Sellers, P. J., 1992: Biophysical models of land surface processes. Chapter 14 in *Climate system modeling*, K. Trenberth, editor, Cambridge University Press, Cambridge.

Sellers, P. J., Y. Mintz, Y. C. Sud, and A. Dalcher, 1986: The design of a Simple Biosphere model (SiB) for use within general circulation models. *J. Atmos. Soc.* **43**, 505-531.

Semazzi, F., and I. Navon, 1986: A comparison of the bounded derivative and the normal mode initialization methods using real data. *Mon. Wea. Rev.* **114**, 2106-2121.

Shapiro, R., 1970: Smoothing, filtering and boundary effects. *Rev. Geophys. Space Phys.* **8**, 359-387.

Shaw, D., P. Lonnberg, A. Hollingsworth, and P. Unden, 1987: Data assimilation: The 1984/85 revisions of the ECMWF mass and wind analysis. *Quart. J. Roy. Meteor. Soc.* **113**, 533-566.

Shukla, J., 1981: Predictability of monthly means. *J. Atmos. Sci.* **38**, 2547-2572.

Shukla, J., J. Anderson, D. Baumhefner, C. Brankovic, Y. Chang, E. Kalnay, L. Marx, T. Palmer, D. Paolino, J. Ploshay, S. Schubert, D. Straus, M. Suarez, and J. Tribbia, 2000: Dynamical seasonal prediction. *Bull. Amer. Meteor. Soc.* **81**, 2593-2606.

Shuman, F., 1957: Numerical methods in weather prediction: I. The balance equation. *Mon. Wea. Rev.* **85**, 329-332.

Shuman, F. G., 1989: History of numerical weather prediction at the National Meteorological Center. *Wea. Forecasting* **4**, 286-296.

Shuman, F. G., and J. B. Hovermale, 1968: An operational six-layer primitive equation model. *J. Appl. Meteor.* **7**, 525-547.

Simmons, A. J., and D. M. Burridge, 1981: An energy and angular momentum conserving vertical finite-difference scheme and hybrid vertical coordinates. *Mon. Wea. Rev.* **109**, 758-766.

Simmons, A. J., and C. Temperton, 1996: Stability of a two-time-level semi-implicit integration scheme for gravity wave motion. *Mon. Wea. Rev.* **125**, 600-615.

Simmons, A. J., B. Hoskins, and D. Burridge, 1978: Stability of the semi-implicit integration scheme for gravity wave motion. *Mon. Wea. Rev.* **125**, 600-615.

Simmons A. J., R. Mureau, and T. Petroliagis, 1995: Error growth estimates of predictability from the ECMWF forecasting system . *Quart. J. Roy. Meteor. Soc.* **121** No. 527, 1739-1771.

Skamarock, W. C., 1989: Truncation error estimates for refinement criteria in nested and adaptive models. *Mon. Wea. Rev.* **117**, 872-886.

Skamarock, W. C., and J. B. Klemp, 1992: The stability of time split numerical methods for the hydrostatic and non-hydrostatic elastic equations. *Mon. Wea. Rev.* **120**, 2109-2127.

Skamarock, W., J. Oliger, and R. L. Street, 1989: Adaptive grid refinement for numerical weather prediction. *J. Comput. Phys.* **80**, 27-60.

Slingo, J. M., 1987: The development and verification of a cloud prediction; model for the ECMWF model. *Quart. J. Roy. Meteor. Soc.* **113**, 899-927.

Slingo, A., 1989: A GCM parameterization for the shortwave radiative properties of water clouds. *J. Atmos. Sci.* **46**, 1419-1427.

Smagorinsky, J., 1956: On the inclusion of moist adiabatic processes in numerical prediction models. *Ber. D. Deutsche. Wetterd.* **5**, 82-90.

Smagorinsky, J., 1983: The beginnings of numerical weather prediction and general circulation modeling: Early recollections. *Adv. Geophy.* **25**, 3-37.

Smagorinsky, J., S. Manabe, and J. L. Holloway, 1965: Numerical results from a nine-level general circulation model of the atmosphere. *Mon. Wea. Rev.* **93**, 727-768.

Smith, L. A., 1997: The maintenance of predictability. In: *Past and present variability of the solar-terrestrial system: measurement, data analysis and theoretical models*. G. C. Castagnoli and A. Provenzale, editors. Proceedings of the International School of Physics Enrico Fermi, Italian Physical Society, IOS Press, Amsterdam, pp. 177-246.

Smith, N. R., 1950: The BMRC ocean thermal analysis system. *Aust. Met. Mag.* **44**, 93-110.

Smith, W., H. Woolf, C. Hayden, D. Wark, and L. McMillin, 1979: The TIROS-N operational vertical sounder. *Bull. Amer. Meteor. Soc.* **60**, 117-197.

Smolarkiewicz, P. K., 1984: A fully multidimensional positive definited advection transport algorithm with small implicit diffusion. *J. Comp. Phys.* **54**, 325-362.

Smolarkiewicz, P. K., and W. Grawoski, 1990: The multidimensional positive definite advection transport algorithm: nonoscillatory option. *J. Comp. Phys.* **86**, 355-375.

Sparrow, C., 1982: *The Lorenz equations: bifurcations, chaos and strange attractors*, Springer-Verlag, New York.

Staff Members of the Institute of Meteorology, University of Stockholm, 1954: Results of forecasting with the barotropic model on an electronic computer (BESK). *Tellus* **6**, 139-149.

Staniforth, A. N., 1987: Formulating efficient finite-element codes for flows in regular domains. *Int. J. Numer. Meth. Fluids* **7**, 1-16.

Staniforth, A. N., 1997: Regional modeling, a theoretical discussion. *Meteor. Atmos. Phys.* **63**, 15-29.

Staniforth, A. N., and J. Côté, 1991: Semi-Lagrangian integration schemes for atmospheric models – a review. *Mon. Wea. Rev.* **119**, 2206-2223.

Staniforth, A. N., and R. W. Daley, 1977: A finite-element formulation for vertical discretization of sigma-coordinate primitive equation models. *Mon. Wea. Rev.* **105**, 1108-1118.

Staniforth, A. N., and R. W. Daley, 1979: A baroclinic finite-element model for regional forecasting with the primitive equations. *Mon. Wea. Rev.* **107**, 107-121.

Staniforth, A. N., and H. L. Mitchell, 1977: A semi-implicit finite-element barotropic model. *Mon. Wea. Rev.* **106**, 439-447.

Staniforth, A. N., and H. L. Mitchell, 1978: A variable-resolution finite-element technique for regional forecasting with the primitive equations. *Mon. Wea. Rev.* **105**, 154-169.

Stauffer, D. R., and N. L. Seaman, 1990: Use of 4-D data assimilation in a limited area mesoscale model. Part 1: Experiments with synoptic scale data. *Mon. Wea. Rev.* **118**, 1250-1277.

Stephens, G. L., 1984: The parameterization of radiation for numerical weather prediction and climate models. *Mon.Wea. Rev.* **112**, 826-867.

Stensrud, D. J., H. E. Brooks, J. Du, M. S. Tracton, and E. Rogers, 1999: Using ensembles for short-range forecasting. *Mon. Wea. Rev.* **127**, 433-446.

Stensrud, D. J., J. W. Bao, and T. T. Warner, 2000: Using initial condition and model physics perturbations in short-range ensembles. *Mon. Wea. Rev.* **128**, 2077-2107.

Strang, G., and G. J. Fix, 1973: *An analysis of the finite element method.* Prentice-Hall, New York.

Stull, R. B., 1988: *An introduction to boundary layer meteorology*, Kluwer Academic Publishers, Dordrecht.

Suarez, M. J., and P. S. Schopf, 1988: A delayed action oscillator for ENSO. *J. Atm. Sci.* **45**, 3283-3287.

Suarez, M. J., A. Arakawa, and D. A. Randall, 1983: The parameterization of the planetary boundary layer in the UCLA general circulation model: formulation and results. *Mon. Wea. Rev.* **111**, 2224-2243.

Sugi, M., 1986: Dynamic normal mode initialization. *J. Met. Soc. Japan* **64**, 623-636.

Sundqvist, H., E. Berge, and J. E. Kristjansson, 1989: Condensation and cloud studies with mesoscale numerical weather prediction model. *Mon. Wea. Rev.* **117**, 1641-1757.

Sundstrom, A., and T. Elvius, 1979: Computational problems related to limited-area modeling. In *Numerical methods used in atmospheric models*, GARP Series Publication No. 17, World Meteorological Organization, Geneva.

Swanson, K., R. Vautard, and C. Pires, 1998: Four-dimensional assimilation and predictability in a quasi-geostrophic model. *Tellus* **50A**, 369-390.

Szunyogh, I., E. Kalnay, and Z. Toth, 1997: A comparison of Lyapunov vectors and optimal vectors in a low resolution GCM. *Tellus* **49A**, 200-227.

Szunyogh, I., Z. Toth, R. E. Morss, S. J. Majumdar, B. J. Etherton, C. H. Bishop, 2000: The effect of targeted dropsonde observations during the 1999 winter storm reconnaissance program. *Mon. Wea. Rev.* **128**, 3520-3537.

Takacs, L., 1985: A two-step scheme for the advection equation with minimized dissipation and dispersion errors. *Mon. Wea. Rev.* **113**, 1050-1065.

Takle, E. S., W. J. Gutowski Jr, R. W. Arritt, Z. Pan, C. J. Anderson, R. Silva, D. Caya, S.-C. Chen, J. H. Christensen, S.-Y. Hong, H.-M. H. Juang, J. J. Katzfey, W. M. Lapenta, R. Laprise, P. Lopez, J. McGregor, and J. O. Roads, 1999: Project to Intercompare Regional Climate Simulations (PIRCS): Description and initial results. *J. Geophys. Res.* **104**, 19443-19462.

Talagrand, O., 1981: A study of the dynamics of four dimensional data assimilation. *Tellus* **33**, 43-60.

Talagrand, O., 1991: The use of adjoint equations in numerical modelling of the atmospheric circulation. In *Automatic differentiation of algorithms: Theory, implementation, and application*, A. Griewank and G. F. Corliss, editors, SIAM, Philadelphia, PA, pp. 169-180.

Talagrand, O., 1997: Assimilation of observations, an introduction. *J. Met. Soc. Japan* Special Issue **75**, 1B, 191-209.

Talagrand, O., and P. Courtier, 1987: Variational assimilation of meteorological observations with the adjoint vorticity equations, Part I. Theory. *Quart. J. Roy. Meteor. Soc.* **113**, 1311-1328.

Tanguay, M., A. Robert, and R. Laprise, 1990: A semi-implicit semilagrangian fully compressible regional forecast model. *Mon. Wea. Rev.* **118**, 1970-1980.

Tarantola, A., 1987: *Inverse problem theory*, Elsevier, Amsterdam.

Tatsumi, Y., 1983: An economical explicit time integration scheme for a primitive model. *J. Meteor. Soc. Japan* **61**, 269-287.

Tatsumi, Y., 1986: A spectral limited-area model with time dependent lateral boundary conditions and its application to a multi-level primitive equation model. *J. Met. Soc. Japan* **64**, 637-663.

Taylor, M., J. J. Tribbia, and M. Iskandarani, 1997a: The spectral element method for the shallow water equations on the sphere. *J. Comp. Phys.* **130**, 92-108.

Taylor, M., T. Loft, and J. Tribbia, 1997b: Performance of a spectral element atmospheric model (SEAM) on the HP Exemplar SPP2000 (personal communication).

Temperton, C., 1976: Dynamic initialization for barotropic and multi-level models, *Quart. J. Roy. Met. Soc.* **102**, 297-311.

Temperton, C., 1984: Variational normal mode initialization for a multi-level model, *Mon. Wea. Rev.* **112**, 2303-2316.

Temperton, C., 1988: Implicit normal mode initialization. *Mon. Wea. Rev.* **115**, 1013-1031.

Temperton, C., 1989: Implicit normal mode initialization for spectral models. *Mon. Wea. Rev.* **117**, 436-451.

Temperton, C., 1991: Finite element methods. In *Numerical methods in atmospheric models* Vol. 1, ECMWF, Shinfield Park, Reading, UK, pp. 103-118.

Temperton, C., and D. Williamson, 1981: Normal mode initialization for a multi-level gridpoint model, Part I: Linear aspects. *Mon. Wea. Rev.* **199**, 729-743.

Teweles, S., and H. Wobus, 1954: Verification of prognostic charts. *Bull. Amer. Meteor. Soc.* **35**, 455-463.

Thepaut, J. N. and P. Courtier, 1991: Four-dimensional variational data assimilation using the adjoint of a multilevel primitive equation model. *Quart. J. Roy. Meteor. Soc.* **117**, 1225-1254.

Thepaut, J. N., R. N. Hoffman, and P. Courtier, 1993: Interactions of dynamics and observations in a 4-dimensional variational assimilation. *Mon. Wea. Rev.* **121**, 3393-3414.

Thiebaux, H. J., 1976: Anisotrophic correlation function for objective analysis. *Mon. Wea. Rev.* **104**, 994-1002.

Thiebaux, H. J., 1985: On approximations to geopotential and wind-field correlation structures. *Tellus* **37A**, 126-31.

Thiebaux, H. J., and M. A. Pedder, 1987: *Spatial objective analysis*, Academic Press, London.

Thiebaux, H. J., H. Mitchell, and D. Shantz, 1986: Horizontal structure of hemispheric forecast error correlations for geopotential and temperature. *Mon. Wea. Rev.* **114**, 1048-1066.

Thompson, P. D., 1961a: *Numerical weather analysis and prediction*, The MacMillan Company, New York.

Thompson, P., 1961b: A dynamical method of analyzing meteorological data. *Tellus* **13**, 334-349.

Thompson, P., 1969: Reduction of analysis error through constraints of dynamical consistency. *J. Appl. Meteor.* **8**, 739-742.

Thompson, P., 1983: A history of numerical weather prediction in the United States. *Bull. Amer. Meteor. Soc.* **64**, 755-769.

Thompson, P. D., 1990: Charney and the revival of NWP. In *The atmosphere – a challenge*, R. Lindzen, E. Lorenz, and G. Platzman, editors, American Meteorological Society, Boston, MA, pp. 93-119.

Thuburn, J., 1997: A PV-based shallow water model on a hexagonal-icosaedral grid. *Mon. Wea. Rev.* **1125**, 2328-2350.

Todling, R. and S. E. Cohn, 1994: Suboptional schemes for atmospheric data assimilation based on the Kalman filter. *Mon. Wea. Rev.* **122**, 2530-2557.

Todling, R., and M. Ghil, 1994: Tracking atmospheric instabilities with the Kalman filter. Part I: Methodology and one-layer results. *Mon. Wea. Rev.* **122**, 183-204.

Tomassini, M., D. LeMeur, and R. Saunders, 1999: Use and impact of satellite atmospheric motion winds on ECMWF analyses and forecasts. *Mon. Wea. Rev.* **127**, 971-986.

Toth, Z., and E. Kalnay, 1993: Ensemble forecasting at NMC: the generation of perturbations. *Bull. Amer. Meteor. Soc.* **74**, 2317-2330.

Toth, Z., and E. Kalnay, 1996: *Ensemble forecasting at NCEP. Seminar Proceedings of a Workshop on Predictability*, Volume 2, ECMWF, Shinfield Park, Reading, UK, pp. 39-60.

Toth, Z., and E. Kalnay, 1997: Ensemble Forecasting at NCEP: the breeding method. *Mon. Wea. Rev.* **125**, 3297-3318.

Toth, Z., E. Kalnay, S. Tracton, R. Wobus, and J. Irwin., 1997: A synoptic evaluation of the NCEP ensemble. *Wea. Forecasting* **12**, 140-153.

Toth, Z., I. Szunyogh, and E. Kalnay, 1999: Response to Notes on the appropriateness of 'Bred Modes' for generating initial perturbations, by R. Errico and R Langland. *Tellus* **51A**, 442-449.

Tracton, M. S., and E. Kalnay, 1993: Ensemble forecasting at NMC: Practical aspects. *Wea. Forecasting* **8**, 379-398.

Treadon, R. E., 1996: Physical initialization in the NMC global data assimilation system. *Meteor. Atmos. Phys.* **60**, 57-86.

Trefethen, L., 1983: Group velocity interpretation of the stability theory of Gustafsson, Kreiss and Sundström. *J. Comp. Phys.* **49**, 199-217.

Trenberth, K. E., editor, 1992: *Climate system modeling*. Cambridge University Press, Cambridge.

Trenberth, K., and J. Olson, 1988: An evaluation and intercomparison of global analyses from the National Meteorological Center and the European Centre for Medium Range Weather Forecasts. *Bull. Amer. Meteor. Soc.* **69**, 1047-1057.

Trevisan, A., and R. Legnani, 1995: Transient error growth and local predictability: a study in the Lorenz system. *Tellus* **47A**, 103-117.

Trevisan, A., and F. Pancotti, 1998: Periodic orbits, Lyapunov vectors and singular vectors in the Lorenz system. *J. Atmos. Sci.* **55**, 390-398.

Tribbia, J., 1981: Non-linear normal mode balancing and the ellipticity condition. *Mon. Wea. Rev.* **109**, 1751-1761.

Tribbia, J., 1984: A simple scheme for higher order non-linear normal mode initialization. *Mon. Wea. Rev.* **112**, 278-284.

Tripoli, G. J., 1992: A nonhydrostatic mesoscale model designed to simulate scale interaction. *Mon. Wea. Rev.* **120**, 1342-1359.

Tripoli, G. J., and W. R. Cotton, 1980: A numerical investigation of several factors contributing to the observed variable intensity of deep convection over South Florida. *J. Appl. Meteorol.* **19**, 1037-1063.

Tsuyuki, T., 1990: Prediction of the 30–60 day oscillation with the JMA Global Model and its impact on extended-range forecasts. *J. Met. Soc. Japan* **68**, 183-201.

Uboldi F, and M. Kamachi., 2000: Time-space weak-constraint data assimilation for nonlinear models, *Tellus* **52A**, 412-421.

Van den Dool, H. M., 1989: A new look at weather forecasting through analogues. *Mon. Wea. Rev.* **117**, 2230-2247.

Van den Dool, H.M., 1994: Searching for analogues, how long must we wait? *Tellus* **46A**, 314-324.

Van Leer, B., 1974: Towards the ultimate conservative difference scheme IV. A new approach to numerical convection. *J. Comp. Phys.* **23**, 276-299.

Vannichenko, N., 1970: The kinetic energy spectrum in the free atmosphere: 1 second to 5 years, *Tellus* **22**, 158-166.

Vannitsem, S., and C. Nicolis, 1997: Lyapunov vectors and error growth patterns in a T21L3 quasigeostrophic model. *J. Atmos. Sci.* **54**, 347-361.

Vautard, R., and B. Legras, 1986: Invariant manifolds, quasi-geostrophy and initialization, *J. Atmos. Sci.* **43**, 565-584.

Vesely, F. J., 2001: *Computational physics – An introduction*, edition. Kluwer Academic/ Plenum Publishers, New York.

Vichnevetsky, R., 1986: Invariance theorems concerning reflection at numerical boundaries. *J. Comput. Phys.* **63**, 268-282.

Vichnevetsky, R., 1987: Wave propagation and reflection in irregular grids for hyperbolic equations. *Appl. Numer. Math.* **2**, 133-166.

Vichnevetsky, R., and L. H. Turner, 1991: Spurious scattering from discontinuously stretching grids in computational fluid flow. *Appl, Numer. Math.* **8**, 289-299.

Wahba, G., and J. Wendelberger, 1980: Some new mathematical methods for variational objective analysis using splines and cross validation. *Mon. Wea. Rev.* **108**, 1122-1143.

Walker, G.T., 1928: World weather, III. *Mem. Roy. Meteor. Soc.* **2**, 97-106.

Wallace, J. M., S. Tibaldi, and A. J. Simmons, 1983: Reduction of systematic forecast errors in the ECMWF model through the introduction of an envelope orography. *Quart. J. Roy. Meteor. Soc.* **109**, 683-717.

Wallace, J. M., E. M. Rasmusson, T. P. Mitchell, V. E. Kousky, E. S. Sarachik, and H. von Storch, 1998: On the structure and evolution of ENSO-related climate variability in the tropical Pacific: lessons from TOGA. *J. Geophys. Res. Oceans* **103**, C7, 14241-14260.

Warner, T. W., and N. L. Seaman, 1990: A real-time mesoscale numerical weather prediction system used for research, teaching and public service at the Pennsylvania State University. *Bull. Amer. Meteor. Soc.* **71**, 792-805.

Warner, T. T., R. A. Peterson, and R. E. Treadon, 1997: A tutorial on lateral boundary conditions as a basic and potentially serious limitation to regional numerical weather prediction. *Bull. Amer. Meteor. Soc.* **78**, 2599-2617.

Washington, W., and C. Parkinson, 1986: *An introduction to three-dimensional climate modelling*, University Science Books, Mill Valley, CA.

Weickman, K. M., G. R. Lussky, and J. E. Kutzbach, 1985: Intraseasonal (30–60) fluctuations of outgoing long-wave radiation and 250 mb streamfunction during northern winter. *Mon. Wea. Rev.* **112**, 941-961.

Wergen, W., 1988: The diabatic ECMWF normal mode initialisation scheme. *Beitr. Phys. Atmosph.* **61**, 274-304.

Wiin-Nielsen, A., 1959: On the application of trajectory methods in numerical forecasting. *Tellus* **11**, 180-196.

Wiin-Nielsen., A., 1991: The birth of numerical weather prediction. *Tellus* **43A-B**, 36-52.

Wilks, D. S., 1995: *Statistical methods in the atmospheric sciences*, Academic Press, New York.

Williamson, D. L., 1968: Integration of the barotropic vorticity equation on a spherical geodesic grid. *Tellus* **20**, 642-653.

Williamson, D. L., 1970: Integration of the primitive barotropic model over a spherical geodesic grid. *Mon. Wea. Rev.* **98**, 512-520.

Williamson, D. L., 1992: Review of numerical approaches for modeling global transport. In *Air pollution modeling and its application,* IX, H. van Dop and G. Kallos, editors, Plenum Press, New York, pp. 377-394.

Williamson, D., and R. Dickinson, 1976: Free Oscillation of the NCAR global circulation model, *Mon. Wea. Rev.* **104**, 1372-1391.

Williamson, D., and A. Kasahara, 1971: Adaptation of meteorological fields forced by updating. *J. Atmos. Sci.* **28**, 1313-1324.

Williamson, D. L., and R. Laprise, 1998: Numerical approximations for global atmospheric general circulation models. In *Numerical modelling of the global atmosphere for climate prediction*. P. Mote and A. O'Neill, editors, Kluwer Academic Publishers, Dordrecht.

Williamson, D. L., and J. G. Olson, 1994: Climate simulations with a semi-Lagrangian version of the NCAR community climate model. *Mon. Wea. Rev.* **122**, 1594-1610.

Williamson, D., and C. Temperton, 1981: Normal model initialization for a multilevel grid-point model, Part II: Non-linear aspects. *Mon. Wea. Rev.* **109**, 745-757.

Winninghoff, F., 1973: Note on a simple, restorative-iterative procedure for initialization of a global forecast model. *Mon. Wea. Rev.* **101**, 79-84.

Wobus, R., and E. Kalnay, 1995: Three years of operational prediction of forecast skill. *Mon. Wea. Rev.* **123**, 2132-2148.

Woollen, J. R., 1991: New NMC operational OI quality control. Preprints, Ninth Conf. on Numerical Weather Prediction, Denver, CO, Amer. Meteor. Soc., pp. 24-27.

Wunch, C., 1978: The North Atlantic general circulation west of 50W determined by inverse methods. *Rev. Geophys. Space Phys.* **16**, 583-620.

Wyrtki, K., 1975: El Niño: the dynamical response of the equational Pacific to atmospheric forcing. *J. Phys. Ocean.* **5**, 572-584.

Xie, P. P., and P. A. Arkin, 1996: Analyses of global monthly precipation using gauge observations, satellite estimates and numerical model predictions. *J. Climate* **9**, 840-858.

Xu, K.-M., and A. Arakawa, 1992: Semi-prognostic tests of Arakawa-Schubert cumulus parameterization using simulated data. *J. Atmos. Sci.* **49**, 2421-2436.

Xu, K. M., and D. A. Randall, 1996: A semiempirical cloudiness parameterization for use in climate models. *J. Atmos. Sci.* **53**, 3084-3102.

Xue, M., K. K. Droegemeier, V. Wong, A. Shapiro, and K. Brewster, 1995: ARPS Version 4.0 User's Guide. (Available from the Center for Analysis and Prediction of Storms, 100 East Boyd Street, Norman, OK, 73019.)

Xue, M., K. K. Droegemeir, and V. Wong, 2000: The Advanced Regional Prediction System (ARPS) – A multiscale non-hydrostatic atmospheric simulation and prediction system. Part I: Model dynamics and verification. *Meteorol. Atmosph. Phys.* **75**, 161-193.

Xue, M., K. K. Droegemeir, V. Wong, A. Shapiro, K. Brewster, F. Carr, D. Weber, Y. Liu, and D. Wang, 2001: The Advanced Regional Prediction System (ARPS) – A multiscale non-hydrostatic atmospheric simulation and prediction system. Part II: Model physics and applications. *Meteorol. Atmosph. Phys.* **76**, 143-165.

Yakimiw, E., 1988: Local error propagation in a global spectral shallow water model: its implication for regional forecast modeling. *Atmopshere-Ocean* **26**, 679-689.

Yakimiw, E., and A. Robert, 1990: Validation experiments for a nested grid-point regional forecast model. *Atmosphere-Ocean* **28**, 466-472.

Yanai, M., S. K. Esbensen, and J. H. Chu, 1973: Determination of bulk properties of tropical cloud clusters from large-scale heat and moisture budgets. *J. Atmos. Sci.* **30,** 611-627.

Yessad, K., and P. Benard, 1996: Introduction of local mapping factor in the spectral part of Meteo-France global variable mesh numerical forecast model. *Quart. J. Roy. Meteor. Soc.* **122**, 1701-1719.

Zapotocny, T. H., D. R. Johnson, and F. M. Beames, 1994: Development and initial test of the University of Wisconsin global isentropic sigma model. *Mon, Wea. Rev.* **122**, 2160-2178.

Zebiak, S. E., and M. A. Cane, 1987: A model El Niño Southern Oscillation. *Mon. Wea. Rev.* **115**, 2262-2278.

Zeigler, C., 1985: Retrieval of thermal and microphysical variables in observed convective storms, Part I: Model development and preliminary testing *J. Atmos. Sci.* **42**, 1487-1509.

Zhang, D.-L., H.-R. Chang, N. L. Seaman, T. T. Warner, and J. M. Fritsch, 1986: A two-way interactive nesting procedure with variable terrain resolution. *Mon. Wea. Rev.* **114**, 1330-1339.

Zhao, Q., and T. L. Black, 1994: Implementation of the cloud scheme in the eta model at NMC. Preprints, 10th Conf. on Numerical Weather Prediction, Portland, OR, Amer. Meteor. Soc., pp. 331-333.

Zhao, Q., and F. H. Carr, 1997: A prognostic cloud scheme for operational NWP models. *Mon. Wea. Rev.* **125**, 1931-1953.

Zhao, Q., T. L. Black, and M. E. Baldwin, 1997: Implementation of the cloud prediction scheme in the Eta model at NCEP. *Wea. Forecasting* **12**, 697-711.

Zhu, Z., J. Thuburn, B. Hoskins, and P. Haynes, 1992: A vertical finite-difference scheme based on a hybrid sigma-theta-p coordinate. *Mon. Wea. Rev.* **121**, 2396-2408.

Zhu, Y, G. Iyengar, Z. Toth, M. S. Tracton, and T. Marchok, 1996: Objective evaluation of the NCEP global ensemble forecasting system. Preprints, 15th Conference on Weather Analysis and Forecasting. Norfolk, VA, Amer. Meteor. Soc., J79-J82.

Ziehmann, C., 2000. Comparison of a single-model EPS with a multi-model ensemble consisting of a few operational models. *Tellus* **52A**, 280-298.

Zou, X., I. M. Navon, and F. X. Le Dimet, 1992: An optimal nudging data assimilation scheme using parameter estimation. *Quart. J. Roy. Meteor. Soc.* **118**, 1163-1186.

Zou, X., Navon I. M. and J. G. Sela, 1993: Variational data assimilation with moist threshold processes using the NMC spectral model *Tellus* **45A**, 370-387.

Zupanski, D., 1997: A general weak constraint applicable to operational 4DVAR data assimilation systems. *Mon. Wea. Rev.* **125**, 2274-2292.

Zupanski, M., 1993: Regional 4-dimensional variational data assimilation in a quasi-operational forecasting environment. *Mon. Wea. Rev.* **121**, 2396-2408.

Zupanski, M., and E. Kalnay, 1999: Principles of data assimilation. In *Global energy and water cycles*, K. A. Browning and R. J. Gurney, editors, Cambridge University Press, Cambridge.

Index

Lightning Source UK Ltd.
Milton Keynes UK
UKOW07f0023071216
289345UK00002B/4/P

9 780521 796293